AF893504

ELI LILLY AND COMPANY
GREENFIELD INTELLIGENCE CENTER
2001 W. MAIN, BOX 708
GREENFIELD, IN 46140

Archives of Toxicology, Supplement 14

Recent Developments in Toxicology: Trends, Methods and Problems

Proceedings of the
European Societies of Toxicology
Meeting Held in Leipzig, September 12–14, 1990

WHEN THIS BOOK IS NO LONGER NEEDED
PLEASE RETURN TO LIB. AGRIC. SERV.

Edited by
Philip L. Chambers, Claire M. Chambers,
Wolf Diether Wiezorek and Siegfried Golbs

With 98 Figures and 65 Tables

Springer-Verlag
Berlin Heidelberg New York
London Paris Tokyo
Hong Kong Barcelona
Budapest

Tox. Ref. Area - GL678 - SS7937-4B $139.00 - (12-16-91) Jars

ELI LILLY AND COMPANY

Claire M. Chambers and Philip L. Chambers
University of Dublin, Department of Pharmacology and Therapeutics, Trinity College, Dublin 2, Ireland

Wolf Diether Wiezorek
Universität Leipzig, Universitätsbereich Medizin, Institut für Pharmakologie und Toxikologie, Härtelstraße 16–18, O-7010 Leipzig, FRG.

Siegfried Golbs
Universität Leipzig, Veterinärmedizinische Fakultät, Institut für Pharmakologie, Pharmazie und Toxikologie, Zwickauer Straße 55, O-7010 Leipzig, FRG.

ISBN 3-540-51422-8 Springer-Verlag Berlin Heidelberg New York
ISBN 0-387-51422-8 Springer-Verlag New York Berlin Heidelberg

Library of Congress Cataloging-in-Publication Data
European Society of Toxicology. Meeting (1990 : Leipzig, Germany) Recent developments in toxicology : trends, methods, and problems : proceedings of the European Society of Toxicology Meeting held in Leipzig, September 12–14, 1990 / edited by Philip L. Chambers ... [et al.]. p. cm. -- (Archives of toxicology. Supplement ; 14) Includes indexes.
ISBN 3-540-51422-8 (alk. paper) : DM 198.00. – ISBN 0-387-51422-8 (alk. paper)
1. Toxicology--Congresses. 2. Target organs (Anatomy)--
Congresses. I. Chambers, P. L. (Philip L.), 1931- . II. Title.
III. Series. [DNLM: 1. Kidney--drug effects--congresses. 2. Liver--drug effects--congresses. 3. Nervous System--drug effects--congresses. 4. Toxicology--congresses. W1AR49GA v. 14 / QV 600 E899r 1990] RA1190.E8 Suppl. vol. 14 [RA1191] 615.9 s--dc20 [615.9] DNLM/DLC for Library of Congress

This work is subject to copyright. All rights are reserved, whether the whole or part of the material is concerned, specifically the rights of translation, reprinting, reuse of illustration, recitation, broadcasting, reproduction on microfilms or in other ways, and storage in data banks. Duplication of this publication or parts thereof is only permitted under the provisions of the German Copyright Law of September 9, 1965, in its version of June 24, 1985, and a copyright fee must always be paid. Violations fall under the posecution act of the German Copyright Law.

© Springer-Verlag Berlin Heidelberg 1991
Printed in Germany

The use of general descriptive names, trade names, trade marks, etc. in this publication, even if the former are not especially identified, is not to be taken as a sign that such names, as understood by the Trade Marks and Merchandise Marks Act, may accordingly be used freely by anyone.

Product Liability: The publisher can give no guarantee for information about drug dosage and application thereof contained in this book. In every individual case the respective user must check its accuracy by consulting other pharmaceutical literature.

Printed and Bookbinding: E. Kieser, 8901 Neusäß
2127/3140/543210 – Printed on acid-free paper

Contents

Neurotoxicology of different noxious compounds

New Aspects and Methods in Toxicopathology

Cardiovascular Toxicology

Toxic Effects on Haemostasis

Toxic Effects on Liver and Kidney

Miscellaneous Toxic Effects

Neurotoxicology of Different Noxious Compounds

Recent Developments in Toxicology:
Trends, Methods and Problems
Arch. Toxicol., Suppl. 14, 3–7 (1991)
© Springer-Verlag 1991

In Vitro Assessment of Antidotes to Organophosphorus Toxicity

S. Das Gupta

Division of Pharmacology and Toxicology, Defence R & D Establishment, Tansen Road, Gwalior-474002. M. P., India

Introduction

In vitro antidotal efficacy against organophosphorus (OP) toxicity was studied employing two different *in vitro* techniques to provide an understanding of both peripheral and central effects.

Poisoning by soman (Pinacolyl methylphosphonofluoridate), an extremely toxic OP compound, usually does not respond to treatment with atropine and a pyridinium oxime (Heilbronn and Tolagen 1965). However HI-6, a bisquaternary oxime has now been designated as the most convenient antidote to soman intoxication (Boskovic 1981). Pretreatment with a carbamate, a reversible acetylcholinesterase (AChE) inhibitor such as pyridostigmine, prevents soman from phosphorylating and thus irreversibly inactivating the active site of AChE (Berry and Davis 1970). This work reports the efficacy of pyridostigmine pretreatment and the therapeutic use of HI-6 at various doses on the recovery of neuromuscular (NM) blockade induced by soman using rat phrenic nerve diaphragm preparation *in vitro* which indicates the role of OP's on the peripheral system. For studying the central effect of OP's the *in vitro* model of the neonatal rat spinal cord was used from which the monosynaptic reflex (MSR) was recorded.

The MSR is a two-neuron arc arising from group 1a afferents and terminating on homonymous alpha-motoneurones. It is a compound action potential recorded in the ventral roots following stimulation of the corresponding dorsal root and is directly proportional to the size of the motoneuron pool (Lloyd 1945). MSR testing can be used to study excitatory as well as inhibitory activity in the spinal cord (Esplin 1972). Very few studies on spinal cord function following toxic insult have been reported. OP agents also affect the MSR both *in vivo* (Swanson and Warnick 1984) and *in vitro* (Yang and Warnick 1983; Yang and Warnick 1984; Das Gupta et al, 1989). These actions were reportedly unrelated to inhibition of AChE since prior inhibition of this enzyme by OP failed to alter the OP's ability to depress or facilitate segmental transmission. Similar observations were reported in the superior cervical ganglia of rats where prior irreversible inhibition of AChE by diisopropylphosphorofluoridate (DFP) did not alter the concentration-dependent depression of synaptic transmission seen with soman (Yarowsky et al 1984). In the present study the antidotal efficacy of different oximes are reported as also is the role of thyrotropin-releasing hormone (TRH) as a possible antidote in OP intoxication using the MSR recorded from neonatal rat spinal cord.

Methods

In vitro Neuromuscular Function Studies: Male Wistar rats weighing (100–200 g) bred in this establishment and maintained on a standard diet (Lipton feed) were used. All animals were provided with water ad libitum. The rat was decapitated and bled out. *In vitro* NM function studies were carried out using the phrenic nerve diaphragm preparation. The left hemidiaphragm was removed as described by Bulbring (1946) and suspended in an organ bath containing Krebs Ringer Solution (composition 119 mM NaCl, 4.7 mM $CaCl_2$, 24.8 mM $NaHCO_3$, 11.1 mM glucose, 1.28 mM KH_2PO_4 and 1.2 mM Mg O_4) which was oxygenated with a mixture of 95% O_2 + 5% CO_2. The temperature was maintained at 37 ± 1°C. The phrenic nerve was stimulated with supramaximal voltage of 0.2 MS duration of 0.2 Hz using a Grass Stimulator Model S-88. The preparation was tetanized every 15 min at 25, 50, 100 and 200 Hz for 3 s (Wolthuis et al 1981, Dube et al 1986). The isometric concentrations of the diaphragm were recorded using a Grass Force Transducer Model FT03 on a Grass Polygraph Model 7-16P-35. The overall percentage of NM function was determined from the four tetanic responses as described by Van Halden et al (1983).

In vitro Recording of MSR from spinal cord of neonatal rat: The housing and preparation of the isolated spinal cord from neonatal rats was similar to that already described (Otsuka & Konishi 1974; Das Gupta et al 1988). The advantage of this *in vitro* spinal cord preparation is that drugs and chemicals can be easily applied to it and the evoked responses readily interpreted. The preparation about to be described can be easily set up and remains responsive for several hours. External variables such as pH and concentration of the applied drug or chemical and the temperature of the preparation can be precisely controlled.

Animals and preparation: Rat pups of either sex (6–9 days old) from female Wistar rats obtained from Charles Rivers Breeding Laboratories Inc (Kingston, NY) were anesthetized with diethyl ether and their spinal columns were removed and placed in a petri dish containing oxygenated physiological solution (124 mM NaCl, 5 mM KCl, 1.3 mM $MgSO_4$, 2.5 mM $CaCl_2$, 1.2 mM KH_2PO_4, 15.0 mM $NaHCO_3$, and 11.0 mM glucose). The spinal cord from the mid-thoracic to the mid-sacral level was removed along with the corresponding dorsal and ventral roots, hemisected in the longitudinal plane and the dura removed. The hemisected cord was then transferred to a small plexiglass bath through which oxygenated physiological solution was constantly perfused at the rate of 2–3 ml/min and maintained at 25 ± 0.5°C.

Stimulation and recording: Suction electrodes were applied to a dorsal and ventral root pair at the L_{3-5} segment. The dorsal root was then stimulated at 0.1 Hz through a stimulus isolation unit controlled by a Digitimer (Medical Systems Model D4030). Stimulating pulses were supramaximal in intensity and had a duration of 0.2 msec. Potential changes in the ventral roots (i.e. the MSR) resulting from dorsal root stimulation were displayed on a Tektronix 5000 series oscilloscope and recorded on a Sony Model SS-HF900 video casette recorder. Control reflex activity was recorded after a stablization period of 1.5–3 h during which the reflex did not change. The spinal cords were then exposed to varying concentrations of the OP compounds for different periods of time or to 1 μM atropine for 15 min and then to TRH with atropine. The MSR was again recorded after completion of each superfusion. In a separate set of experiments the hemisected cord was exposed to different concentration of oximes for 30 min or to sarin for 2 h or to sarin followed by an oxime.

Analysis: Taped responses were subjected to waveform analysis using a MINC/ DECLAB 23 computer. Reflex activity was quantified by measuring the area under the curve. Data were statistically analyzed using analysis of variance and Student's "t" test.

Estimation of cholinesterase: Spinal cord cholinesterase was determined by Ellman's method (Ellman et al 1961).

Results and Discussion

Soman given in a concentration 100 nM and 1 μM produced a dose-dependent tetanic fade in all the frequencies studied. Repeated washing with Krebs-Ringer solution could not produce complete recovery even 90 min after its administration. Pyridostigmine (1×10^{-7}M) *per se* did not produce any significant effect on tetanic tension. However a dose and time dependent reversal of pyridostigmine pretreatment on the NM blockade produced by soman (1 μM) was observed on all the four frequencies studied, being highly pronounced at lower frequencies. Therapeutic administration of HI-6 (1.5 mM) after soman (1 μM) and washing after 30 min resulted in a complete recovery of soman-induced tetanic fade.

Pyridostigmine (1×10^{-7}M) pretreatment was observed to produce complete reversal of tetanic fade produced by different concentrations of soman. This protection is attributed to the beneficial partial inhibition of AChE by the carbamate which can spontaneously breakdown regenerating the active enzyme. The partial recovery at higher frequency may be attributed to more release of acetylcholine quanta at the synaptic junction (Jeevaratnam et al 1990). HI-6 induced recovery of soman induced tetanic fade may be due to either recovery of inhibited AChE activity or direct pharmacological effects or blockade of nicotinic-ganglionic receptors (Kirsch et al 1981).

The OP compounds DFP and sarin depressed the MSR. The concentrations by which it depressed 50% were 100 μM for DFP and 10 nM for sarin. Simultaneous superfusion of the spinal cords with thyrotropin-releasing hormone (TRH) resulted in a reversal of the depression. The potentiating effect of TRH was not affected by atropine (1 μM) although atropine easily reversed the OP-induced depression (Das Gupta et al 1988). This finding represents a possible use of TRH as an adjunct in the therapy of OP toxicity. The three oximes viz. pralidoxime, trimedoxime and diethyxime (0.1–10 μM) did not have any effect on the MSR per se although these oximes (0.1–10 μM) along with sarin (0.1 μM) for 30 min resulted in a dose-dependent reversal of the depression which is not related to regeneration of spinal AChE. Thus, the therapeutic efficacy of oximes may reside in their interaction with AChE receptors and not in the dephosphorylation of inactivated AChE (Warnick et al 1989).

Acknowledgements: The author thanks Brig. (Dr.) K. M. Rao, Director, Defence R & D Establishment, Gwalior for his keen interest in this study; Dr. Jordan E. Warnick, Department of Pharmacology and Experimental Therapeutics, University of Maryland, School of Medicine, Baltimore, Maryland, USA for permitting the use of the data of the monosynaptic reflex of neonatal rat spinal cord which was carried out in his laboratory; and also Mr. Mitthan Lal for secretarial assistance.

References

Berry WK, Davis DR (1970) The use of carbamates and atropine in the protection of animal against poisoning by 1, 2, 2-trimethylpropyl methyl-phosphonofluoridate. Biochem Pharmacol 19:927–934

Boskovic B (1981) The treatment of soman poisoning and its perspectives. Fundam Appl Toxicol 1:203–213

Bulbring E (1946) Observations on the isolated phrenic nerve diaphragm of the rat. Br J Pharmacol 1:38–61

Das Gupta S, Bass KN, Warnick JE (1989) Interaction of reversible and irreversible cholinesterase inhibitors on the monosynaptic reflex in neonatal rats. Toxicol Appl Pharmacol 99:28–36

Das Gupta S, Deshpande SB, Warnick JE (1988) Segmental synaptic depression caused by diisopropylphosphorofluoridate and sarin is reversed by thyrotropin-releasing by hormone in the neonatal rat spinal cord. Toxicol Appl Pharmacol 95:499–506

Dube SN, Ghosh AK, Jeevaratnam K, Kumar D, Das Gupta S, Pant BP, Batra BS, Jaiswal DK (1986) Studies on the efficacy of diethyxime as an antidote against organophosphorus intoxication in rat. Japan J Pharmacol 41:267–271

Ellman GL, Courtney DK, Andres V (Jr), Featherstone RM (1961) A new and rapid colorimetric determination of acetylcholinesterase activity. Biochem Pharmacol 7:88–95

Esplin DW (1972) In: Synaptic system models in experimental models of epilepsy. Purpura DP, Penry JK, Tower DB, Woodbury DM, Walter RD (eds) Raven Press. N Y P-223

Heilbronn E, Tolagen B (1965) Toxogonin in sarin, soman and tabun poisoning. Biochem Pharmacol 14:73–77

Jeevaratnam K, Dube SN, Das Gupta S (1990) Dose dependent efficacy of pyridostigmine pretreatment in soman poisoning. Asia Pacific J Pharmacol 5:(2), 145–149

Kirsch D, Hauser W, Weger N (1981) Effect of bispyridinium oxime HGG-12 and HGG-42 and ganglion blocking agents in synaptic transmission and NADPH fluorescence in the superior cervical ganglion of the rat after soman poisoning *in vitro*. Fund Appl Toxicol 1:169–176

Lloyd DPC (1945) On the relationship between discharge zone and subliminal fringe in a motoneuron pool supplied by a homogenous presynaptic pathway. Yale J Biol Med 18:117

Otsuka M, Konishi S (1974) Electrophysiology of mammalian spinal cord in vitro. Nature (London) 252:733–734

Swanson KL, Warnick JE (1984) Tabun facilitates and depresses spinal reflexes in cat and neonatal rat spinal cords. Neurosci Abst 10:417

Van Halden HPM, Van der Weil HJ, Wolthuis OL (1983) Therapy of organophosphorus poisoning: the marmoset as model for man. Br J Pharmacol 78:579–589

Warnick JE, Das Gupta S, Bass KN, Deshpande SB (1989) Mechanism of organosphorphate and carbamate induced potentiation and depression of the monosynaptic reflex. Abstr Soc Toxicol 9:148

Wolthuis OL, Berends F, Meeter E (1981) Problems in the therapy of soman poisoning. Fund Appl Toxicol 1:183–192

Yarowsky P, Fowler JC, Taylor G, Weinreich D (1984) Noncholinesterase actions of an irreversible acetylcholinesterase inhibitor on synaptic transmission and membrane properties in autonomic ganglia. Cell Mol Neurobiol 4:351–366

Yang QZ, Warnick JE (1983) Effect of sarin and soman on spinal reflexes in the cat. Neurosci Abstr 9:230

Yang QZ, Warnick JE (1984) Antagonism of organophosphate induced depression of reflex activity in the neonatal rat spinal cord. Neurosci Abstr 10:417

Recent Developments in Toxicology:
Trends, Methods and Problems
Arch. Toxicol., Suppl. 14, 8–14 (1991)
© Springer-Verlag 1991

Primary Cultures of Adult Mammalian Sensory Neurons and Other *in Vitro* Systems of Use in Neurotoxicological Studies

R. A. Smith

Department of Anatomy, University of Glasgow, Glasgow G12 8QQ, Scotland

Introduction

There has been a significant trend in recent years, from not only within scientific and medical communities but also from a wide spectrum of social, ethical, political and commercial concerns, to seek reliable alternatives to whole animal experimentation in toxicological research. The impetus for adapting and developing methodologies to reduce, if not replace, the numbers of animals used is as great in studying substances potentially toxic to the complex nervous system as for agents to other target organs. Stringent validation is required before acceptance of *in vitro* methods in the predictive screening of new compounds, and this is well recognised by workers in the field (e.g. Atterwill & Steele 1987; Shahar & Goldberg 1987; Mehlman et al 1989). *In vitro* approaches do exist, which are capable of generating data on the mechanisms underlying acute toxicity. These permit precisely defined dosing in controlled environments, which is not possible in most *in vivo* studies, and are relatively quick and inexpensive to carry out. It is in mechanistic studies that cell and tissue culture can be of most value.

A range of techniques may be used to culture tissues from virtually any region of nervous system. Each has advantages, but also some limitations, so that the balance of these must be carefully considered. Further, the endpoints employed in assessing toxic insult are somewhat diverse. Established neural cell lines, for example human and animal neuroblastomas (Ericsson and Walum 1984), can give pertinent data, although extrapolation to the original cell type remains open to some debate for transformed cells in respect to genetic, physiological, biochemical and differentiated states. Organotypic explants of fetal spinal cord and sensory ganglia in long term cultures have proved useful in neurotoxicity studies (Spencer et al 1978). Elegant fetal brain reaggregate cultures are another system of great potential for studying effects on the developing nervous system (Honeggar and Richelson 1977; Kontur et al 1987), whilst short term toxicity screening of cultured hippocampal slices of adult brain represents an integrated system of differentiated mammalian origin (Greene et al 1985).

Primary cultures of PNS and CNS cells can be initiated from both fetal and adult sources. They are proving of increasing value in neurotoxicological research (e.g. Khera and Whalen 1988). Primary adult mouse sensory neuron cultures have the advantage of being postmitotic, differentiated neurons which may be maintained for at least six weeks and offer reproducible cultures without the complex interactions of some *in vitro* systems (Smith and McInnes 1986). Cultures have been monitored following drug treatments (Smith et al 1987), and after alcohol and aldehyde exposures (Smith et al 1990). As with other *in vitro* studies several endpoints are chosen to observe toxic insult: cell viability, membrane integrity, neurite retention, ultrastructural changes and protein leakage.

Materials and Methods

Cultures are prepared from adult male CBA mice following dispersal by collagenase incubation and by gentle mechanical means (Smith and McInnes 1986). They are maintained in a supplemented Dulbecco's medium as either (a) cultures where non-neuronal cells (NNC's) are allowed to proliferate; or (b) NNC-reduced cultures where 5 μM cytosine arabinoside is added to the medium to eliminate dividing NNC's. In this case, to enhance survival, neurons were maintained on substrata coated with either fibronectin, laminin or polylysine (at concentrations of 10 μg/ml), or with substrate-attached material (microexudates) derived from monolayer primary cultures of mouse hepatocytes and from the fibroblastic cell line 3T6, as previously described (Smith & Orr 1987, Orr & Smith 1988).

Ethanol, propanol, acetaldehyde and acrolein treatments have been carried out in serum-free Dulbecco's medium after 8 div (days in vitro), with dishes taped to minimise loss of the volatile agents as discussed by Smith et al (1990). Cell viabilities were observed, in triplicate, following treatments of one to 48 hours. Cells fixed with 3% glutaraldehyde and postfixed with 1% osmium tetroxide were prepared for scanning electron microscopy (EM) to assess surface damage and changes in neurite outgrowth (Smith et al 1990). Other cultures were screened by transmission EM following ethanol and acetaldehyde treatments.

Results and Discussion

Dorsal root ganglion (DRG) neurons and non-neuronal cells (NNC's) adhere to the culture substratum within 24 hours of dispersal. Large refractile, spherical neurons may by identified by phase contrast microscopy, whilst the NNC's become flattened and stellate. By 7 div the density of NNC's due to proliferation is such that it is difficult to monitor by phase contrast microscopy the neurite extension and growth by the neurons. In order to overcome this problem the effect of cytosine arabinoside (CA) which selectively kills mitotic NNC's was studied. With the reduction in NNC outgrowth however, the

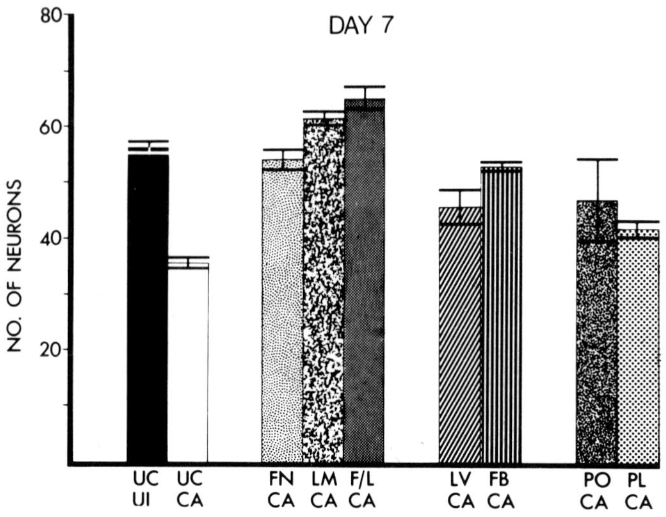

Fig. 1. Neuronal counts (x ± s.e.m.) in a standard reticle field of cultures maintained on uncoated (UC) plastic in uninhibited (UI) or 5 μM cytosine arabinoside (CA) supplemented Dulbecco's medium; and compared with cultures in CA-medium on the following coated substrata: Fibronecton (FN); Laminin (LM); Combined fibronectin and laminin (F/L); Hepatocyte microexudate (LV); Fibroblast exudate (FB); Polyornithine (PO) and Polylysine (PL)

number of surviving neurons also decreased. A study was carried out to investigate whether coating the plastic dishes would create a more conducive substratum, and therefore would prevent a reduction in neuronal numbers when NNC proliferation was inhibited. Fibronectin and laminin, both individually and in combination, were found to be most effective in retaining neuronal numbers in the presence of CA at levels comparable, if not higher, than when NNC proliferation occurred (Fig. 1). Microexudates, prepared from a fibroblast cell line and from primary hepatocytes, also enhanced neuronal maintainence in the presence of CA, as did polylysine and polyornithine, although not as effectively as the fibronectin and laminin alone (Orr and Smith 1988). It is suggested that 5 μM CA does not directly kill postmitotic neurons, as reported by Martin et al (1990), after 4 days with 50 μM CA, but that neuronal survival is less on uncoated substrata since a physical or trophic effect of NNC's is lacking. This is restored and neuronal numbers are retained with fibronectin, laminin and to a lesser extent the other coating agents. Furthermore fibronectin and laminin stimulated neurite regeneration (Orr and Smith 1988), a characteristic used in subsequent alcohol and aldehyde studies (Smith et al 1990). As serum may absorb the alcohols and aldehydes and reduce potency, treatments are carried out in serum-free Dulbecco's media, on cultures transferred along with matched controls to serum-free conditions at 8 div. Certainly 50% reductions in cell survival were observed with ethanol much more rapidly in the absence of serum (Smith et al 1990), and at lower doses than were necessary with continuous administration for 12 div if serum was a component of the media (Scott et al 1986).

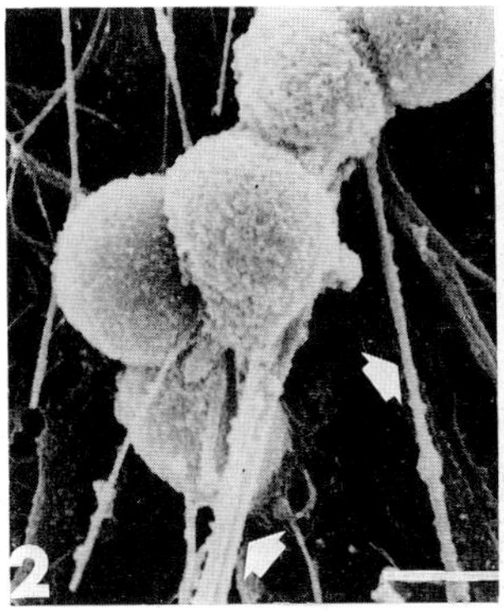

Fig. 2. Scanning electron micrograph of untreated neurons in NNC-reduced culture at 8 div. Note presence of neurites (arrows). Bar = 10 μm

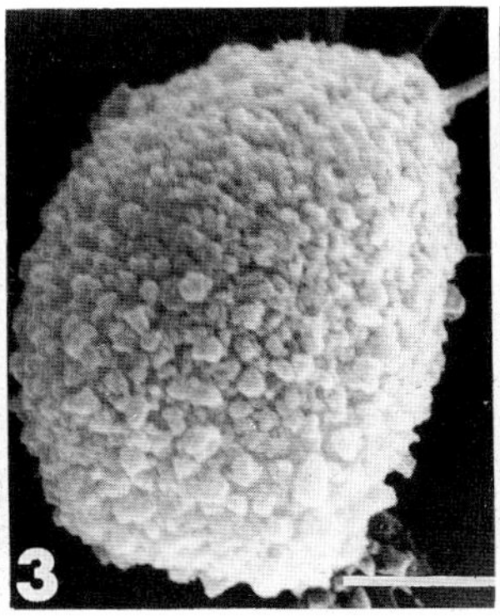

Fig. 3. Neuron with blebbed perikaryon and loss of neurite after 1 h treatment with 1 M ethanol. Bar = 5 μm

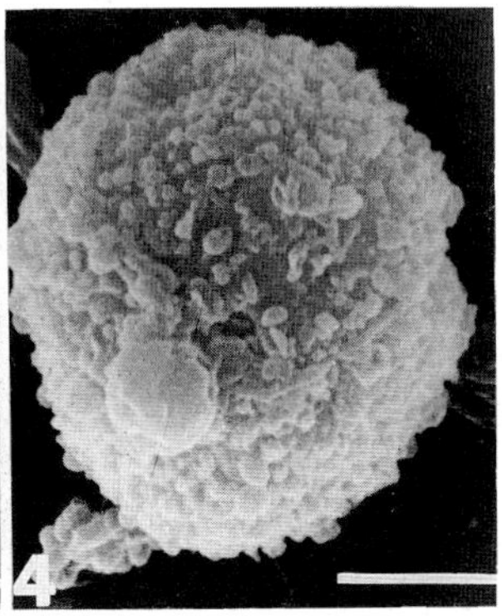

Fig. 4. Neuronal blebbing following treatment with 1 mM acetaldehyde for 6 h. Bar = 5 μm

As previously reported by Smith et al in 1990 the order of potency, obtained by using the endpoints of cell viability and by observing effects on neurite retention and on surface morphology, was acrolein > acetaldehyde > propanol > ethanol. Acrolein had an EC_{50} concentration of 600 μM after 24 hours, acetaldehyde of 1 mM after 24 hours, propanol of 100 mM after 24 hours and ethanol of 500 mM after 48 hours as evidenced by cell viability studies. Neurons showed structural changes, evident by scanning electron microscopy, with shorter exposures (1–6 hours) and lower concentrations. Control cultured neurons have a perikaryon which is roughened by the presence of fine microvilli (Orr and Smith 1988; Panesse et al 1990) and from which neurites extend (Fig. 2). Treated neurons however have blebbed and pitted cell bodies and the neurites, where present, are shortened and thickened (Figs. 3 and 4). Neuronal effects were noted both in NNC-reduced and in uninhibited co-cultures. With propanol and acetaldehyde neurons appeared more sensitive than NNC's. Scott and co-workers (1986) reported LD_{50}'s of 386 mM and 452 mM ethanol for cultured adult mouse and fetal neurons respectively after 12 day exposures. They further found that toxicity was greater for non-neuronal cells with corresponding LD_{50}'s of 206 mM and 152 mM for adult and fetal NNC's. In a recent study of embryonic rat DRG cultures 86–172 mM ethanol was considered to affect Schwann cell proliferation rather than cell survival, with cell morphology remaining unaltered (Mithen et al 1990). Nor did these workers observe ultrastructural changes in DRG neurons cultured in media containing 344 mM ethanol for 35 days. This contrasts with the increasing accumulations of myelin figures and lipofuscin granules in adult neurons following chronic exposure to 386 mM ethanol for 12 days (Scott et al 1986).

Similar ultrastructural changes were noted in primary adult mouse DRG neurons treated with ethanol and acetaldehyde, but using shorter exposures and lower concentrations. Control neurons are characterized by a large pale

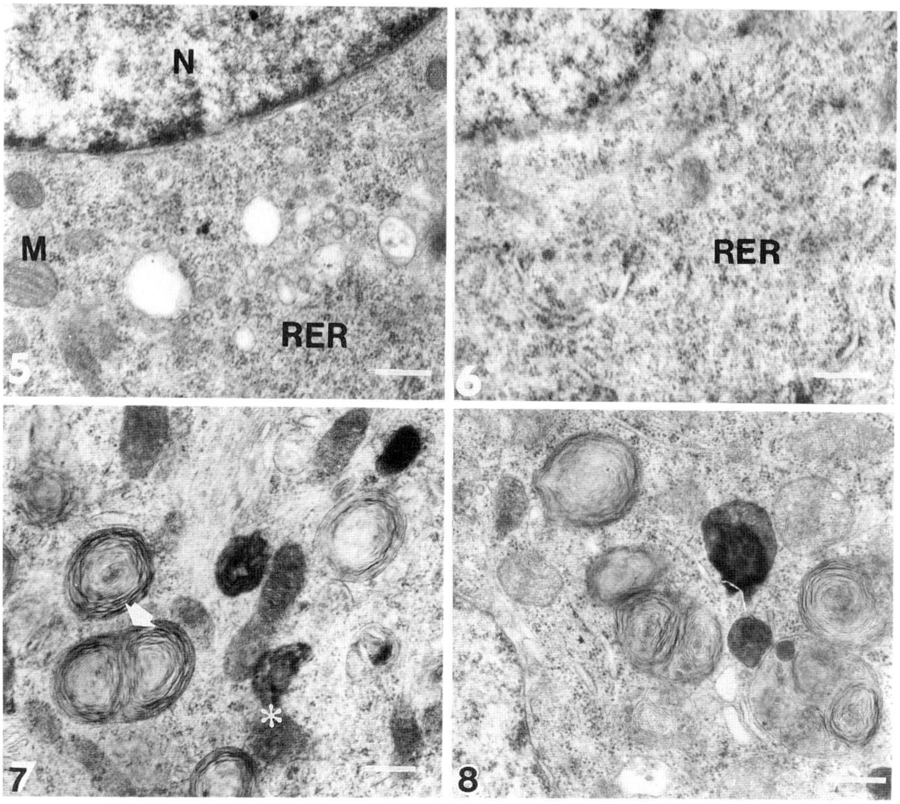

Fig. 5. TEM of untreated 8 div neuron. N.B.: Nucleus (N); mitochondria (M), rough endoplasmic reticulum (RER). Bar = 0.5 μm

Fig. 6. TEM of neuron after 24 h exposure to 50 mM ethanol. No discernible changes from untreated cell are evident. Bar = 0.5 μm

Fig. 7. TEM of neuron treated with 100 mM ethanol for 24 h. Note myelin-like figures (only occasionally seen in untreated cells) (arrowed), and electron dense granules (asterisk) in cytoplasm. Bar = 0.5 μm

Fig. 8. TEM of neuron after 24 h exposure to 300 mM ethanol with an increased accumulation of myelin-like figures. Bar = 0.5 μm

nucleus surrounded by cytoplasm densely packed with large amounts of rough endoplasmic reticulum, perinuclear-located golgi complexes and elongated mitochondria (Fig. 5). Exposure to 50 mM ethanol did not alter this control morphology (Fig. 6). However, whorl-like myelin figures, together with aggregations of lysosomes and electron-dense bodies, were evident with 100 mM doses for 24 hours (Fig. 7). These structures are dose dependent, numbers increasing at higher ethanol concentrations (Fig. 8). Whether these represent a common response of sensory neurons to alcohol and aldehyde exposures remains to be elucidated.

As well as cell viability and morphological criteria others have employed various functional endpoints to assess toxic insult *in vitro*. For PNS neurons, mainly of fetal origin, these include: electrophysiological properties (Oakes and Pozos 1982; Scott and Edwards 1981; White et al 1990); and GABA responses (Skerritt et al 1984). Calcium uptake changes were screened for neural cell lines (Greenberg et al 1987); whilst electrophysiological studies have also been undertaken on cultured CNS neurons (e.g. Skerritt et al 1984; Greene et al 1985).

Enzyme leakage has been quantitatively assessed to monitor injury of cultured fetal mouse cortical neurons (Koh et al 1990). Protein leakage, analysed by electrophoresis of test media, is currently under consideration as an endpoint in this system of cultured adult sensory neurons (Smith, unpublished observations). Future studies are also planned to determine whether specific DRG neuronal subsets, which can be characterised for their neuropeptide and transmitter content by immunostaining and other cell markers (Schoenen et al 1989), are differentially affected by alcohol and aldehyde treatments adding to the endpoints for observation in this system.

Obvious caution is necessary in extrapolation from *in vitro* findings to possible *in vivo* effects, particularly for substances that may require metabolic activation or where the blood-brain barrier may be an important limiting consideration. *In vitro* systems do however permit controlled dosing in defined and regulated culture conditions. Elimination of complex interactions with other body systems can be an advantage of *in vitro* methods for mechanistic studies of toxic action. Although some systems can be cultured for long periods, for example organotypic explants and established neural cell lines, in general *in vitro* methods are, at present, more appropriate for acute toxicity studies. As techniques improve and more is known of the basic growth requirements of isolated nervous tissue, one might predict that cell culture will also be of use in chronic neurotoxicity studies. The work of Smith and co-workers has focussed on the primary culture of adult neurons from the PNS, and has screened these for the effects of alcohols and aldehydes. A larger number of studies by others use fetal nervous tissue, particularly when concerned with regions of the CNS; with refinement however this limitation should eventually diminish. In summary *in vitro* methods add to an understanding of the mechanisms of toxic action. They offer alternative approaches in neurotoxicological investigations hence complementing *in vivo* studies.

Acknowledgements. The technical assistance of Mrs. T. Wubetu is gratefully acknowleged for the transmission electron microscopic studies.

References

Atterwill CK, Steele CE (1987) In vitro methods in toxicology. C.U.P., Cambridge

Ericsson A-Ch, Walum E (1984) Cytotoxicity of cyclophosphamide and acrylamide in glioma and neuroblastoma cell lines cocultured with liver cells. Toxicol Letters 20:251–256

Greenberg DA, Carpenter CL, Messing RO (1987) Ethanol-induced component of Ca uptake in PC12 cells is sensitive to Ca channel modulating drugs. Brain Res 410:143–146

Greene RW, Haas HL, Hermann A (1985) Effects of caffeine on hippocampal pyramidal cells in vitro. Br J Pharmac 85:163–169

Honeggar P, Richelson E (1977) Kainic acid alters neurochemical development in fetal rat brain aggregating cell cultures. Brain Res 138:580–584

Khera KS, Whalen C (1988) Detection of neuroteratogens with an in vitro cytotoxicity assay using primary monolayers cultured from dissociated fetal rat brains. Toxic *in Vitro* 2:257–273

Koh JY, Goldberg MP, Hartley DM, Choi DW (1990) Non-NMDA receptor-mediated neurotoxicity in cortical culture. J Neurosci 10:693–705

Kontur PJ, Hoffmann PC, Heller A (1987) Neurotoxic effects of methamphetamine assessed in three-dimensional reaggregate tissue cultures. Dev Brain Res 31:7–14

Martin DP, Wallace TL, Johnson EM (1990) Ara-C kills postmitotic neurons in a fashion resembling trophic factor deprivation: Evidence that a deoxycytidine-dependent process is required for NGF signal transduction. J Neurosci 10:184–193

Mehlman MA, Pfitzer EA, Scala RA (1989) A report on methods to reduce, refine and replace animal testing in industrial toxicology laboratories. Cell Biol Toxicol 5:349–358

Mithen FA, Reiker MM, Birchem R (1990) Effects of ethanol on rat Schwann cell proliferation and myelination in culture. In Vitro Cell Dev Biol 26:129–139

Oakes SG, Pozos, RS (1982) Electrophysiologic effects of acute ethanol exposure. I. Alterations in the action potentials of DRG neurons in dissociated culture. Dev Brain Res 5:243–249

Orr DJ, Smith RA (1988) Neuronal maintenance and neurite extension in non-neuronal cell-reduced culture is dependent on substratum coating. J Cell Sci 91:555–561

Pannese E, Ledda M, Conte V, Procacci P, Matsuda S (1990) SEM observations of the perikaryal projections of rabbit spinal ganglion neurons after enzymatic removal of connective tissue and satellite cells. Cell Tiss Res 260:167–173

Schoenen J, Delree P, Leprince P, Moonen G (1989) Neurotransmitter phenotype plasticity in cultured dissociated adult rat DRG: An immunocytochemical study. J Neurosci Res 22:473–487

Scott BS, Edwards BAV (1981) The effect of chronic ethanol exposure on the electric membrane properties of DRG neurons in cell culture. J Neurobiol 12:379–390

Scott BS, Petit TL, Lew J (1986) Differential survival of fetal and adult neurons and non-neuronal cells exposed chronically to ethanol in cell culture. Neurotoxicol 7:81–90

Shahar A, Goldberg A (1987) Model systems in neurotoxicolog-Alternative approaches to animal testing. AR Liss, New York

Skerritt JH, Werz MA, McLean MJ, Macdonald RL (1984) Diazepam and its anomalous p-chloro-derivative Ro 5-4864: Comparative effects on mouse neurons in culture. Brain Res 310:99–105

Smith RA, McInnes IB (1986) Phase contrast and electron microscopical observations of adult mouse DRG cells maintained in primary culture. J Anat 145:1–12

Smith RA, Orr DJ (1987) The survival of adult mouse sensory neurons in vitro is enhanced by natural and synthetic substrata, particularly fibronectin. J Neurosci Res 17:265–270

Smith RA, Orr DJ, McInnes IB (1987) A refined primary culture system of adult mouse sensory neurons for neurotoxicity testing. In Shahar A, Goldberg A (eds) Model systems in neurotoxicology-Alternative approaches to animal testing. Alan R Liss Inc, New York, pp 69–75

Smith RA, Orr DJ, Haetzman ML, MacPherson N, Storey ND (1990) The response of primary cultured adult mouse sensory neurons to ethanol, propanol, acetaldehyde and acrolein treatments. Virchows Arch (B) 58:323–330

Spencer PS, Peterson ER, Madrid R, Raine CS (1978) Effects of thallium salts on neuronal mitochondria in organotypic cord-ganglia-muscle combination cultures. J Cell Biol 58:79–95

White G, Lovinger DM, Weight FF (1990) Ethanol inhibits NMDA-activated current but does not alter GABA-activated current in an isolated adult mammalian neuron. Brain Res 507:332–336

Recent Developments in Toxicology:
Trends, Methods and Problems
Arch. Toxicol., Suppl. 14, 15–24 (1991)
© Springer-Verlag 1991

Animal Models in Behavioural Toxicology and Teratology

R. LEDERER, J. ELSNER, and G. ZBINDEN

Institute of Toxicology, Swiss Federal Institute for Technology and University of Zurich, CH-8603 Schwerzenbach, Switzerland

Introduction

Behavioural toxicology and teratology is generally classified as being a subdiscipline of neurotoxicology. While the nervous system is indeed the most important organ involved in behaviour, there are others, and one should regard the study of behaviour rather as a holistic approach to functional aspects of biology. In this respect, the problem of laboratory animal behaviour serving as model for human behaviour is an important one and has to be solved.

A pragmatic approach to this question which is generally and implicitly adopted is to avoid any attempt of extrapolation from animal behaviour to man. If any behavioural change induced by an investigated substance is observed in a laboratory animal, one assumes that the chances are high that there will be some behavioural effects in man, even if these may be quite different in quality and quantity.

Another approach is a more rational one. Starting from known fundamental mechanisms of human psychopathies, behavioural aspects of animals may be sought which are based on the same underlying mechanisms. Toxicity and teratogenicity are always the expression of a disturbance of basic biological functions at the biochemical and cellular level. If species specific expressions of these functions are investigated, the resulting observable effects on a particular organism may be quite different from those in another organism. If, however, fundamental aspects of such functions are considered, then the effects should be quite comparable across species. As examples of such an approach, possible rat models for minimal brain dysfunction are presented.

Minimal Brain Dysfunction in Humans

Minimal brain dysfunction (MBD) is one of the most common behavioural syndromes encountered in present-day pediatrics. Worldwide 10 to 12% of children are classified as falling into various degrees of MBD. Boys suffer from this syndrome about five times more commonly than girls (Toelle 1985). It is characterized by hyperactivity, decreased attention span, learning difficulties, perceptual problems and impulsivity. Children suffering from this behavioural dysfunction have difficulties in the tactile-kinesthetic system (Affolter and Stricker 1980) and with fine motor skills. Although hyperactivity disappears in most children at the time of puberty, decreased attention span, perceptual and emotional problems persist (Shaywitz et al 1977; Lansdell 1980).

The etiology of MBD is not well-known. Possible causal factors could be genetic, neurobiologic, neurotoxic and psychosocial. Oxygen deficiency during birth and some infectuous diseases are the most well documented causes of MBD (Remschmidt 1987). Metabolic studies point to possible alterations in monoamine metabolism, especially dopamine. Examinations of the cerebrospinal fluid in most of these children showed a significantly decreased level of homovanillic acid, the major dopamine metabolite (Shaywitz et al 1978; Lansdell 1980).

Possible Animal Models of Minimal Brain Dysfunction

Methylmercury is a known human behavioural teratogen which induces, at high exposure levels, the congenital Minamata disease and, at lower levels, an increased incidence of mental retardation (Harada 1978). However, no formal study was ever performed on the implications of low-level methylmercury exposure with the incidence of MBD. Therefore, one still has to rely on animal experimentation to estimate this risk. Created in the context of an interlaboratory study on methods in behavioural teratology (EIBT; Elsner 1986; Schreiner et al 1986; Suter and Schön 1986) the following exposure scheme was applied in three different experiments. Wistar rat dams were subjected to methylmercury chloride in the drinking water (1.5 and 5 mg/l) from two weeks before pairing until the end of lactation, resulting in an approximate daily methylmercury exposure of 0.2 and 0.6 mg/kg. This exposure did not affect the weight and the weight gain of the dams and offspring and no overt toxicity or teratogenicity could be detected (data not shown). The offspring of dams were subjected at specific ages to different behavioural testing procedures (Elsner 1986; Elsner et al 1990; Elsner 1991).

Ultrasonic Vocalization of Neonatal Rat Pups as a Model of Baby Cry Features

The human baby cry reflects the capacity of the nervous system to the activated, and also its ability to inhibit or to modulate this activation (Parmelee 1962). It seems plausible that cry characteristics may mirror the brain integrity of the human neonate. Lester (1987) showed that several acoustic features (the average and the rate of change of the fundamental frequency and of the first formant, average amplitude) of baby cries are strongly correlated with Bayley Scales of Infant development at 18 months and with the verbal and perceptual behaviour at five years of age expressed by the McCarthy General Cognitive Index.

Early information about the integrity of the nervous system may also be obtained by a microanalysis of ultrasonic vocalization of neonatal rats (Adams et al 1983; Elsner et al 1990). In the first two or three weeks of life, young rats elicit ultrasounds with a frequency range between 40 and 55 kHz. Ultrasonic vocalizations are produced in distressful situations like cold stress, i.e. lower than the nest temperature of 33°C. The ontogeny of cold stress vocalizing is related to the maturation of the animal's thermoregulation. In the first few days after birth, rat pups are not able to regulate their body temperature and are very resistant to cold (Okon 1972; Allin and Banks 1971). During this period they do not emit many calls. In the period between the onset of the thermoregulation, between 4 and 8 days of age, and its full development at about 10 to 14 days of age, the rat pups start the most intensive cold stress vocalizations (Davis 1982).

The ultrasonic vocalization of 5 to 15 day old rat pups perinatally treated with methylmercury was recorded for one minute every other day. Each rat pup was isolated in a glass beaker, placed in a glass desiccator standing in a thermostatically controlled water bath at 20°C. A directed condenser microphone (Bruel and Kjaer 4135) was fixed 10 cm above the bottom of the beaker. The preconverted frequency signal (positive zero-crossings per second) was recorded and analysed by an LSI 11 computer.

The arithmetic means of the number of calls/min as a function of age are represented in Fig. 1a. The age evolution of the exposed rats exhibited a developmental lag. No increase after day five was detected in the calling rate of the control rats (in contrast with most published reports on rodents, Oswalt and Meier 1975, Cuomo et al 1987). A peak in the calling rate of exposed offspring rats occured, however, on day 9 for the low dose group and on day 11 for the high dose group. In addition, the exposed animals exhibited a lower overall calling rate than the controls. The call durations are represented in Fig. 1b. They were significantly decreased in exposed rats during the first three recording days. The power density of the frequency modulation, represented in Fig. 1c, was lower in the exposed rats on day 7. Other variables calculated from the vocalization frequency (mean frequency, base-interval and several distributions) did not show any exposure effects.

These data imply that methylmercury exposure induced not only a dose-dependent developmental lag, but also a general reduction in the calling rate and call duration. It is interesting to compare the observed poorer frequency modulation of methylmercury exposed rat pups with the human data of Lester (1987), who found a correlation between low change of the first formant of neonate cries and low McCarthy Verbal scores at 5 years of age.

Modified setup. An improved procedure for recording and analyzing ultrasonic vocalizations was developed recently. Restraining tubes were constructed which have an anatomical form with flat bottoms fitting the shape of the rat pups. The tubes are made of double glass walls permitting precise control of the glass temperature by circulating water. This makes it possible to vary the temperature and to study the influence of temperature on call incidences. To avoid interferring noise, the glass tube, which is fixed on a wooden frame,

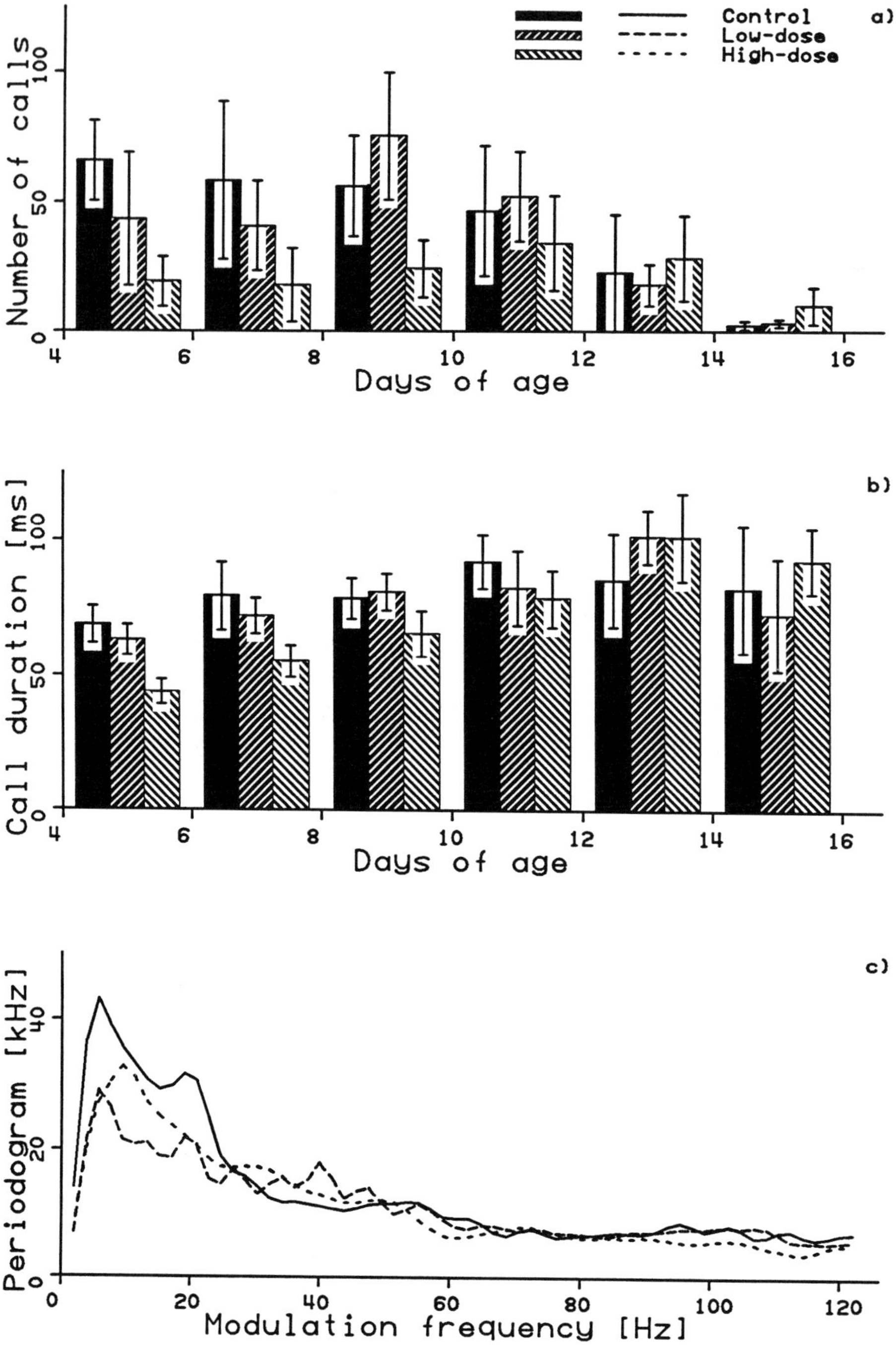

Fig. 1 a–c. **a** calling rate (calls per minute), **b** call duration (milliseconds) as functions of day since birth, and **c** periodogram of the frequency modulation at day 7 evaluated from the vocalization frequency recordings of three groups perinatally exposed to methylmercury (adapted from Elsner et al 1990)

slides into a box of sound absorbing material. The microphone is mounted at the front of the box allowing a constant distance to the pup's snout. Differently sized tubes are used as the pups grow. The frequency and amplitude of the recorded calls are converted simultaneously with a sampling rate of 100 Hz to digital values by separate electronic circuits and stored in a Macintosh computer. This new compact procedure allows the precise recording and microanalysis of both frequency and amplitude of ultrasonic calls in a highly standardized environment. An experiment was first done in this new setup with rats prenatally exposed to nicotine, showing that rats called at a still substantial rate after day 15, an exposure effect in the number of calls having been visible only at day 19 (data not shown).

Locomotion Analysis as Model of Human Hyperkinesis

When MBD was first described, it was mainly characterized by hyperkinesis. Consequently, measures of activity of laboratory animals were, and still are, widely used both in behavioural toxicology and teratology as animal models of hyperkinesis. Several automated setups are proposed for this purpose, ranging from simple infrared activity monitoring (Tamborini et al 1989), through automated open-fields from different manufacturers and complex mazes to the detailed study of activity pattern sequences (Wootten et al 1985; Paulus et al 1990). Most developments in this field are motivated by the recognition of the fact that human hyperkinesis is not as much a simple increase in plain activity magnitude but more an alteration of structural and sequential aspects of activity.

Locomotor behaviour can also be studied in the wheel-shaped activity monitor, a structured environment in which local and locomotor activity including qualitative aspects of path choices are assessed over a complete day (Elsner 1986). Six alleys emerge from a center and are connected by a circular alley. Infrared optical gates are located in each segment and are connected to a digital computer which stores each break and release of a gate together with its event time for later microanalysis.

Rats select their paths across this environment according to two main, conflicting incentives. One is the motivation to explore the available space by continually altering path choices, the other is the conservative patrolling of familiar paths. Rats pre- and perinatally exposed to methylmercury tended to avoid exploration as mirrored by strongly increased incidences of repeated paths along the alleys (Fig. 2). These rats had also an increased tendency to move along the circular alley (data not shown). These observations, together with a slightly increased locomotor activity (data not shown) find again their analogy in the MBD symptomatology (Elsner 1986; Elsner et al 1988).

Operant Behaviour as Model of Human Attention Deficit

Following the progresses in the behavioural analysis of MBD, this syndrome was relabeled in 1980 as "attention deficit disorder" (DSM-III 1980). Distracta-

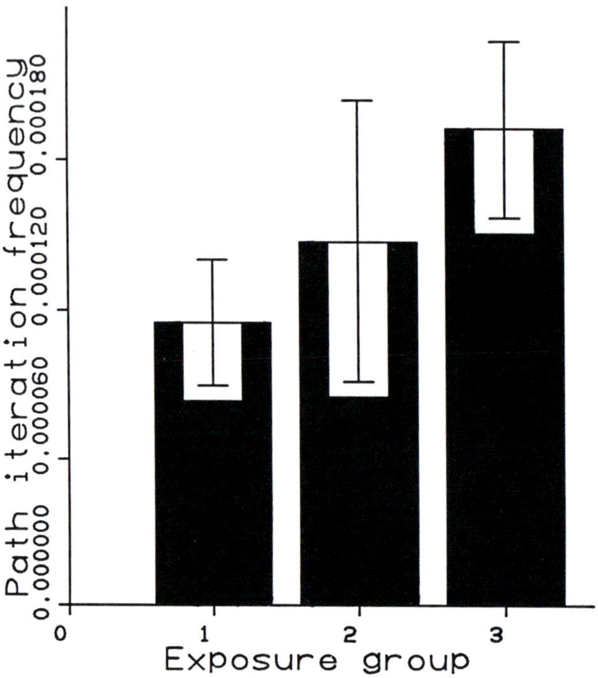

Fig. 2. Path iteration frequency for path length 7 of three groups perinatally exposed to methylmercury: control is group 1, low-dose is group 2, and high-dose is group 3 (adapted from Elsner 1986)

bility seems to be at the origin of altered activity patterns and hyperkinesis in MBD. These attentional problems are usually detected only in the more demanding situations at school and are the causes for learning disabilities for these otherwise normally intelligent children. Moreover, these characteristics of MBD are highly unstable, being almost absent on certain days and very marked on others.

Operant conditioning provides a powerful tool for the assessment of cognitive aspects of animal behaviour. The possibility to detect learning deficits is only one aspect of this experimental method. In addition, by using techniques of microanalysis of behaviour, several fundamental aspects of instrumental behaviour, such as attention spans, may be investigated.

Rats perinatally exposed to methylmercury were trained to lever-press for water reinforcements. In discrete trials, signalled by the illumination of the glass-levers, the rats were subjected to three learning phases. Firstly, both of two levers were reinforced when pressed during a trial; secondly, only one of the two levers was illuminated and reinforced during a trial (visual discrimination); finally, both levers were illuminated during each trial and the rats had to press the left lever in the first, the right lever in the second and alternatively in the subsequent trials in order to earn each time a reinforcement (spatial alternation).

The behaviour of the different exposure groups was identical in the first two easier training phases. In the third, most difficult phase, short attention spans (number of trials responded in a row) of the perinatally exposed rats were much more frequent than in the controls (Fig. 3), while the performance

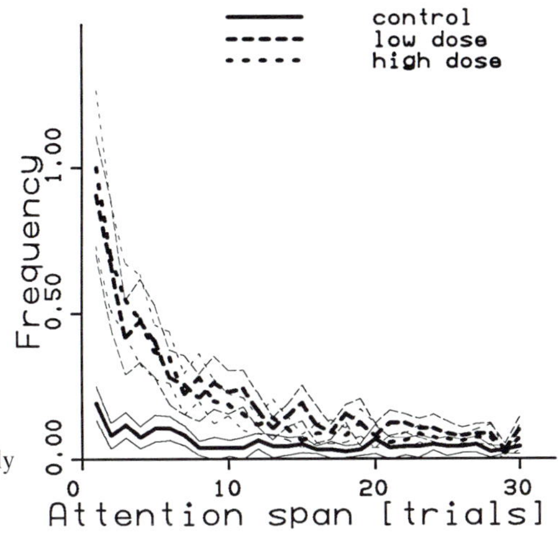

Fig. 3. Mean frequency distribution of attention spans of three groups perinatally exposed to methylmercury (from Elsner 1986)

(number of correct lever presses) was unaltered. Moreover, the variability from session to session of the observed attentional deficit was very high. Thus, all observed aspects of the operant behaviour of methylmercury exposed rats were again in close analogy to above-mentioned characteristics of MBD (Elsner 1986).

Differential Response Force as a Model of Human Tactile-Kinesthetic Perception Deficits

Affolter and Stricker (1980) showed that children with MBD-type characteristics have clearcut deficiencies in the tactile-kinesthetic perception. The tactile-kinesthetic system is the one which relates an organism to reality and, according to Piaget (1970), is the necessary prerequisite for the cognitive development of a human being. Thus, a deficit in the tactile-kinesthetic system may, presumably, be the cause of the complex and unstable symptomatology of MBD.

An operant conditioning technique was developed for the specific assessment of rat tactile-kinesthetic performance (Elsner 1991). A force sensitive lever was used for training perinatally methylmercury exposed rats to press, in discrete trials, during 1 second within a force band of 60 to 80 g (differential reinforcement of response force) without delivering any additional feedback information other than the rat's own tactile kinesthetic perception. In the preliminary training steps in which the force-band was wider, no treatment effect was observed. In the last, most difficult phase, however, the number of correct responses of the exposed rats was only half that of the controls (Fig. 4). Moreover, the failing responses of the exposed rats were mainly due to premature excess above the upper threshold of 80 g and the total number of responses exceeding 80 g was dose-dependantly increased.

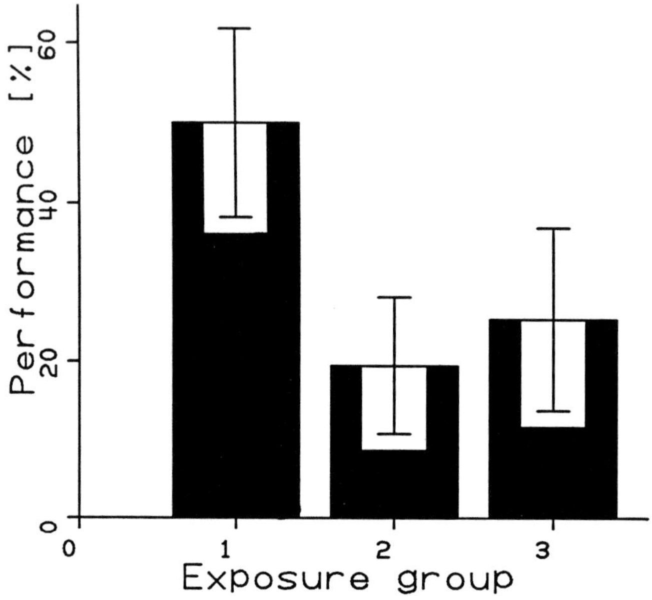

Fig. 4. Performance of three groups perinatally exposed to methylmercury in the schedule of differential reinforcement of response force (60–80 g 1 sec) in the first session after training in successively narrower force bands: control is group 1, low-dose is group 2, and high-dose is group 3 (adapted from Elsner 1991)

Again, a close analogy was found in this experiment with the human condition. The tactile-kinesthetic system of methylmercury exposed rats appeared to be deficient, including a tendency to coarse motoricity, as it is observed in children suffering from MBD.

Discussion

These experiments show that by applying appropriate models, the symptomatology of the minimal brain dysfunction syndrome can be simulated in animals. However, not all indicated analogies are well documented. Increased MBD-incidence due to methylmercury has never been investigated in humans; baby cry features have not been studied in methylmercury-exposed children; no systematic testing of the tactile-kinesthetic system has been performed in children suffering from MBD. Nevertheless, the presented tools of behavioural microanalysis result in a symptomatology in rats perinatally exposed to methylmercury which is strongly reminiscent of the psychopathological descriptions of MBD, while presenting no overt behavioural or other clinical abnormalities.

As another possible animal model of MBD at the neurochemical level, some authors have explored 6-hydroxydopamine (6-OHDA) brain-lesions (Shaywitz et al 1976; Thieme et al 1980). When this neurotoxicant was administered intracisternally or intraventriculary to developing rats after pretreatment with desmethylimipramine (DMI), the resulting effect was a rapid and profound depletion of brain dopamine. Several studies which recorded motor activity during the second and third weeks of rat pups treated with 6-OHDA as neonates found a significantly increased activity. However, when the rat pups reached puberty, the hyperactivity was no longer apparent (Shaywitz et al 1976; Erinoff et al 1979; Fobes and Olds 1981; Concannon and Schechter 1982; Concannon et al 1983).

Thus, the observed hyperactivity in rats with depleted dopaminergic neurons appears to correspond to that of children suffering from MBD. The other important symptoms of MBD, such as attention and learning behaviour, tactile-kinesthetic perception and emotionality have, however, not yet been studied intensively in this model (Oke and Adams 1978; Archer et al 1988). In an effort to test further biochemical processes of this syndrome and to explore the behavioural implications of dopaminergic lesions in more detail, animals treated with 6-hydroxydopamine and desmethylimipramine are presently being subjected to the described behavioural testing methods.

References

Adams J, Miller DR, Nelson DJ (1983) Ultrasonic vocalizations as diagnostic tools in studies of developmental toxicity. Neurobehav Toxicol Teratol 5:29–34

Affolter F, Stricker E (1980) Perceptual processes as prerequisites for complex human behavior, Hans Huber, Bern

Archer T, Danysz W, Fredriksson A, Jonsson G, Luthman J, Sundstroem E, Teiling A (1988) Neonatal 6-hydroxydopamine-induced dopamine depletions: Motor activity and performance in maze learning. Pharmacol Biochem Behav 31:357–364

Allin JT, Banks DM (1971) Effects of temperature on ultrasound production by infant albino rats. Dev Psychol 4:149–156

Concannon JT, Schechter MD (1982) Openfield behavior in dopamine depleted rat pups and their mothers. Pharmacol Biochem Behav 18:285–287

Concannon JT, Braughler JM, Schechter MD (1983) The development of brain biogenic amines, cyclic nucleotides and hyperactivity in 6-OHDA treated rat pups. Pharmacol Biochem Behav 18:477–482

Cuomo V, Cagiano R, Renna G, De Salvia MA, Racagni G (1987) Ultrasonic vocalization in rat pups: Effects of early postnatal exposure to SCH 23390 (a DA1-receptor antagonist) and sulpiride (a DA2-receptor antagonist). Neuropharmacology 26:701–705

Davis JM (1982) Ethological approaches to behavioral toxicology. In: Mitchell CL (ed) Nervous System Toxicology, Raven Press, New York, pp 29–44

DSM-III (1980) Diagnostic and Statistical Manual of Mental Disorders, Third Edition, American Psychiatric Association, Washington, DC

Elsner J (1982) PASTOR, a new schedule programming language. Behav Res Meth Instr 14:254–263

Elsner J (1986) Testing strategies in behavioral teratology: III. Microanalysis of behavior. Neurobehav Toxicol Teratol 8:573–584

Elsner J, Hodel B, Suter KE, Oelke D, Ulbrich B, Schreiner G, Cuomo V, Cagiano R, Rosengren LE, Karlsson JE, Haglid KG (1988) Detection limits of different approaches in behavioral teratology, and correlation of effects with neurochemical parameters. Neurotoxicol Teratol 10:155–167

Elsner J, Suter D, Alder S (1990) Microanalysis of ultrasound vocalization of young rats: Assessment of the behavioral teratogenicity of methylmercury. Neurotoxicol Teratol 12:7–14

Elsner J (1991) The tactile-kinesthetic system of rats as an animal model for minimal brain dysfunction. Arch Toxicol, in press

Erinoff L, Macphail RC, Heller A, Seiden L (1979) Age-dependent effects of 6-hydroxydopamine on locomotor activity in the rat. Brain Res 164:195–205

Fobes JL, Olds ME (1981) Effects of neonatal 6-hydroxydopamine treatment on catecholamine levels and behavior during development and adulthood. Psychopharmacology 73:27–30

Harada M (1978) Congenital minimata disease: intrauterine methylmercury poisoning. Teratology 18:285–288

Lansdell HC (1980) Theories of brain mechanisms in minimal brain dysfunctions. In: Rie HE et al (eds) Handbook of Minimal Brain Dysfunctions. Wiley and sons, New York, pp 117–146

Lester BM (1987) Developmental outcome prediction from acoustic cry analysis in term and preterm infants. Pediatrics 80:529–534

Oke AF, Adams N (1978) Selective attention dysfunctions in adult rats neonatally treated with 6-hydroxydopamine. Pharmacol Biochem Behav 9:429–432

Okon E (1972) Factors affecting ultrasound production in infant rodents. J Zoology 168:139–148

Oswalt G, Meier G (1975) Olfactory, thermal, and tactual influences in infantile ultrasonic vocalization in rats. Dev Psychobiol 8:129–135

Parmelee AH (1962) Infant crying and neurologic diagnosis. J Pediatr 61:801–802

Paulus MP, Geyer MA, Gold LH, Mandell AJ (1990) Application of entropy measures derived from the ergodic theory of dynamical systems to rat locomotor behavior. Proc Natl Acad Sci USA 87:723–727

Piaget J (1970) Piaget's theory. In: Mussen PH (ed) Carmichael's manual of child psychology, 3rd ed, 1: John Wiley & Sons, Inc., New York, pp 703–732

Remschmidt H (1987) Kinder- und Jugendpsychiatrie, Georg Thieme, Stuttgart

Schreiner G, Ulbrich B, Bass R (1986) Testing strategies in behavioral teratology: II. Discrimination learning. Neurobehav Toxicol Teratol 8:567–572

Shaywitz BA, Yager DR, Klopper JH (1976) Selective brain dopamine depletion in developing rats: An experimental model of minimal brain dysfunction. Science 191:305–307

Shaywitz BA, Cohen DJ, Bowers MB (1977) CSF monoamine metabolites in children with minimal brain dysfunction: Evidence for alteration of brain dopamine. J Pediatr 90:67–71

Shaywitz SE, Cohen DJ, Shaywitz BA (1978) The biochemical basis of minimal brain dysfunction. J Pediatr 92:179–187

Suter KE, Schön H (1986) Testing strategies in behavioral teratology: I. Testing battery. Neurobehav Toxicol Teratol 8:561–566

Tamborini P, Sigg H, Zbinden G (1989) Quantitative analysis of rat activity in the home cage by infrared monitoring. Application to the acute toxicity testing of acetanilide and phenylmercuric acetate. Arch Toxicol 63:85–96

Thieme RE, Dijkstra H, Stoof JC (1980) An evaluation of the young dopamine-lesioned rat as an animal model for minimal brain dysfunction (MBD). Psychopharmacology 67:165–169

Toelle R (1985) Psychiatrie, Springer, Berlin

Wender PH (1971) Minimal brain dysfunction syndrome, Wiley, New York

Wootten V, Brown DR, Callahan BG, Vetrano K, Wadman P, Melia J, Mulligan T, Schatz A (1985) Behavioral and biochemical alterations following in utero exposure to methylmercury. Neurobehav Toxicol Teratol 7:767–773

Recent Developments in Toxicology:
Trends, Methods and Problems
Arch. Toxicol., Suppl. 14, 25–29 (1991)
© Springer-Verlag 1991

Cytotoxic Brain Oedema as a Model in the Examination of Cerebroprotective Substances

K. Andreas

Institute of Pharmacology and Toxicology, Medical Academy "Carl Gustav Carus", Dresden, Lingnerplatz 1, O-8010 Dresden, FRG

Introduction

Hexachlorophene (HCP) or 2.2'-methylene bis (3, 4, 6-trichlorphenol) formerly used as germicide, given in the diet of animals, or absorbed through the skin of newborn animals or humans, produces severe oedema in the white matter of the central nervous system (Kimbrough 1971 and 1973; Lampert et al 1973; Mullick 1973; Towfighi 1980). In accordance with Klatzo's definition (Klatzo 1967) the hexachlorophene-induced brain oedema is characterized as cytotoxic oedema. We intend to use this brain oedema as a model for testing cerebroprotective substances. Therefore rats, put on a diet containing high concentrations of HCP, were examined and the recovery of the rats, after cessation of HCP-exposure, was observed with respect to the water, sodium and potassium content of the brain, the histology of the cerebellum and the motor behaviour. With special respect to the question of suitability as a model for testing of cerebroprotective substances, the oedema reducing effect of substances used to manage brain oedema or of potential therapeutic value was tested.

Material and Methods

Hexachlorophene 99% was obtained from Aldrich-Chemie GmbH u. Co. KG Steinheim/Albuch v. FRG. Source of mannitol-solution (10 per cent) and piracetam was the Pharmaceutisches Kombinat GERMED, GDR. Dexamethasone was obtained from KRKA, Novo Mesto, Yugoslavia.

130 adult (160 g) male Wistar rats (Shoe: Wist/1, VEB Versuchstierproduktion Schönwalde) were given a liquid milk diet containing hexachlorophene and pellet food ad libitum during 12 days, thereafter the HCP exposure was cut off. The treated rats received a mean dose of 100 mg/kg HCP per day. Control rats received liquid diet without HCP. During the development of the oedema and in the period of restitution the brain water content was measured in one

hemisphere and expressed as percent of wet mass, the concentration of sodium and potassium in the tissue was measured flame-photometrically. For the purpose of histology the cerebellum was fixed in formaldehyde, the cryo-sections were stained with cresylviolet (Fluka Chemie AG, Switzerland). The functional disorder of the treated animals was characterized in a simple behavioural test. The animals were put on a horizontally placed small ladder with 1.20 m length and 8 cm width and a distance of 2.5 cm between the rungs. The number of falls betweens the rungs during 4 minutes was counted as expression of motor disturbance. The numeric data were statistically evaluated by using Student's t-Test. The treatment with mannitol 20 mg/kg, dexamethasone 16 mg/kg, or piracetam 300 mg/kg started immediately after cessation of HCP pretreatment (see above). These substances were daily administered intravenously. The control animals received the same volume (1% of body volume) of sodium chloride (0–9% NaCl) solution.

Results and Discussion

The brain water content increased during the hexachlorophene treatment and was significantly higher in comparison with the control values from the 8th day of HCP-exposure. The sodium concentration increased, and the potassium concentration decreased (Table 1). These alterations indicate a disturbance of the cellular ion homoestasis. The return to normal of the water and electrolyte values is shown in Table 2. Already from the sixth day vacuoles in the white matter of the cerebellum were visible. The vacuolation increased and developed to a vacuolic network by the 12th day. 6 weeks after cessation of HCP treatment the alterations had disappeared. As an expression of function motor disturbance the HCP-treated rats showed, in the ladder test, on an average, 40 failures during 4 minutes in comparison with 5 failure at most in control animals. The HCP-treated rats recovered quickly after the cessation of treatment and the number of failures decreased (see Fig. 1, controls). An additional treatment with mannitol 20 mg/kg iv. (Fig. 1 A), dexamethasone 16 mg/kg iv.

Table 1. Brain content of water, sodium, and potassium during hexachlorophene treatment

	Day of treatment						
	Control	2	4	6	8	10	12
water (per cent ww)	75.6 (0.3)	76.0 (0.4)	76.9 (0.3)	76.3 (0.5)	77.1 (0.4)	77.1 (0.5)	79.7 (0.3)
sodium (mmol/kg ww)	49.8 (0.6)	53.7 (2.9)	57.9* (4.2)	57.8* (3.0)	56.8* (1.2)	63.7* (2.3)	57.2* (3.2)
potassium (mmol/kg ww)	102.2 (0.9)	103.4 (3.1)	104.2 (1.5)	103.1 (2.9)	96.9* (2.0)	91.9* (3.4)	82.9* (1.5)

Values are mean ± SD of ten rats in each group.
Student's t-test: * = $p<0.05$ (comparison to control)

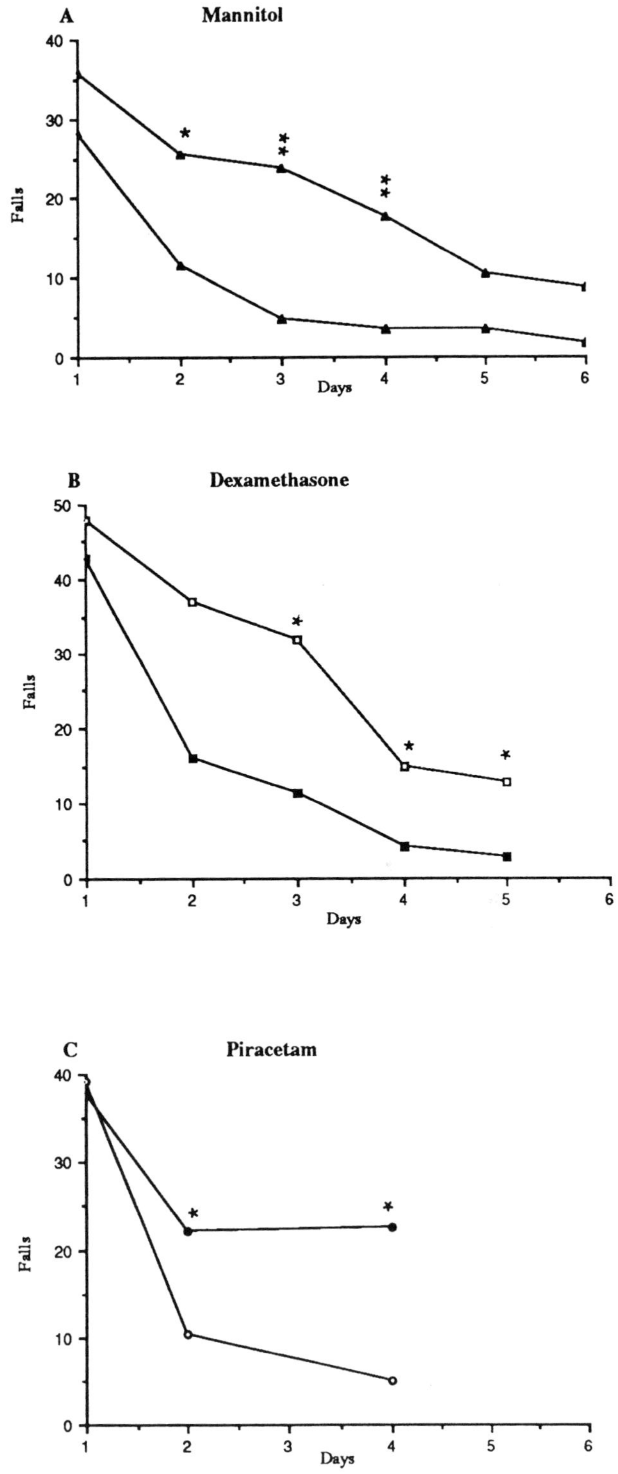

Fig. 1. "Ladder test" in HCP-pretreated rats (100 mg/kg daily in liquid diet during 2 weeks). after **a)** mannitol 20 mg/kg iv. (5 d), **b)** dexamethasone 16 mg/kg iv. (5 d), or **c)** piracetam 300 mg/kg iv. (4 d). Control animals received sodium chloride solution (0.9 per cent). Ordinate: Failure during 4 minutes. Abscissa: Time after cessation of HCP-pretreatment.
n = 10, * = $p<0.05$,
** = $p<0.02$

Table 2. Brain contant of water, sodium, and potassium after cessation of hexachlorophene treatment

	Day after cessation of treatment						
	6	8	10	14	18	41	48
water (per cent ww)	78.7* (0.4)	78.2* (0.4)	75.9 (0.2)	74.9 (0.2)	75.6 (0.2)	74.8 (0.5)	75.6 (0.2)
sodium (mmol/kg ww)	54.9* (1.5)	57.8* (1.6)	53.1* (1.0)	48.3 (0.5)	50.7 (0.4)	49.7 (0.4)	50.2 (0.7)
potassium (mmol/kg ww)	99.1* (2.2)	97.9* (2.3)	97.4* (0.8)	100.9* (0.6)	100.7* (1.1)	104.2 (0.7)	103.5 (0.9)

Values are mean ± SD of ten rats in each group.
Student's t-test: * = $p<0.05$ (comparison to control Table 1)

(Fig. 1 B), or piracetam 300 mg/kg iv. (Fig. 1 C) resulted in acceleration of recovery. The failure rate differed significantly on several days in comparison between mannitol, dexamethasone or piracetam treated rats and the control animals pretreated with HCP alone. The results of the experiments confirm that the HCP-pretreatment is a proper neurotoxicological model to test substances for use in brain of oedema management. The course of development and recovery from the oedema can be observed biochemically, histologically and by simple behavioural tests. The HCP-oedema can be influenced by substances with potential therapeutic value. Mannitol and dexamethasone are usual in therapy of brain oedema (Baethmann et al 1982; Klein and Schmidt 1982). Piracetam is a nootropic substance with cerebroprotective effects caused by its membranotropic efficacy (Schmidt 1982). Its clinical employment is not yet common with respect to brain oedema. The efficacy of other nootropics should be examined.

References

Baethmann A, Oettlinger W, Moritake K, Chaussy L, Jesch F (1982) Metabolic effects of corticosteroids in central nervous tissue. In: Hartmann A, Brock M (eds) Treatment of cerebral oedema, Springer, Berlin Heidelberg New York, pp 24–30

Kimbrough RD (1973) Review of recent evidence of toxic effects of hexachlorophene. Pediatrics 51:391–397

Kimbrough RD, Gaines TB (1971) Hexachlorophene effects on the rat brain: study of high doses by light and electron microscopy. Arch Environmental Health 23:114–120

Klatzo I (1967) Presidential adress: neuropathological aspects of brain oedema. J Neuropath exp Neurol 26:1–14

Klein HJ, Schmidt K (1982) Hyperosmolar solutions and diuretics in the treatment of brain oedema. In: Hartmann A, Brock M (eds) Treatment of cerebral oedema, Springer, Berlin Heidelberg New York, pp 81–99

Lampert PW, O'Brien J, Garrett R (1973) Hexachlorophene encephalopathy. Acta Neuropathol 23:326–335

Mullick FG (1973) Hexachlorophene toxicity: Human experience at the AFIP. Pediatrics 51:395–401

Schmidt J (1982) Zur Pharmakologie der Nootropika. In: Oehme P, Löwe H, Göres E (eds) Beiträge zur Wirkstofforschung, vol 17, Akademie-Industrie-Komplex Arzneimittelforschung, Berlin, pp 1–175
Towfighi I (1980) Hexachlorophene. In: Spencer PS, Schaumburg HH (eds) Experimental and Clinical Neurotoxicology, Williams and Wilkins, Baltimore London, pp 440–455

Recent Developments in Toxicology:
Trends, Methods and Problems
Arch. Toxicol., Suppl. 14, 30–32 (1991)
© Springer-Verlag 1991

Neurochemical Characterization of Hexachlorophene Induced Brain Oedema

K. Andreas, M. Blaschke, E. Bergsträsser, H.-D. Fischer, and J. Schmidt

Institute of Pharmacology and Toxicology, Medical University "Carl Gustav Carus" Dresden; O-8010 Dresden, Lingnerplatz 1

Introduction

Hexachlorophene induces severe generalized oedema of the cytotoxic type in white matter of CNS producing paralytic symptoms which disappear spontaneously within some weeks after termination of drug treatment. The aim of the present study was to provide an additional neurochemical characterization of hexachlorophene oedema at the paralytic climax of its occurrance. The parameters used were

(i) the kinetic constants calculated on the Ca^{2+}-dependence of K^{+}-stimulated dopamine release from striatum slices, which represents a very sensitive probe for synaptic membrane function, and

(ii) the activity of high affinity Ca^{2+}-ATPase in a crude brain synaptosomal fraction, possibly connected with release mechanisms.

Materials and Methods

Animals and experimental design: Male albino rats (strain Shoe: WIST/1), 180–200 g, allowed free access to pellet food and a liquid diet, were used. In experiments with hexachlorophene, rats were treated for 10 days with a liquid diet containing the drug (purchased from Aldrich-Chemie, Steinheim, FRG) in a concentration which resulted in a dose of 100 mg/kg/d. 48 hours after termination of hexachlorophene treatment the animals were killed.
Dopamine release (Fischer et al 1982): Striatum slices of 200 μM thickness were loaded with ^{14}C-dopamine (rDA, 0.03 mM, spec.act. 2.2 GBq/mM,

---

Abbreviations: ATPase = Adenosine-5'-triphosphatase; CNS = central nervous system; EGTA = 1,2-Bis-(2-aminoethoxyethane)-N,N,N',N'-tetraacetic acid; FER = fractional efflux rate; rDA = radiolabeled dopamine

Radiochem. Centre Amersham) by incubation at 37°C for 30 mins. After washing and transfer (about 10 mg of tissue wet weight) to a chamber (0.5 ml) the slices were superfused (0.5 ml/min) at 37°C with Krebs' solution (pH 7.4) containing 4.7 mM KCl. K^+-stimulation was performed by increasing KCl-concentration in the superfusion fluid to 14 mM or 28 mM for 30 sec (balanced by decrease in Na^+-concentration). In order to test the Ca^{2+}-dependence, $CaCl_2$-concentration in the superfusion fluid, $[Ca^{2+}]_s$, was changed from zero to 1.8 mM gradually by doubling 0.225 mM. The stimulus-related evoked overflow of rDA (fractional efflux rate, FER) was expressed as a percentage of tissue radioactivity at the onset of every stimulation period. Ca^{2+}-dependence was analysed according to enzyme kinetic calculations resulting in two parameters:
(i) FER_{max}, a virtual maximal amount of rDA released by a definite K^+-stimulus in presence of an optimal $[Ca^{2+}]_s$ and
(ii) K_{Ca} which signifies $[Ca^{2+}]_s$ necessary to obtain half-maximal FER by any K^+-stimulus.

The numerical values of the two constants were extracted graphically from reciprocal plots of 1/FER vs. $1/[Ca^{2+}]_s$.

High affinity Ca^{2+}-ATPase (Blaschke et al 1988): A crude synaptosomal fraction was prepared by fractionated centrifugation (10,000·g sediment) of brain (without cerebellum) tissue homogenate (7% w/v in Tris·HCl, pH 7.4). ATP hydrolysis (37°C, 30 min) was determined in the presence of 0.2 mM EGTA (Fluka AG, Buchs, Swiss), 0.6 mM strophanthin (Sigma Chemie GmbH, Deishofen, FRG) and increasing $CaCl_{2+}$-concentrations (.75 to 170 μM). Enzyme activity was calculated as μM inorganic phosphate formed per mg protein and hour.

Statistics: Student t-test was used.

Results and Discussion

FER_{max} (Table 1) which depends essentially on the extent of depolarization (Gregerson and Selmanoff 1987) and which reflects the size of the vesicular rDA pool, is apparently increased in hexachlorophene oedema. Obviously an additional artifical compartment is opened. The release from this pool seems to be K^+-sensitive but not Ca^{2+}-dependent in the same extent like the functional one. The considerable rise of K_{Ca}, an estimation of stimulus-release coupling

Table 1. Quasi-kinetic parameters of Ca^{2+}-dependence of K^+-stimulated dopamine release from rat striatum slices (n = number of individual results used for calculation)

Treatment	Stimulus	n	FER_{max}(%)	K_{ca}(mM)
(Control)	14 mM KCl	12	1.77	0.64
(Control)	28 mM KCl	15	2.87	
Hexachlorophene	14 mM KCl	10	5.03	5.11
Hexachlorophene	28 mM KCl	11	6.36	

Table 2. Dependence on Ca^{2+}-concentration of high affinity Ca^{2+}-ATPase activity (mean ± SD, number of experiments in parenthesis) in a crude synaptosomal rat brain fraction

Treatment	ATPase activity ($\mu M\ P_i$/mg prot./h)		
	in presence of 75 $\mu M\ CaCl_2$	110 $\mu M\ CaCl_2$	170 $\mu M\ CaCl_2$
(Control)	0.57±0.03 (22)	0.80±0.04 (19)	1.15±0.04 (22)
Hexachlorophene	0.52±0.04 (28)	0.74±0.03[b] (25)	0.93±0.04[a] (27)

[a] Significantly different from control group: $p<0.01$
[b] Significantly different from control group: $p<0.05$.

quality, reflects an impairment of this function. However, the artificial compartment mentioned above will contribute to this result, showing it at least in part as a sham effect.

Reduced high affinity Ca^{2+}-ATPase activity (Table 2) refers to altered kinetic conditions of this functionally important enzyme (Lin and Way 1984) which is connected with neuronal Ca^{2+} homoiostasis.

References

Blaschke M, Fischer H-D, Schmidt J (1988) Zur peroxidativen Schädigung der synaptosomalen hochaffinen Ca^{2+}-ATPase und ihrer pharmakologischen Beeinflußbarkeit. Biomed Biochim Acta 47:887–893

Fischer H-D, Rudolph E, Schmidt J (1982) Relation between Na^+, K^+-ATPase activity and dopamine release from rat striatum slices. Acta Biol Med Germ 41:471–476

Gregerson KA, Selmanoff M (1987) Rapid release of ^3H-dopamine from median eminence and striatal synaptosomes. J Neurochem 48:1222–1230

Lin SCC, Way EL (1984) Calcium transport in and out of brain nerve endings in vitro. The role of synaptosomal plasme membrane Ca^{2+}-ATPase in Ca^{2+}-extrusion. Brain Res 298:225–234

Recent Developments in Toxicology:
Trends, Methods and Problems
Arch. Toxicol., Suppl. 14, 33–37 (1991)
© Springer-Verlag 1991

The Effects of Organophosphorous Compounds on the Central Nervous System of Rats

I. Dési, L. Nagymajtényi, R. Lorencz, and Z. Molnár

Department of Hygiene and Epidemiology, Albert Szent-Györgyi University Medical School, 6720 Hungary, Szeged, Dóm tér 10

The mechanism of the organophosphorous compounds, the inhibition of the enzyme cholinesterase causing a cholinergic crisis and special clinical signs of intoxication are well known (Hayes 1982, Klaassen et al 1986). The electrophysiological changes of the central and peripheral nervous system produced by these compounds have been investigated only by a few authors (Hyde et al 1978; Solima et al 1982; Dési 1983).

In previous acute and subchronic experiments it was found that dimethoate, DDVP and parathion-methyl caused significant changes of the EEG and functioning of the peripheral nerve (Nagymajtényi et al 1988).

The aim of the present studies was to investigate the pathophysiological processes of the EEG and the somatosensory, visual and auditory evoked potentials in acute experiments giving large doses of the compounds to the rats.

Methods

In the experiments male Wistar rats (250–300 g) in groups of 6 were used.

The doses of the pure compounds (2.0 LD_{50} ip) were as follows: dimethoate (0,0-dimethyl-S-methylcarbamoylmethyl phosphorodithioate; 98,2%; VEB Chemiekombinat Bitterfeld, GDR) 550 mg/kg; dichlorvos (DDVP) (2,2-dichlorovinyl dimethyl phosphate; 98,6%; Universal Laboratories, Hungary) 8 mg/kg; parathion-methyl (0,0-dimethyl 0-4-nitrophenyl phosphorothioate; 90%; VEB Chemiekombinat Bitterfeld, GDR) 5 mg/kg. As an antidote atropine (0,1% EGIS Pharmacological Works Hungary) (8 mg/kg) was used given to the rats 5 minutes before their treatment with the chemicals. The control animals were treated with physiological salt solution.

The EEG was registered with silver electrodes placed frontally and contralateral parieto-occipitally on the skull of the rats. The EEG was recorded before the treatment and every 20 minutes after the injection of the compounds. The analysed parameters of EEG were: mean amplitude, mean frequency, power density of the frequency wave bands, an index formed by the proportion of the slow (delta + theta) and fast ($beta_1$ + $beta_2$) wave bands.

For the somatosensory stimulation two electrodes were inserted into the whiskery part of the face of the rat. The parameters of the stimuli were: 0.2 msec, 2 V, 0.5 Hz. For the visual stimulation flashes were used (0.2 msec, 0.5 Hz) conducted from the stimulator to the opened eye of the rat by means of fibreoptics. The parameters of the sound clicks used for auditory stimulation were 0.2 msec, 40 dB, 0.5 Hz. The evoked potentials were registered with monopolar silver electrodes put on the primary somatosensory, visual and auditory centres. They were recorded before and every 5 minutes after the administration of the chemicals. 25 potentials were averaged by a signal averager program and the latencies and peak to peak amplitudes were measured.

As the initial latency and amplitudes of the evoked potentials were different, the data were standardised. The initial latencies and amplitudes were taken for basis (1.00) and the other data were compared to these.

The cholinesterase activity in the cortex and white matter was measured by the method of Meinecke and Oettel (1966).

Results

All the rats showing the typical signs of intoxication died within 45–70 minutes because of the gross toxic effect of the large doses of the chemicals.

The EEG data are shown in Fig. 1. The tendency of the changes caused by the compounds are practically the same: significant decrease of the mean amplitude and significant increase of the mean frequency. The EEG index decreased because of the changes of the power density of the different wave bands (Fig. 1 A).

The changes of EEG of the rats given not only chemicals but atropine, also were practically the same as those of the animals treated with the compounds (Fig. 1 B).

The tendency of the changes of the somatosensory evoked potentials was also the same in the case of all agents. The latencies increased continuously and at the end of the experiments they were significantly longer. The amplitudes decreased faster and the differences compared with the basic values were more marked (Fig. 2 A). The antidote could not influence the toxic effect of the compounds on the somatosensory evoked potentials. The extension of the latency and the shortening of the amplitude were the same when the animals were given not only dimethoate but also the antidote (Fig. 2 B).

The alteration of the visual and auditory evoked potentials caused by the compounds was the same and atropine could not prevent the toxic effect.

The data obtained by measuring the activity of cholinesterase showed that the inhibition of the enzyme was not more than 30–35%.

Discussion

On the basis of the results it was possible to verify that the large doses of the organophosphorous compounds produced a well marked alteration of the func-

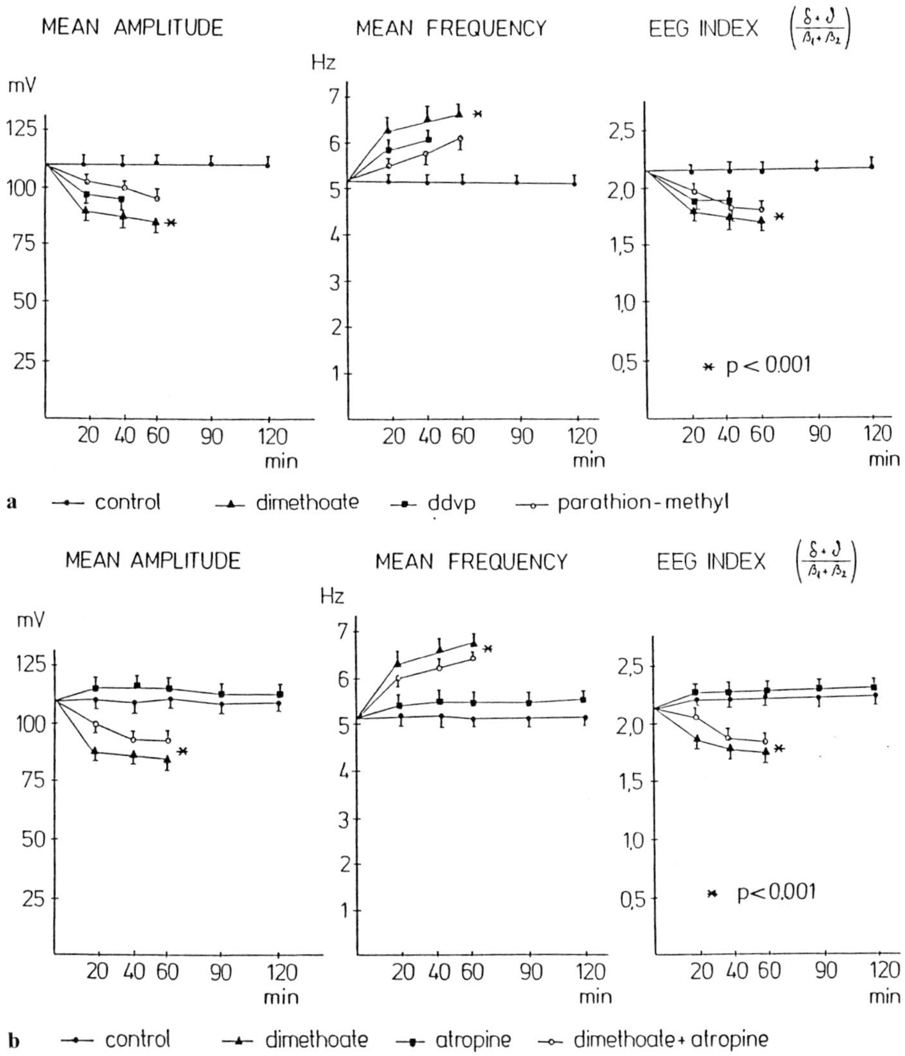

Fig. 1a, b. **a)** Effect of organophosphates on the EEG; **b)** Effect of dimethoate and atropine on the EEG

tioning of the central nervous system. The significant decrease of the mean amplitude and EEG index and the increase of the frequency represented the elevated activity of the brain. It should also be noted that the antidote, atropine, could not influence the toxic effect of the compounds.

There were significant changes of the somatosensory, visual, and auditory evoked potentials. Although the divergencies of the various evoked potentials were different it was noted that the functioning of the investigated tracts was considerably damaged by the compounds and the antidote could not reduce this toxic effect.

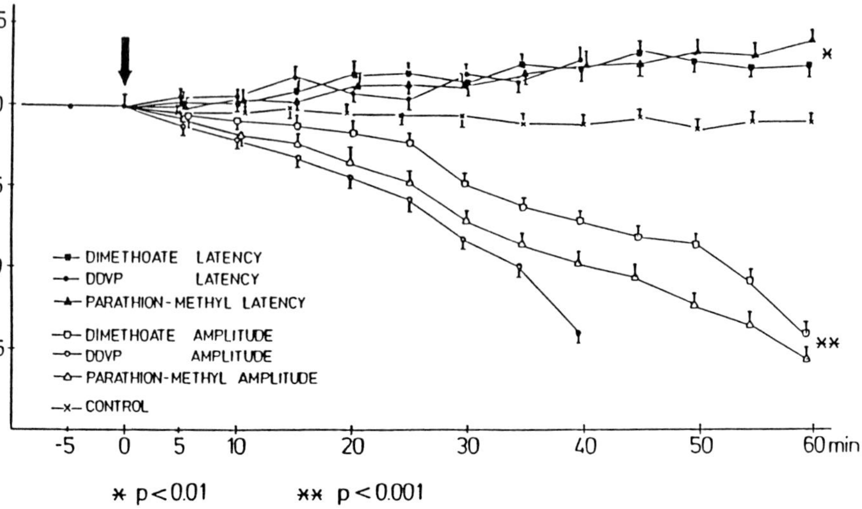

a

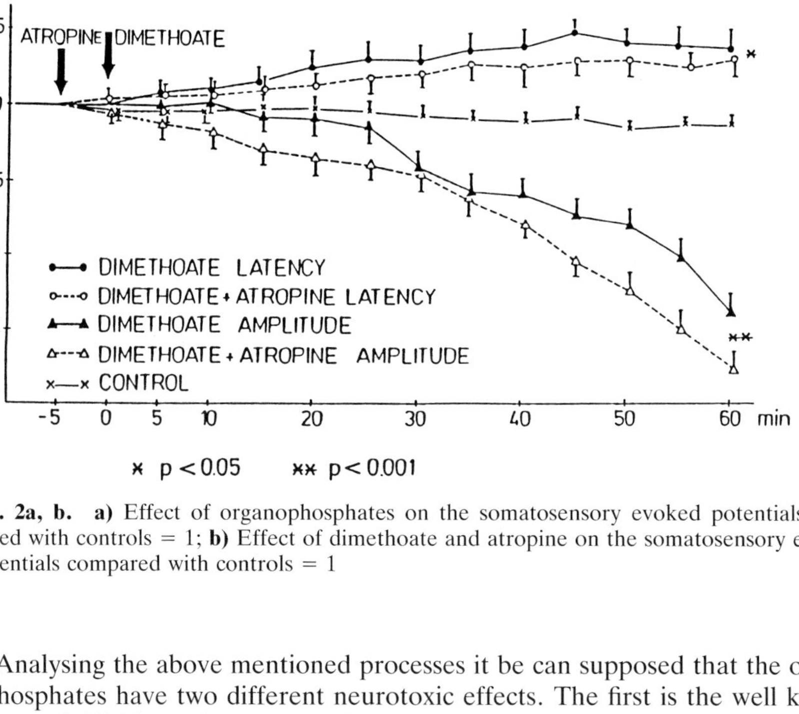

b

Fig. 2a, b. **a)** Effect of organophosphates on the somatosensory evoked potentials compared with controls = 1; **b)** Effect of dimethoate and atropine on the somatosensory evoked potentials compared with controls = 1

Analysing the above mentioned processes it be can supposed that the organophosphates have two different neurotoxic effects. The first is the well known enzyme inhibition. The second might be a direct effect, since there are several changes which could not be explained solely as a result of the enzyme inhibition and also important was that the changes were not prevented by the large dose of antidote. If this is true the present concept of the mechanism and the treatment of the intoxication of the organophosphorous compounds should be changed.

References

Dési I (1983) Neurological investigation of pesticides in animal experiments. Neurobehav Toxicol Teratol 5:503–515

Hayes WJ (1982) Toxicology of pesticides. Williams and Wilkins, Baltimore

Hyde KM, Crandall JC, Kortman KE, McCoy WK (1978) EEG, ECG and respiratory response to acute insecticide exposure. Bull Environm Contam Toxicol 19:47–55

Klaassen CD, Amdur MO, Doull J (1986) Toxicology. The basic science of poisons. Macmillan Publishing Company, New York

Nagymajtényi L, Dési I, Lorencz R (1988) Neurophysiological markers as early signs of organophosphate neurotoxicity. Neurotox Terat 10:429–434

Meinecke KH, Oettel H (1966) Mikromethode zur Bestimmung der Acetyl-cholinesterase-Aktivität in Erythrozyten und Plasma von Mensch und Tier. Arch Toxicol 21:321–332

Solima SA, Farmer J, Curley A (1982) Is delayed neuropathy a property of all organophosphorous compounds? Toxicology 23:267–279

Recent Developments in Toxicology:
Trends, Methods and Problems
Arch. Toxicol., Suppl. 14, 38–40 (1991)
© Springer-Verlag 1991

The Effects of n-Hexane and its Metabolites on Erythrocyte and Synaptosome Membrane Acetylcholinesterase *in Vitro*

S. RANTANEN, S. HYPPÖNEN, I. AHONEN, and H. TÄHTI

Department of Biomedical Sciences, University of Tampere, P.O. Box 607, Tampere, and Regional Institute of Occupational Health in Tampere, Finland

Introduction

Organic solvents may exert their anaesthetic effect on central nervous tissue by affecting the cell membrane integral enzymes, which regulate the ion transport through the membrane (Korpela and Tähti 1986). In the present study, the effect of n-hexane and its metabolites 2,5-hexanedione, 2-hexanol and 2-hexanone on membrane acetylcholinesterase (AChE) were studied.

Material and Methods

Human peripheral erythrocytes and male rat brain synaptosomes (Sprague Dawley, Department of Biomedical Sciences, University of Tampere) were used as membrane models (Korpela and Tähti 1988). The enzyme activities were determined after 30 min incubation of membrane preparations in incubation admixtures with the following added concentrations of solvents (p.a. Merck): 3 mM, 6 mM, 9 mM, 20 mM and 30 mM at 37°C. The enzyme activities were determined by the method of Ellman et al 1961, and expressed as per cents of control activity. All measurements were made as triplicates.

Results and Discussion

The AChE activity was affected by n-hexane more in erythrocyte than in synaptosome preparations (Fig. 1), which confirms the previous results on aromatic hydrocarbons (Korpela and Tähti 1988), and is due to the different compositions of these membrane models. 2-Hexanone and 2,5-hexanedione, the neurotoxic metabolites of n-hexane inhibited AChE in synaptosomal mem-

---

Abbreviations: AChE = acetylcholinesterase

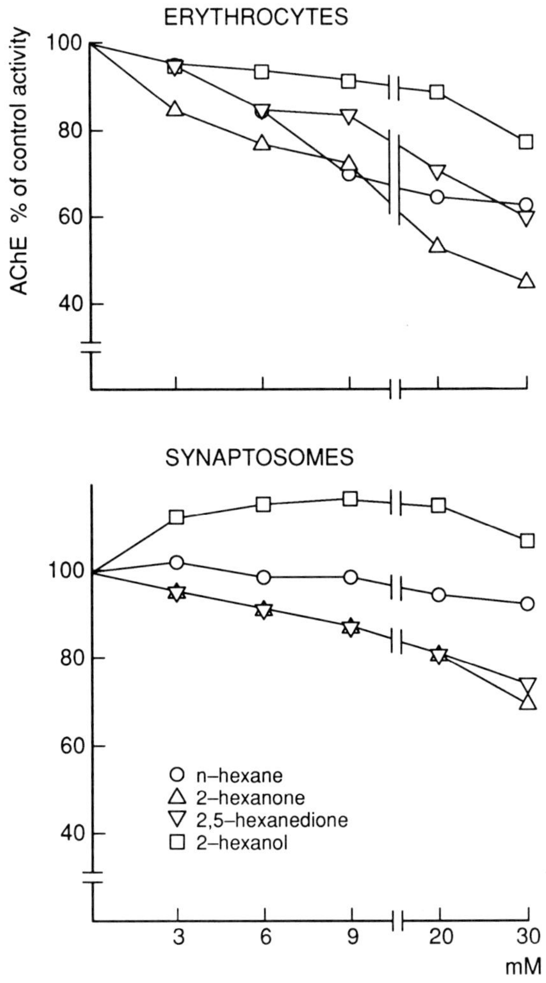

Fig. 1. The effects of n-hexane and its metabolites 2-hexanone, 2,5-hexanedione and 2-hexanol on human erythrocyte and rat synaptosome membrane AChE

brane more than in erythrocyte membrane. 2-hexanol, however, caused a slight enzyme activation in synaptosomal membrane in accordance with previous results on alcohols (Korpela and Tähti 1986). Despite the slight differences in the effects of n-hexane and its metabolites on synaptosome and erythrocyte membranes, the peripheral erythrocyte membranes may be used when the neurotoxicity of n-hexane and its metabolites are evaluated.

Acknowledgements. This study was financially supported by the Finnish Ministry of Agriculture and Forestry.

References

Ellman GL, Courtney KD, Andres V, Featherstone RM (1961) A new and rapid colorimetric determination of acetylcholinesterase activity. Biochem Pharmacol 7:88–95

Korpela M, Tähti H (1986) The effect of selected organic solvents on intact human red cell membrane acetylcholinesterase in vitro. Toxicol Appl Pharmacol 85:257–262

Korpela M, Tähti H (1988) Human and rat erythrocyte membrane as a model for studying the effects of organic solvents on membrane-bound acetylcholinesterase *in vitro*. In Vitro Toxicol 2:135–137

New Aspects and Methods in Toxicopathology

Recent Developments in Toxicology:
Trends, Methods and Problems
Arch. Toxicol., Suppl. 14, 43–49 (1991)
© Springer-Verlag 1991

Assessment of Toxicopathological Effects in Ageing Laboratory Rodents

G. J. Krinke, L. P. A. Perrin, and R. Hess

Ciba-Geigy AG, Basel, Switzerland

In the ageing laboratory rodents used for the testing safety of chemicals, toxicopathological effects which may occur, must be assessed against the background of natural ageing changes. The following issues are of particular concern: normal "age-related" data, consistency and relevance of pathology data, detection of subtle changes, age-related susceptibility to toxicants.

Normal "Age-Related" Data

During life the functional and morphological properties of organs and tissues undergo natural, age-related changes. Consequently, all data acquired from laboratory animals reflect their actual age. As with advancing age the frequency of non-neoplastic and proliferative lesions increases, the data acquired from aged animals may be biased by the presence of spontaneous disease. Therefore, it is important to discriminate natural ageing changes, expected to occur with little variation in most individuals of comparable age, from age-related disease affecting some, but not all individuals.

Examples of such a situation are afforded by the assessment of spleen pathology in two recent long-term experiments. In one of these studies, carried out on mice, the statistical analysis indicated in males a significant, dose-related decrease of spleen weight. However, macroscopic observation indicated that enlarged spleens occurred more frequently in the control animals.

Histopathology showed that hemoblastic lesions, such as malignant lymphoma, leukemia and mast cell tumors accounted for most of the instances of splenic enlargement. Therefore, the individuals affected by such diseases were excluded from the data used for calculation of the mean weight of spleen: this correction resulted in quite a flat distribution of the mean weight of spleen among the groups, indicating that the original values in the treated animals were closer to the "normal" values, than those observed in the untreated control animals (Fig. 1). Obviously, the spontaneous occurrence of hemoblastic diseases was supressed in male animals by exposure to the test agent in this experiment.

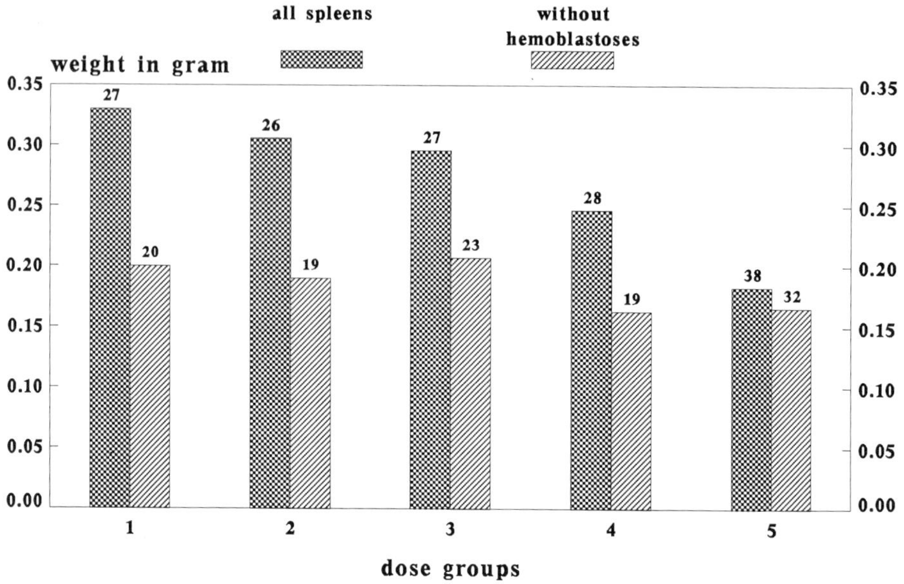

Fig. 1. Diagram of mean weight of the spleen in five groups of male mice used in a long-term carcinogenicity study. The uncorrected values show a dose-related decrease (group 1 = control, group 5 = high dose). After correction by removal the values of spleens affected by hemoblastic diseases, this trend is no more apparent. The number of available individual data is indicated above each column

In the other study, which was a 2-year study in rats, the hematology data indicated a dose-related decrease in number of erythrocytes as well as a decline of hematocrit values. Body weight, spleen-to-brain weight ratio, and frequency of extramedullary hematopoiesis in the spleen were decreased in a dose-related manner. Concomitantly the exposure to the test agent resulted in the dose-related prolongation of survival, associated with a decrease in frequency of "masses" occuring on the body surface. It was therefore of interest to find out whether the decreased number of red blood cells was associated with the decreased number of superficial masses. Since in the female animals mammary tumors accounted for most of the superficial masses, the relationship, between the presence of extramedullary hematopoiesis in the spleen and the presence of mammary tumors (of any kind) in the same individuals, was analysed in groups of control female animals from different long-term studies. The analysis of the association of mammary tumors with extramedullary hematopoiesis in spleen showed in six of seven instances more frequent hematopoiesis in tumor-bearing animals than in the animals without mammary tumors (Fig. 2).

Therefore, it was presumed that in this study a certain decrease in hematopoietic activity could be attributed to the decline of neoplastic lesions in the treated animals.

Deep understanding of the pathology of animals subjected to morphological examination is an indispensable prerequisite for the overall evaluation of

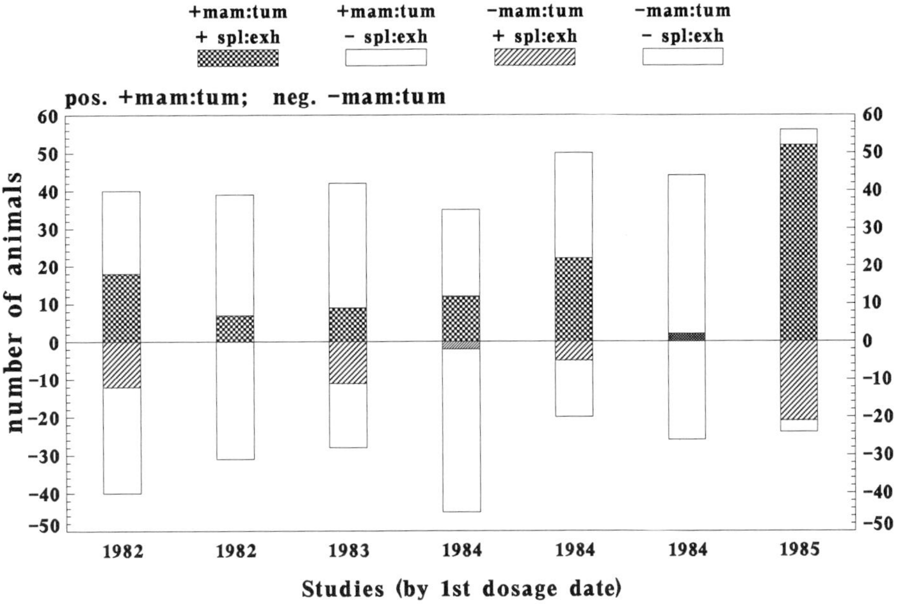

Fig. 2. Diagram of distribution of splenic extramedullary hematopoiesis in relationship to mammary tumors in control female rats from 7 different long-term carcinogenicity experiments. The hematopoiesis appears to be more frequent in the tumor-bearing animals. In more recent studies, the spontaneous tumor rate is increased

experiments, in particular for appropriate interpretation of statistical results (Haseman 1990). Special attention should be paid to the possibility that treated animals may be in better health than the respective control groups.

Consistency and Relevance of Pathology Data

The nomenclature of rodent pathology has been developed for the purpose of extrapolation of the findings in rodents to other species, in particular to humans. This utilitarian approach resulted in the application of terms used in human pathology to rodents even in instances where it may not be appropriate. So, for instance, proliferative lesions in the rat brain may be roughly differentiated in 3 categories: diffuse, circumscribed and meningeal. Most of these lesions hardly compare with the well differentiated human tumors such as astrocytomas, oligodendrogliomas, ependymomas or meningiomas. In certain rat strains, practically all meningeal tumors are granular cell tumors, but most pathologists would not dare to use the term "granular cell meningioma" because it is not established in the terminology of other species (Russel and Rubinstein 1989).

Another example is provided by the adrenal medullary proliferative lesions in the rat very frequently consisting of small fast growing cells, exhibiting invasive growth and producing distant metastases, so that they resemble sympaticoblastomas rather than pheochromocytomas. The latter term, however, is preferred by many pathologists since it pertains to the condition expected in other species. The example of rat adrenal medullary proliferative lesions also demonstrates the futility of attempting to discriminate "neoplastic" from "hyperplastic" proliferative lesions: the difference between focal hyperplasia, benign and malignant tumor is arbitrary because there is obviously a continuous progression from initial to advanced lesions.

The nomenclature used in rodent pathology suffers from the lack of unanimity especially when the lesions are specific to rodents and virtually unknown in other species. For example a proliferative cardiac lesion in rats has been named using a number of different, partly contradictory terms, such as endocardial thickening, neurinoma, or Anitschkow cell sarcoma. The histogenesis of this lesion is uncertain, and its biological behavior is in doubt (Landes et al 1988).

Toxicological pathologists have become increasingly aware of the necessity to consistently use standard pathology terms. The Society of Toxicological Pathologists in the USA has initiated an endeavour to develop a widely accepted standard nomenclature. Concurrently, a Pathology Registry Group, under the auspices of the Fraunhofer Institute of Toxicology and Aerosol Research in Hanover, FRG, is developing a systematized nomenclature of proliferative lesions occurring in laboratory rats (Mohr et al 1990). The use of standard terminology enables the analysis of the character and frequency of pathological changes occurring in ageing control animals in long-term studies from various laboratories. By that means, a spectrum of age-related lesions characteristic for control animals of different breeds can be established. It is hardly conceivable, however, that the reporting of "normal" age-related pathology can be dispensed with by using a data bank, as these data may undergo changes in the course of time, even in one laboratory using the same animal strain. For instance, the data presented in Figure 2 indicate that the number of control animals bearing mammary tumors appears to increase in the more recent studies.

Certain age-related changes in rodents seem to be devoid of relevance for humans, either because the affected organs are not present in humans forestomach, Harderian gland) or because the changes, such as harderization of the lacrimal gland, are particular to rodents. This change of the rat lacrimal/exorbital gland is age-related and sex-dependent. A low degree of harderization is present in young rats of both sexes; with advancing age its extent increases in males but not in female animals and this increase may be reversed by castration. The reason for harderization of rat lacrimal gland is unknown; it bears no relevance to humans (Krinke and Krinke 1990).

Detection of Subtle Changes

The limited resolving power of routine histopathological examination carried out on haematoxylin and eosin stained paraffin sections does not allow detection of subtle changes at the cellular and subcellular level. Owing to the frequent occurrence of neuropsychiatric disorders in ageing man, the age-

Table 1. Age-related changes described in the rat brain

Atrophy of brain
Loss of synaptic contacts
Atrophy of neurons
Intracellular inclusions
Deposition of heavy metals
Deposition of lipofuscin
Loss of neurons
Demyelination
Hypertrophy of astrocytes
Activation of microglia
Decline of vascularization
Spongiform encephalopathy

related pathology of the rat nervous system is of particular interest. For methodological reasons quantitative changes, such as loss of synaptic contacts or loss of neurons have been neglected. Also other age-related changes, such as deposition of lipofuscin, or demyelination are rarely detected in long-term toxicological experiments (Table 1).

Obviously, this is due not only to the technical confines, but also to the fact that during the first two years of life the rat brain seems to be permanently growing and shows little signs of atrophy other than local compression by expansive intracranial lesions. The continuous growth of rat brain is reflected by the steady increase in brain weight (Fig. 3).

While the design of long-term toxicological experiments in rodents is satisfactory with regard to their main purpose, i.e. bioassay of carcinogenicity, they are unsuitable for the assessment of age-associated pathology of the brain (Krinke and Eisenbrandt 1990).

Susceptibility to Toxicants

Despite the conflicting results of numerous investigations dealing with differences between young and old individuals, there is sufficient evidence for increased susceptibility to intoxication in the aged rodent, at least for certain chemical agents. Old rodents are less able than young animals to metabolize xenobiotics; they show a decline of the level and the inducibility of the renal and hepatic microsomal mixed function oxidase system, involved in phase I metabolism. The capacity for conjugation of phase I metabolites in phase II appears to be decreased as well. Decreased hepatic or renal function in ageing rodents may impair the excretion of chemical agents and may contribute to their increased toxicity. In the ageing rats the presence of chronic progressive nephropathy is a common cause of renal dysfunction.

Slow elimination rate is known to account for species differences in susceptibility to intoxication. For example chickens, that are more sensitive to delayed neurotoxicity of organophosphorous compounds than rats, eliminate tri-o-cresyl phosphate more slowly than rats (Abou-Donia et al 1990). Unfortunately

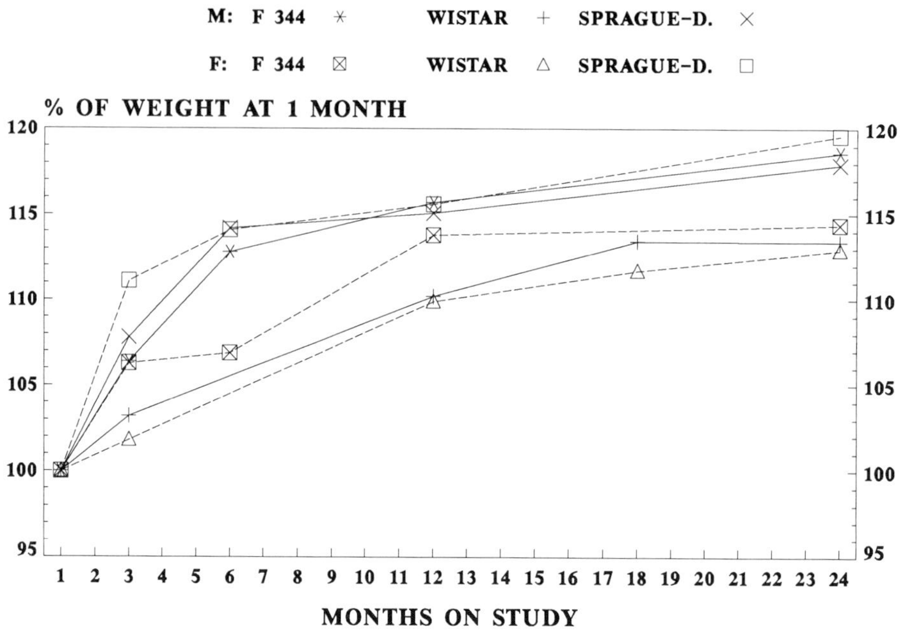

Fig. 3. Diagram of the mean brain weight in control groups of male and female rats of three different strains in relationship to animal age. The initial age after one month on study (about 2 months of virtual age) was taken as 100% to enable this comparison. This diagram includes data kindly provided by Dr. D. L. Eisenbrandt, and Dr. R. B. Beems

the elimination rate was not compared in young and old chickens: a faster elimination could be expected in young chickens, that are practically insensitive to delayed neurotoxicity. Another example is afforded by pyridoxine which is more toxic to dogs than rats and is eliminated more slowly from dogs than rats.

In a recent study comparing general toxicity of doxorubicin in young and old rats, the higher toxicity in old rats was tentatively attributed to decreased urinary elimination of this agent in old animals. Ageing rats also exhibit an increased sensitivity to acetaminophen-induced nephrotoxicity.

Neurotoxicity of 6-aminonicotinamide shows a dramatic change with age in the distribution and severity of lesions. While in the young animals the lesion affects the phylogenetically early developed structures such as the spinal gray matter and brain stem with limbic structures, in the old rats the phylogenetically later developed cerebral cortex is involved. The mechanism controlling increased susceptibility to the effects of 6-aminonicotinamide is unknown: it might be based on age-dependent changes of enzymatic activities and might be involved in presenile and senile human neurodegenerative disease. The spongiform encephalopathy induced with 6-aminonicotinamide in old rats resembles human spongiform lesions such as Creutzfeld-Jacob disease, as well as spongiform encephalopathy known as "mad cow" disease in cattle and scrapie in sheep. Thus this disease complex may reflect a common underlying metabolic defect (Horita et al 1980). Spongiform encephalopathy can be experimentally induced in rodents infected with a murine neurotropic retrovirus: its pathogenesis is explained by abortive replication of the virus in neurons. It has been proposed, that retroviral etiology may be instrumental in such human degenerative diseases as Alzheimer's and Creutzfeld-Jacob (Sharpe et al 1990).

Spongiform encephalopathy of yet unknown etiology occurs in aged laboratory mice and rats and sometimes distributes in treated animals in a dose-related manner. Genetic spongiform encephalopathy occurring in "zitter" rats shows a very similar morphology and age-dependent progression (Gomi et al 1990).

The problem of toxicity related to ageing certainly will deserve the increased attention of the scientific community in the years to come.

References

Abou-Donia MB, Suwita E, Nomeir AA (1990) Absorption, distribution, and elimination of a single oral dose of (^{14}C)tri-o-cresyl phosphate in hens. Toxicology 61:13–25

Gomi H, Inui K, Taniguchi H, Yoshikawa Y, Yamanouchi K (1990) Edematous changes in the central nervous system of zitter rats with genetic spongiform encephalopathy. J Neuropath Exp Neurol 49:250–259

Haseman JK (1990) Use of statistical decision rules for evaluating laboratory animal carcinogenicity studies. Fundamental and Applied Toxicology 14:637–648

Horita N, Ishii T, Izumiyama Y (1980) Ultrastructure of 6-aminonicotinamide (6-AN)-induced lesions in the central nervous system of rats. II. Alterations of the nervous susceptibility with aging. Acta Neuropathol (Berl) 49:19–27

Krinke GJ, Krinke AL (1990) Changes with age in the rat lacrimal gland. In: Capen CC, Dungworth D, Mohr U (eds) ILSI Monograph on Pathology of Aging Animals, Volume 1: Rat, in preparation

Krinke GJ, Eisenbrandt DL (1990) Changes with age in the rat brain. In: Capen CC, Dungworth D, Mohr U (eds) ILSI Monograph on Pathology of Aging Animals, Volume 1: Rat, in preparation

Landes CH, Ruefenacht HJ, Naylor DC, Krinke GJ (1988) Rat endomyocardial disease: a neural origin? Exp Pathol 34:65–69

Mohr U et al (1990) Tumor registry data base. Suggestions for a systematized nomenclature of pre-neoplastic and neoplastic lesions in rats. Exp Pathol 38:1–18

Russel DS, Rubinstein LJ (1989) Pathology of tumours of the nervous system. Edward Arnold, London, Melbourne, Auckland, 1012 pp

Sharpe AH, Hunter JJ, Chassler P, Jaenisch R (1990) Role of abortive retroviral infection of neurons in spongiform CNS degeneration. Nature 346:181–183

Recent Developments in Toxicology:
Trends, Methods and Problems
Arch. Toxicol., Suppl. 14, 50–54 (1991)
© Springer-Verlag 1991

New Microtechniques in Toxicopathology

K. Augsten, and J. Güttner

Institute of Microbiology and Experimental Therapy, Jena, Beutenbergstraße 11, O-6900 Jena, Germany

The contact with or uptake of xenobiotics and/or infectious agents by men and animals may cause alterations of tissues, organs or the whole organism. On the other hand, living material requires a defined content of trace elements. Changes of these often unknown, optimal contents of trace elements are connected with deviations from the normal state.

Commonly, the morphologic or histologic evidence of the influence of foreign materials or trace elements on the organism are evaluated routinely by means of light microscopy. Although these changes can be dramatic, they may uniformly consist of the same, non-specific exudative, necrotising, granulomatous, or fibrotic reactions, in spite of different etiology. Therefore, the value of light microscopy for the reliable etiological identification is often limited. For these reasons methods are needed which allow to detect ions of the foreign material *in situ* or trace element(s) within a given biological compartment. To detect and localize these materials or elements in certain cellular or subcellular constituents, microanalytic methods are used. The prefix "micro" has a double meaning, firstly to characterize the volume, in the order of some μm^3 or parts of them, and secondly the amount or content of a substance or element in μg or less. The purpose of this paper is to characterize and evaluate physically essential, destructive and nondestructive micromethods and to give some examples to demonstrate the usefulness of the scanning electron microscope connected with an energy dispersive spectrometer.

The commercial microanalytical techniques and their essential characteristics to evaluate the content of elements in microvolumes are described in Tables 1 and 2. Optical emission spectrometry (FAES: Flame Atomic Emission Spectroscopy, FIAS: Flow Injection Atomic Spectroscopy, ICP-AES: Inductively Coupled Plasma Atomic Emission Spectroscopy) uses spectra emitted by free atoms or ions generated within a source such as flame, spark and arc or inductively coupled plasma. The monochromator isolates light from various wavelengths corresponding to emission lines of the different elements in the analyte, and differentiates this light from the background radiation.

Table 1. Comparison of microanalytical techniques; non-microscopical; a) destructive; b) non-destructive

Technique	Mass limit	Concentration limit	Elemental range	Sample Volume	Sample State
Emission (FAES; FIAS)[a)]	10^{-9}–10^{-12} g	10 ppm	all	0.4 ml	fluid; suspension
Absorption (FAAS; GF-AAS)[a)]	10^{-9}–10^{-13} g	<0.1 ppm	most metals	0.05 ml	fluid; solid
Inductively coupled plasma (ICP-AES; ICP-MS)[a)]	10^{-9}–10^{-11} g	<10 ppb	75 elements	0.1–10 ml	fluid; suspension
Ion-selective electrodes[b)]	$<10^{-10}$ g	10^{-5}–10^{-10} M	43 elements (free ions)	5–20 μl (intra-cellular)	fluid
Neutron activation analysis[b)]	10^{-7}–10^{-15} g	<0.1 ppm	$Z > 5$	cellular, tissue	solid; fluid

FAES – Flame Atomic Emission Spectroscopy; FIAS – Flow Injection Atomic Spectroscopy; FAAS – Flame Atomic Absorption Spectroscopy; GF-AAS – Graphite Furnace Atomic Absorption Spectroscopy; ICP – AES: Inductively Coupled Plasma – Atomic Emission Spectroscopy; ICP – MS: Mass Spectroscopy

Atomic absorption spectroscopy (FAAS: Flame Atomic Absorption Spectroscopy, GF-AAS: Graphite Furnace-Atomic Absorption Spectroscopy) based on the absorption of the emission line of the hollow cathode lamp through an atomic vapour in a flame or graphite furnace tube. Both methods are commonly used techniques in the analysis of trace elements and of small volumes, however, they are again destructive. The following methods in Table 1 (ion selective electrodes, neutron activation analysis) need special sensors and also electrical equipments or a neutron source.

But the understanding of the role of elements or ions in processes such as cellular injury or influence on the growth potential requires the knowledge of total cell elements and the content of elements/ions in the cytoplasm. Besides, the comprehension needs the identification of the sites of release, uptake and binding of those ions/elements within the cells and the correlation with pathological lesions such as changes in cell metabolism or pathological findings. Correlative structural-functional studies, basing on the static probe method of X-Ray Microanalysis (XRMA or EPMA), Electron Energy Loss Spectroscopy (EELS), Particle Induced X-ray Emission (PIXE), Laser Microprobe Mass Analysis (LAMMA), Light- or Electron (LM- or EM) histochemistry combined with biochemical and electrophysiological techniques, have focussed on the role of, e.g., mitochondria, sarcoplasmic reticulum, vacuoles, lysosomes, or mucous granules.

The applications of EPMA, XRMA, and LAMMA are varied in diagnostic pathology, physiology, and anatomy. In the pathology the usefulness of these

Table 2. Comparison of microanalytical techniques; microscopical; a) destructive; b) non-destructive

Technique	Mass limit (g)	Concentration limit (ppm)	Elemental range	Spatial resolution
X-ray microanalysis (XRMA, EPMA)[b)]	10^{-19}	50	Z > 4	40 nm
Electron energy loss spectroscopy (EELS)[b)]	10^{-21}	0.01	Z > 3	0.5 nm
Auger electrons[b)]	10^{-11}–10^{-16}	1000	Z > 4	0.1 μm
Secondary ion mass spectroscopy (SIMS)[a)]	10^{-20}	0.01	all	0.5 μm
Particle induced X-ray emission (PIXE)	10^{-16}	1	Z > 4	2–10 μm
X-ray fluorescence[b)]	10^{-11}	30	Z > 4	50 μm
X-ray-absorption[b)]	10^{-14}	100	Z > 12	10 μm
Cathodoluminescence (CL)[b)]	10^{-11}	20	compounds	90 nm
Autoradiography[b)]	10^{-10}–10^{-15}	0.05	all	0.1 μm
Laser microprobe mass analysis (LAMMA)[a)]	10^{-20}	0.1	all	0.5 μm
EM histochemistry	10^{-18}–10^{-21}	0.01	compounds	ultrastructural
LM histochemistry	10^{-12}–10^{-14}	0.1	compounds	0.25 μm

techniques can be roughly grouped into several broad categories on analyses of xenobiotic particles, metal deposits, endogeneous particulates, soluble ions, and materials in forensic science (Baker et al 1985). In the following, the performing of EPMA together with SE-imaging on liver necrosis is described more in detail (Augsten and Güttner 1990). Liver necrosis of mice that had suffered from Tyzzer's disease, infectious ectromelia as well as necrosis of unknown origin were investigated. The approach to the analysis is to study sections carefully by LM prior to EPMA. That means, the current "gold standard" in morphological pathology around the world is the H&E stained paraffin section. Therefore the liver was processed for routine histology by fixation in 4% neutralized formalin for two days and embedded in paraffin; subsequently following serial sections, 4–5 μm thick, were mounted on glass slides and Al-foil or C-discs. After deparaffination the sections were, respectively, stained with H&E or sputtered with a thin Ni-layer. Al-foils were than mounted onto carbon planchets and specimens were observed and analyzed at a magnification of 2000 x, an accelerating voltage of 20 kV, a 200 seconds effective collection time and a beam current of 0.2 nA with a JEOL JSM 35 scanning electron microscope (Tokyo, Japan) attached with an energy dispersive spectroscopy capability (KEVEX, USA). Five animals per group were used and spectra from each animal from five randomly chosen necrotic areas were compared with six spectra generated from neighbouring areas without light-microscopically changed structures in the same slide. The background and line intensity of the spectra were calculated using a Quantex computer program from KEVEX. The ratio of the peak intensity (P) to the background intensity (B) of the same energy (P/B) or P for Si, S, Cl and Ca was used for comparison of preserved and necrotic tissue (Statham and Pawley 1978).

The EPMA of impaired liver shows more or less characteristic patterns of peak intensities, depending on aetiology and time. The Ca-peak at 3.69 keV consistently increased in foci of acute necrosis of various origins and dropped to normal values when the necrosis was being replaced by granulation tissue. With Tyzzer's disease the intensity of the P-peak (12.4 counts/s) and S-peak (10.6 counts/s), respectively, decreased in both the acute and subacute phase of the disease. In ectromelia liver the P-peak was in the necrotic and preserved area (nearly) equal (3–4 counts/s) but the S-peak dropped from 5.8 counts/s to 4.4 counts/s.

Additionally, in both diseases in mice, a Si-peak (4.3 counts/s) was detected which was not present in necrosis of unknown origins (2.3 counts/s). If the necrotic area became too large the spectra of histologically preserved liver tissue of ectromelia disease was equal to the spectra of necrosis, suggesting that the preserved tissue appeared also to be damaged. The alterations found in the composition of histologically equivalent necrosis may be a first step to characterize the necrosis in the course of pathomechanism in more detail. To our knowledge, there have not been descriptions of necrosis from Tyzzer's disease and ectromelia, respectively, which have detected Si within the damaged tissue. The reason for this results is still unknown.

The summarized microprobe techniques and their results often have diagnostic, therapeutic and/or legal implications. They are no longer solely research tools and supply new informations and increase knowledge about fundamental processes within the cells and tissues.

References

Abraham JL (1980) Biomedical microanalysis – Putting it to work now in diagnostic pathology. Scanning Electron Microscopy IV:171–178

Augsten K, Stein G (1988) Scanning electron microscopy and x-ray microanalysis investigation of aluminium deposition in samples from hemodyalized patients. Trace Elem. Med. 5:55–59

Augsten K, Güttner J (1990) Use of x-ray microanalysis to characterize liver necrosis of mice, rats and guinea-pig. Proc XII. Internat. Congress Electron Microscopy, Seattle (USA), 350–351

Baker D, Kupke KD, Ingram P, Roggli VL, Shelburne JD (1985) Microprobe analysis in human pathology. Scanning Electron Microscopy II:659–680

Boyde A, Reid SA (1983) Simple collector for cathodoluminscence in the SEM made from aluminium foil. J Microscopy 132:239–242

Burns MS (1982) Applications of SIMS in biological research: a review. J Microscopy 127:237–258

Chandler JA (1987) X-ray microanalysis in the electron microscope Vol. 5, Part II of the Series "Practical Methods in Electron Microscopy", 317–547; Ed by A. M. Glauert

Chong NS, Norton NL, Anderson JL (1990) Multielement trace metal determination by electrodeposition, scanning electron microscopic x-ray fluorescence, and inductively coupled plasma mass spectrometry. Analytical Chemistry 62:1043–1050

Dymott TC (1989) Spectrometer design for inductively coupled plasma. Optical emission spectrometry. Intern. Laboratory, March, 26–37

Echlin P (1977) Microanalysis in biology and medicine. Nature 270:102–103
Egerton RF, Egerton M (1983) An electron energy-loss bibliography. Scanning Electron Microscopy I:119–142
Galle P, Berry JP, Escaig F (1983) Secondary ion mass microanalysis: Applications in biology. Scanning Electron Microscopy II:827–829
Hantsche H (1989) Comparison of basic principles of the surface-specific analytical methods: AES/SAM, ESCA (XPS), SIMS and ISS with x-ray microanalysis, and some applications in research and industry. Scanning 11:257–280
Janssen AP, Venables JA (1979) Scanning Auger microscopy – an introduction for biologists. Scanning Electron Microscopy II:259–278
Martin BW (1980) A microprobe based on particle-induced x-ray emission (PIXE) – a powerful tool for microanalyses in minerals and cells. Scanning Electron Microscopy I:419–438
Schmidt PF (1984) Localization of trace elements with the laser microprobe mass analyzer (LAMMA). Trace Elem Med 1:13–20
Shelburne JD, Estrada H, Hale M, Ingram P, Tucker JA (1989) Correlative microscopic and microprobe analysis in pathology. Proc. 47 Annual Meeting of the Electron Microscopy Society of America, 900–901
Statham P, Pawley J (1978) A new method for particle x-ray microanalysis based on peak background measurements. Scanning Electron Microscopy I:469–478
Zielkowski R, Bächmann K (1978) Instrumentelle Multianalyse in biologischen Matrices: Ein Vergleich von NAA, RFA und AAS. Fresenius Z. Anal. Chem. 290:143–144

Recent Developments in Toxicology:
Trends, Methods and Problems
Arch. Toxicol., Suppl. 14, 55–57 (1991)
© Springer-Verlag 1991

Tumor Registry Data Base: Systematized Nomenclature of Preneoplastic and Neoplastic Lesions in Rat Gonads

S. Qureshi, and R. A. Ettlin

Sandoz Pharma Ltd., Drug Safety Assessment/Toxicology, CH-4002 Basel, Switzerland

Introduction

In order to improve the evaluation and interpretation of results from carcinogenicity studies with rodents, pathologists from one research institute and from several chemical and pharmaceutical companies in West Germany and Switzerland are collaborating to set up a data base for preneoplastic and neoplastic lesions. For this purpose the nomenclature applied to various lesions was systematized. This paper illustrates some of the criteria employed by giving specific examples of lesions of the male and female rat gonads.

Methods

Groups of several toxicopathologists were assigned the task of describing the light microscopic features of proliferative lesions of various organs and organ systems and of defining diagnostic criteria. Further aspects covered were biological behavior of the tumors, histogenesis and differential diagnosis. Extensive literature searches were conducted and the key references listed. Special attention was paid to define rules for the differentiation of hyperplasias from benign tumors and benign tumors from their malignant counterparts. When appropriate, macroscopic features, electron-microscopic characteristics and (immuno)histochemical staining properties were also covered. The dossiers were reviewed by other pathologists involved in the Registry Data Bank Project and from academia. Publication of an illustrated reference manual is planned.

Parallel to these activities, histopathology seminars and panel sessions on histological slides were organized in order to test the diagnostic criteria for their practical usefulness. The experience gained was incorporated into the manuscripts.

Table 1. Proposed nomenclature for proliferative lesions of the male and female rat gonad

TESTIS	**OVARY**
teratoma, benign	teratoma, benign
teratoma, malignant	teratoma, malignant
	adenoma *(cystic, mucinous, papillary, serous, tubulostromal)*
	adenocarcinoma *(cystic, mucinous, papillary, serous, tubulostromal)*
TESTIS, LEYDIG CELL	**OVARY, SEX CORD STROMA**
hyperplasia	hyperplasia, sex cord-stromal *(granulosa cell-type, theca cell-type, Sertoli cell-type, mixed)*
tumor, benign	granulosa cell, tumor, benign
tumor, malignant	granulosa cell, tumor, malignant
	luteoma, benign
	luteoma, malignant
	theca cell, tumor, benign
	theca cell, tumor, malignant
TESTIS, SERTOLI CELL	
tumor, benign	Sertoli cell, tumor, benign
tumor, malignant	Sertoli cell, tumor, malignant
TESTIS, GERM CELL	**OVARY, GERM CELL**
seminoma, malignant	carcinoma, yolk sac
	choriocarcinoma

Results and Discussions

The proposed nomenclature for proliferative lesions of the male and female rat gonad are given in Table 1. Proliferative lesions of the endocrinologically active cell compartments (Leydig cells, sex cord stromal cells) appear to represent a continuous spectrum from small collections of hyperplastic cells to large tumors. The diagnostic criteria for separating hyperplasias and tumors must therefore be arbitrary. In this system, for example, proliferations up to a mass diameter of 3 seminiferous tubules are designated as Leydig cell hyperplasia, and beyond that as Leydig cell tumor. This is in contrast to the definition of the (American) Society of Toxicologic Pathologists, who chose the diameter of one seminiferous tubule to distinguish hyperplasias and tumors. The Registry Data Bank Group found that most trophic Leydig cell alterations of control animals are below the limit of 3 tubular diameters. Thus, compounds which induce a neoplastic growth are more easily picked up. The lesion size was found to be the best, albeit not perfect, criterion available for routine use. Cytometric investigations performed by the authors (Ettlin et al, submitted) have shown distinct cellular differences (nuclear size, density, shape and chromatin texture) between hyperplastic and neoplastic lesions. These changes are, however, too subtle to serve as criteria during qualitative histological evaluation.

Similarly, the term ovarian sex cord-stromal hyperplasia was introduced in this nomenclature in accordance with Montgomery and Alison (1987) and MacLachlen (1987), as such lesions, which are frequently seen in old rats, exhibit a morphological spectrum including granulosa, theca, Sertoli, luteal and Leydig cells in varying quantities. In the absence of other easily applicable criteria, the lesion size was again used to distinguish hyperplasia from neoplasia, the latter being by definition larger than 2–3 mm in diameter. The size of a lesion depends upon the section plane. However, as the assessment of the carcinogenic risk is ultimately a statistical exercise, and as the position of the section plane is, at the microscopical level, random, the size criterion is a valid one.

The creation of a common nomenclature was found to be time-consuming. However, data banks with histological diagnoses not based on a common, well-defined nomenclature and not reviewed by experienced pathologists are considered to be of very limited use. Furthermore, the group believes that the registration of chemicals and drugs is facilitated by using a well-defined and recognized terminology.

References

Ettlin RA, Qureshi S, Perentes E, Stirnimann P, Greiner B, Hauser R, Prentice DE, Oberholzer M (submitted) Morphometrical investigations of trophic Leydig cell alterations. Path Res Pract

MacLachlan NJ (1987) Ovarian disorders in domestic animals. Environ Health Perspect 73:27–33

Mohr U, Bader R, Ernst H, Ettlin R, Gembardt C, Harleman JH, Hartig F, Jahn W, Kaliner G, Karbe E, Kaufmann W, Krieg K, Krinke G, Küttler K, Landes C, Mettler F, Morawietz G, Notman J, Püschner H, Qureshi S, Reznik G, Rittinghausen S, Tuch K, Urwyler H, Weisse G, Weisse I, Zehnder J (1990) Tumor Registry Data Base: Suggestions for a systematized nomenclature of pre-neoplastic and neoplastic lesions in rats. Exp Pathol 38:1–18

Montgomery CA, Alison RH (1987) Nonneoplastic lesions of the ovary in Fischer 344 rats and B63F_1 mice. Environ Health Perspect 73:77–90

Recent Developments in Toxicology:
Trends, Methods and Problems
Arch. Toxicol., Suppl. 14, 58–62 (1991)
© Springer-Verlag 1991

Mortality and Neoplastic Lesions in Untreated Sprague-Dawley Rats

J. Kaspareit-Rittinghausen[1], and F. Deerberg[2]

[1] Squibb-von Heyden, Abt. Toxikologie, 8400 Regensburg 13, Donaustaufer Str. 378, FRG
[2] Central Institute for Laboratory Animal Breeding, 3000 Hannover, Hermann-Ehlers-Allee 57, FRG

Introduction

Data on tumour types and frequencies of rodent species obtained from longevity studies can be helpful tools for the interpretation of carcinogenicity studies, since conclusions concerning carcinogenicity of a particular substance require the knowledge of incidences of naturally occuring neoplasms of the animals under study. In the present paper a report is given on the mortality and the most important neoplastic lesions of untreated Han:SPRD rats kept in a life-span study from weaning to their natural death.

Material and methods

A total of 394 Han:SPRD rats, 200 males and 194 females, were maintained in groups of five (separated according to sex) in polycarbonate cages type No. IV (base area 1750 cm^2), on softwood bedding, from weaning until their natural death. They were kept in a barrier-type animal quarter at 22 ± 1°C room temperature, with 55 ± 5% relative humidity, 15 mm H_2O atmospheric pressure, 12:12 hours light-dark sequence (6 a.m.–6 p.m.) and an air exchange rate of 20 times per hour. All rats were fed an autoclaved (120°C for 5 min) commercial (Han:MR3, Fa. Eggersmann, Rinteln), cereal-based diet (supplemented with vitamins and minerals) ad libitum. Acidified (pH 2.5) and pasteurized (95°C for 15 s) tap water was continually available. The breeding colony is examined routinely and the rats were known to harbour no murine viral infections (Reo virus, Sendai virus, PVM, LCM, Theiler virus, Adeno-virus, Kilham, virus, Toolan's H-1 virus) and to be free of endo- and ectoparasites, protozoa as well as specific pathogenic bacteria, mycoplasma and fungi.

All rats were checked daily. A complete necropsy was performed on those animals found dead or moribund. Dead rats were not replaced by new animals.

After fixation in 10% formalin tissue samples from almost all organs and from all pathological alterations were embedded in paraffin. Tissue sections (4 μm) were stained routinely with haematoxylin-eosin. Selected tumour sections were stained with Azan according to Heidenhain and with reticulin stain according to Gomori. Representative tumour tissue was embedded in hydroxyethyl-methacrylate and 1 μm thick sections were stained with toluidine blue. Statistical calculation was performed by chi-square analysis.

Results

Mortality of Han:SPRD rats

The first female died during the 7th month and the first male during the 12th month of age. Thereafter, mortality increased age-dependently in both sexes, with only slight differences between them. The oldest female reached an age of 38 months and the oldest male an age of 39 months. The mean life-span of males was 29.6 ± 5.0 months and that of females 27.6 ± 7.2 months ($p<0.001$). The highest mortality occurred during the third year of life in both sexes.

Total tumour incidence and frequent neoplasms of Han:SPRD rats

The tumour incidences of male and female Han:SPRD rats are presented in Table 1. In 91.5% of the males and 93.5% of the females one or more

Table 1. Mortality and tumour frequency in male and female. Han:SPRD rats according to age (n/%)

Sex	Age (months)				
	1–12	13–24	25–36	37–48	Total
Mortality:					
Males	2/1.0	24/12.0	162/81.0	12/6.0	200/100
Females	4/2.1	52/26.8	129/66.5	9/14.6	194/100
Tumour frequency:					
Males	2/1.0	19/ 9.5	151/75.5	11/5.5	183/91.5
Females	2/1.0	45/23.2	125/64.4	9/4.6	181/93.3

Table 2. Frequency of multiple tumours (n/%) in tumour-bearing Han:SPRD rats

Sex	Number of tumours						
	1	2	3	4	5	6	7
Males	37/18.5	57/28.5	43/21.5	26/13.0	17/8.5	3/1.5	—
Females	39/20.1	48/24.8	47/24.2	26/13.0	16/8.2	4/2.1	1/0.5

Table 3. Types of tumours in organs frequently affected by neoplastic lesions in Han:SPRD rats

	Tumour incidence			
	n		%	
Animals examined	males 200	females 194	males 100	females 100
Type of tumour				
Mammary gland				
Fibroadenoma		109		56,2
Adenocarcinoma		14		7.2
Thyroid				
Follicular adenoma	1	2	0.5	1.0
Follicular carcinoma	2	1	1.0	0.5
C-cell adenoma	108	85	54.0	43.8
C-cell carcinoma	9	12	4.5	6.2
Pituitary gland				
Adenoma	93	81	46.5	42.0
Adenocarcinoma	13	6	6.5	3.1
Endocrine pancreas				
Islet-cell adenoma	36	16	18.0	8.2
Islet-cell carcinoma	4		2.0	
Adrenal glands				
Cortical adenoma	8	26	4.0	13.0
Cortical carcinoma	4	1	2.0	0.5
Phaeochromocytoma	46	30	23.0	15.5
Preputial gland				
Papilloma	1		0.5	
Adenoacanthoma	1		0.5	
Squamous cell carcinoma	30		15.0	
Clitoral gland				
Papilloma		2		1.0
Squamous cell carcinoma		16		8.3

neoplasms were found. Sixty-four different types of tumours were observed in males and fifty-three in females. The great majority of tumour-bearing males (73.0%) and females (72.8%) had multiple neoplasms (Table 2). Mammary gland, endocrine organs and preputial/clitoral gland were the organs most often affected by tumours. The types of tumours found in these organs are presented in Table 3.

The most common tumour of females were neoplasms of the mammary gland (63.4%). Most of them were benign fibroadenomas (56.2%), while only 7.2% were classified histologically as adenocarcinomas. Tumours of the endocrine system were frequent in animals of both sexes. In 38.3% of the animals even multiple occurrence of endocrine neoplasms was observed. The endocrine organ affected by the highest incidence of tumours was the thyroid. In 60.0% of the males and 51.5% of the females uni- or bilateral neoplasms of the thyroid gland were present. Almost all tumours originated from interfollicular calcitonin-producing C-cells of the gland. Pituitary gland tumours were found in 53.0% of the males and 45.1% of the females in the Han:SPRD stock. Less frequently adrenal glands and islet-cells of the pancreas were affected by neoplasms. Phaeochromocytomas were observed in 23% of male rats and 15.5% of females, while tumours of the adrenal gland cortex occurred in 13.5% of the females, but only in 6.0% of the males. Tumours that originated from the islet-cells of the pancreas could be found in 20.5% of male rats and 8.2% of females. The only significant sex-difference existed between the number of islet-cell tumours ($p<0.001$) of males and females. Tumours of the preputial and clitoral gland were observed with the uncommonly high incidence of 16.0% in males and 9.3% in females. The tumours of the preputial gland were classified histologically as papilloma (1), squamous cell carcinomas (30) and adenoacanthoma (1), that of the clitoral gland as papillomas (2) and squamous cell carcinomas (16).

Discussion

In the present study the mean life-span of female rats was shorter than that of males. The only cause this sex-difference can attributed to is the early removal of many females from the study in consequence of the presence of large, superficially ulcerated mammary gland tumours. Tumours were observed in more than 90% of animals of both sexes. Tumour incidences in life-span studies on SPRD rats comparable to the present study (60% in males, 95% in females; respectively 77% in males, 85% in females) were described by Muraoka et al (1977) and Löhrke et al (1982). Mammary gland neoplasms were the most common tumours in female Han:SPRD rats as well as in females of other studies on Sprague-Dawley rats. Incidences of mammary gland tumours observed previously varied between 26.0% and 59.0% in aged Sprague Dawley rats (Thompson et al 1961; Morii and Fujii 1973; Prejean et al 1973; Löhrke et al 1982). The observation of a relatively high portion of adenocarcinomas of the mammary gland in Han:SPRD rats, confirms the findings of previous studies (Prejean et al 1973; Löhrke et al 1982).

Other organs often affected by neoplasms belonged to the endocrine system. The highest incidence of neoplasms (originating nearly completely from the C-cells) was found in the thyroid. A comparable high incidence of C-cell neoplasms of the thyroid have also been observed after examination of serial sections of this organ by Thompson and Hunt (1963). They were also the most frequent endocrine tumours in the studies of Suzuki et al (1979) and Löhrke et al (1982) on aged Sprague-Dawley rats. In addition a general predisposition of SPRD rats for endocrine tumours of pituitary, adrenal glands and islet-cells can be concluded from the results of the present study and from several previous studies on this strain (Prejean et al 1973; Cohen et al 1978; Suzuki et al 1979; Anver et al 1982).

There are only few reports on preputial or clitoral gland neoplasms in rats. Tumours found included adenomas (Coleman et al 1977), cystadenomas (Reznik and Reznik-Schüller 1980) as well as adenomas and carcinomas (Goodman et al 1979; Goodman et al 1980; Solleveld et al 1984). Naturally occurring squamous cell tumours of this gland were incidental findings in rats up to now (Anver et al 1982; Goodman et al 1979). The tumours observed with the unusual high incidence of 16% in males and 9.3% in females of the Han:SPRD outbred stock, were almost completely squamous cell neoplasms. No bite wounds were observed in the peripreputial region that could be responsible for a tumour development caused by chronic inflammation. Since preputial and clitoral gland neoplasms of squamous cell origin have not been described in other Sprague-Dawley rats before, they must be considered as a strain-specific feature of the Han:SPRD stock.

References

Anver MR, Cohen BJ, Lattuada CP, Foster JS (1982) Age-associated lesions in barrier-reared male Sprague-Dawley rats: A comparison between Hap:(SD) and Crl:COBS$^{(R)}$CD$^{(R)}$(SD) stocks. Exp Aging Res 8:3–24

Cohen GL, Anver MR, Ringler DH, Adelmann RC (1978) Age-associated pathological changes in male rats. Fed Proc 37:2848–2850

Coleman GL, Barthold SW, Osbaldison OW, Foster SJ, Jonas AM (1977) Pathological changes during aging in barrier-reared Fischer 344 rats. J Geront 32:256–278

Goodman DG, Ward JM, Squire RA, Chu KC, Linhart MS (1971) Neoplastic and nonneoplastic lesions in aging F 344 rats. Toxicol and Appl Pharmacol 48:237–248

Goodman DG, Squire RA, Paxton MB, Reichhardt WD, Chu KG, Linhardt MS (1980) Neoplastic and nonneoplastic lesions in aging Osborne-Mendel rats. Toxicol and Appl Pharmacol 55:433–447

Löhrke H, Hesse B, Goerttler K (1982) Spontaneous tumours in male and female specified pathogen-free Sprague-Dawley rats (outbred stock Sut:SDT). Z Versuchstierkd 24:225–230

Morii S, Fujii T (1973) Spontaneous tumours in Sprague-Dawley rats. Exp Animals 22:127–138

Muraoka Y, Itoh M, Yamashita F, Hayashi Y (1977) Spontaneous tumours in aged SD-JCL rats. Exp Pathol 26:13–22

Prejean JD, Peckham JC, Casey AE, Gridwold DP, Weisburger EK, Weisburger JH (1973) Spontaneous tumours in Sprague-Dawley rats and Swiss mice. Cancer Res 33:2768–2773

Reznik G, Reznik-Schüller H (1980) Pathology of the clitoral and preputial glands in aging F 344 rats. Lab Anim Sci 30:845–851

Solleveld HA, Haseman JK, McConnel EE (1984) Natural history of body weight gain, survival and neoplasia in F 344 rats. J Natl Cancer Inst 72:929–940

Suzuki H, Mohr U, Kimmerle G (1979) Spontaneous endocrine tumours in Sprague-Dawley rats. J Cancer Res Clin Oncol 95:184–196

Thompson SW, Huseby RA, Fosc MA, Davis CL, Hunt RD (1961) Spontaneous tumours in the Sprague-Dawley rat. J Natl Cancer Inst 27:1037–1057

Thompson SW, Hunt RD (1963) Spontaneous tumours of the Sprague-Dawley rat: Incidence rates of some types of neoplasms as determined by serial sections versus single section techniques. Ann NY Acad Sci 108:832–845

Recent Developments in Toxicology:
Trends, Methods and Problems
Arch. Toxicol., Suppl. 14, 63–67 (1991)
© Springer-Verlag 1991

Investigations on the Sensitivity of Several Biochemical and Histological Methods after Liver Intoxication Induced by Pentachlorophenol and Carbon Tetrachloride

CH. ELLINGER, W. SEFFNER, and F. SCHILLER

Research Institute for Hygiene and Microbiology, Bad Elster, 9933, GDR

Introduction

The connection between injury to energy metabolism, lipid peroxidation and cytotoxicity may be a target in the search for sensitive parameters for detecting different types of hepatocellular damage. The objective of the present study was to define more precisely the availability of thiobarbituric acid (TBA) reactivity, a method for studying lipid peroxidation, and of polarographic determination of oxidative phosphorylation in this field. Carbon tetrachloride (CCl_4) was used as an experimental model agent to initiate lipid peroxidation after the metabolic conversion to the trichloromethyl radical (Recknagel et al 1989). Pentachlorophenol (PCP) which acts as an uncoupler of oxidative phosphorylation (Koch and Axmann 1987) was also used.

Materials and Methods

Analytical-grade pentachlorophenol was obtained from Fluka AG (Buchs). Carbon tetrachloride (analytical grade) was purchased from Laborchemie Apolda. All other chemicals were commercial products of highest available purity.

Male, approximately 3 month old rats (F1 from female WELS/Fohm and male BDIX) were maintained as described by Heinecke (1988). 54 animals were used for the acute CCl_4 and 36 for the subchronic PCP experiment. 0.16 mg/kg b.w. CCl_4 dissolved in peanut oil to 4 ml volume were injected i.p. Control 1 received oil under the same conditions and control 2 remained untreated. The rats were killed 24 hr post dosing. For subchronic PCP treatment, the PCP group received the substance in drinking water (250 mg/l) with 1% ethanol for 90 days. Controls were administered drinking water with 1% ethanol. In both experiments rats were fasted overnight and then killed by decapitation. Blood was collected, allowed to coagulate and centrifuged.

Serum was immediately used for determination of aspartate- and alanine aminotransferases (Bergmeyer and Bernt 1970), leucine nitranilidase (Appel 1970) and γ-glutamyl transpeptidase (Szasz 1970) by standard enzymatic procedures. Livers were immediately removed, washed (Sucrose, 0.25 mol/l) and weighed. Pieces from the lateral median lobe of each liver were fixed in 4% formalin and processed by usual histological techniques. Slices of 5 μm were stained with hematoxylin-eosin. Frozen slices were stained with Sudan III for lipids. Mitochondria were immediately isolated as described by Steinbrecht and Kunz (1970). Respiratory activity of mitochondrial suspensions was measured as oxygen consumption recorded polarographically by a Clark electrode operated at 25°C in a closed and magnetically stirred chamber with a test mixture as described by Küster et al (1976). Products from lipid peroxidation were detected in liver homogenate using the TBA assay according to Ohkawa et al (1979). The protein content was determined with the Biuret method (Gornall et al 1949), and ethylmorphine-N-demethylation was measured according to Klinger and Müller (1977) in 10.000 g supernatant.

Assays were performed either in duplicate or triplicate for each animal. The data were analysed by multiple comparisons with LSD test (HP: General statistics on computer HP 9845 T).

Table 1. CCl_4- and PCO-induced alterations of serum enzyme activities and other biochemical parameters. Values represent means and standard deviation for groups of 12 (acute experiment) or 18 (subchronical experiment) rats

Parameter	Acute Experiment			Subchronical Experiment	
	Control 1	Control 2	CCl_4	PCP	Control
TBA-reactivity (Δ A)	0.31±0.05	0.34±0.05	0.41±0.07*	0.23±0.03	0.25±0.03
Ethylmorphine-N-demethylation[1)]	0.40±0.14	0.35±0.06	0.26±0.08	0.59±0.20	0.43±0.13
Alanine aminotransferase[2)]	2.87±2.5	3.30±3.00	15.7±12.6*	5.51±3.30	4.22±2.61
Aspartate aminotransferase[2)]	7.75±3.61	7.46±4.51	18.10±8.6*	9.92±6.21	5.78±2.13
Leucine nitranilidase[2)] ($\times 10^{-3}$)	7.81±2.90	6.42±4.13	15.1±11.2*	7.90±5.01	8.53±6.14
γ-glutamyl-transpeptidase[2)]	9.33±8.01	5.71±5.15	11.0±7.3*	5.32±2.91	5.61±4.52
ADP/O-ratio (1)	1.91±0.16	2.08±0.21	1.10±0.53*	0.81±0.56*	2.11±0.28

* statistically significant ($p<0.05$)

[1)] nmol formaldehyde/mg protein x min

[2)] μmol/s

Results and Discussion

CCl_4 treatment

CCl_4 caused liver damage at the microscopic level characterised by lipid infiltration, necroses, vacuolar degenerations and cellular infiltration as already reported for similar experiments (Kniepert and Seffner 1989). There was no change in liver weight. The increase of serum enzyme activities (Table 1) and a decrease of the ASAT/ALAT ratio indicate a liver disease. The decrease of microsomal ethylmorphine-N-demethylation and of the mitochondrial ADP/Oxygen ratio as well as the decrease of the respiratory control index (RCI) indicate an inactivation based on destruction of cytochrome P-450 and the respiratory chain, both of which haem containing.

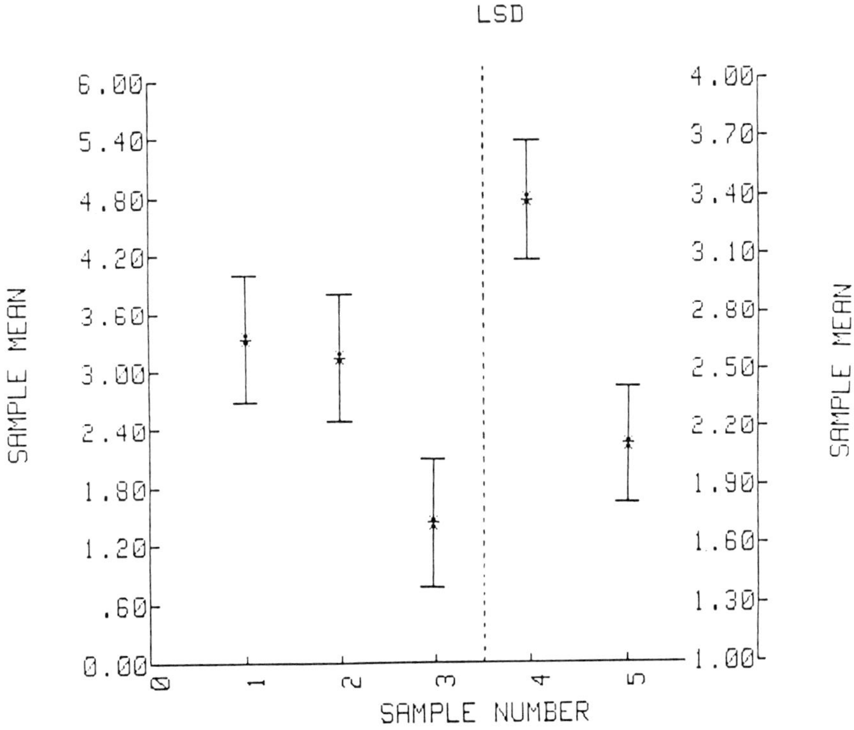

Fig. 1. Sample means and least significant difference (LSD) of respiratory control indices of liver mitochondria from 12 (acute experiment) or 18 (subchronical experiment) rats.

Sample number

1 control 1: oil
2 control 2: untreated
3 0.16 g/kg CCl_4
} acute experiment

4 control: 1% ethanol
5 250 mg/l PCP in drinking water with 1% ethanol
} for 90 days

Compared with the histological results the increase of TBA activity is not very marked, although it represents one of the initial steps of liver injury caused through lipid peroxidation. It is possible, that a correlation between malondialdehyde development and cytotoxicity could be expected only in a small range under specific circumstances, since the TBA method does not provide an endpoint for the reaction chain. In mice TBA-values gradually rose to a maximum at 24 hr postdosing (Kim and LaBella 1987). Development of lipid peroxidation by rats may be different from mice. That is why the 24 hr results reported, emphasize the necessity for several analytical indices of lipid peroxidation measured at different time intervals. In similar experiments, reported by Kim et al (1990), corn oil shows significant effects on hepatotoxicity of CCl_4. Comparison of control 1 and 2 could not confirm their results for peanut oil.

PCP treatment

PCP is strongly lipophylic and a powerful uncoupler of oxidative phosphorylation (Duxbury and Thompson, 1987). But its liver toxicity is low-grade compared with CCl_4. So histological changes could not be detected. The expected increase in microsomal activity (Schiller and Seffner 1988) was not significant. However, marked differences in oxidative phosphorylation (RCI, ADP/O ratio) indicate liver changes.

The two parameters could be assessed as very sensitive parameters for detecting PCP induced liver changes. Owing to its high yield of energy transducing activity, mitochondrial oxidative phosphorylation is responsible for supplying of ATP requirement of cells. That is the reason disturbances in this area are initiating mechanisms of cytotoxicity.

References

Appel W (1970) Peptidasen. In: Methoden der enzymatischen Analyse (Bergmeyer HU ed). Vol 1 Berlin: Akad Verl:916–921

Bergmeyer HU, Bernt E (1970) Glutamat-Oxalacetat-Transaminase UV-Test mit Malat-Dehydrogenase als Indikator-Enzym. In: Methoden der enzymatischen Analyse (Bergmeyer HU ed). Vol 1 Berlin: Akad Verl:685–690

Bergmeyer HU, Bernt E (1970) Glutamat-Pyruvat-Transaminase UV-Test mit Lactat-Dehydrogenase als Indikator-Enzym. In: Methoden der enzymatischen Analyse (Bergmeyer HU ed). Vol 1 Berlin: Akad Verl:717–723

Duxbury CL, Thompson JE (1987) Pentachlorphenol Alters the Molecular Organization of Membranes in Mammalian Cells. Arch Environ Contam Toxicol 16:367–373

Gornall AG, Bardawill CJ, David MM (1949) Determination of serum proteins by means of Biuret reaction. J Biol Chem 177:751–766

Heinecke H (1988) Definierte Tiermodelle in der Mikrobiologie. Schr Reihe Gesundh Umw Bad Elster 4 Vol 5:2–27

Kim RS, LaBella FS (1989) Criteria for determination of lipid peroxidation in tissues: estimation in liver of mice intoxicated with carbon tetrachloride. Can J Physiol Pharmacol 65:1503–1506

Kim HJ, Odend'hal S, Bruckner V (1990) Effect of Oral Dosing Vehicles on the Acute Hepatotoxicity of Carbon Tetrachloride in Rats. Toxicol Appl Pharmacol 102:34–49

Klinger W, Müller D (1977) Ethylmorphin-N-demethylation by liver homogenate of newborn and adult rats. Acta biol med germ 36:1149–1159

Kniepert E, Seffner W (1989) Histological and histometrical changes of liver damage in rats after treatment with ethanol and carbon tetrachloride. Exp Pathol 36:185–188

Koch R, Axmann C (1987) Pentachlorphenol. Eigenschaften-Toxizität-Umweltverhalten-Gefährlichkeit. Schrreihe Gesundh Umw Bad Elster 3 Vol 5:3–69

Küster U, Bohnensack R, Kunz W (1976) Control of oxidative phosphorylation by the extramitochondrial ATP/ADP ratio. Biochem Biophys Acta 440:391–402

Ohkawa H, Ohishi N, Yagi K (1979) Assay for lipid peroxides in animal tissues by thiobarbituric acid reaction. Anal Biochem 95:351–358

Recknagel RO, Glende EA, Dolak JA, Waller RL (1989) Mechanisms of carbon tetrachloride toxicity. Pharmac Ther 43:139–154

Schiller F, Seffner W (1988) Untersuchungen zur Kombinationswirkung halogenierter Kohlenwasserstoffe und chlorierte Phenole auf einige Parameter des Leberstoffwechsels und des Immunsystems im subchronischen Tierversuch. Schrreihe Gesundh Umw Bad Elster 4 Vol 5:48–62

Steinbrecht I, Kunz W (1970) Anwendung der "Cycling"-Technik zur stichprobenartigen quantitativen Bestimmung des Reduktionsgrades des NAD- und NADP-Systems von Rattenlebermitochondrien bei kontinuierlich registrierenden Messungen. Acta biol med germ 25:731–747

Szasz G (1970) γ-Glutamyl-Transpeptidase. In: Methoden der enzymatischen Analyse (Bergmeyer HU ed). Vol 1 Berlin: Akad Verl:733–738

Recent Developments in Toxicology:
Trends, Methods and Problems
Arch. Toxicol., Suppl. 14, 68–70 (1991)
© Springer-Verlag 1991

The Glutathione System in Rat Liver Chronically Injured by Thioacetamide

M. Kretzschmar[1], H. Franke[2], G. Machnik[3], T. Zimmermann[2], R. Dargel[2], and W. Klinger[1]

[1] Institute of Pharmacology and Toxicology,
[2] Institute of Pathological Biochemistry and
[3] Institute of Pathological Anatomy of Medical Faculty, Friedrich Schiller University Jena, Löbderstraße 1, 6900-Jena

Introduction

Glutathione (reduced form: GSH, oxidized form: GSSG) serves a broad spectrum of functions in the liver (for review see Meister and Anderson 1983; Kretzschmar and Klinger 1990). GSH is transported into the circulation through the sinusoidal membrane by a carrier transport system (Kaplowitz et al 1985). This fact is important for supplying extrahepatic tissues with GSH or cysteine (Bannai and Tataishi 1986). Recently, Burgunder and Lauterburg (1987) demonstrated that liver cirrhosis in man is associated with decreased GSH plasma levels. The authors concluded from these data that GSH synthesis in cirrhotic liver is decreased.

The thioacetamide (TAA)-induced cirrhosis-like liver lesion in rats was found to be similar to major morphological and biochemical features of the human disease (Zimmermann et al 1987). Hence this model was used to investigate the relationship between intrahepatic glutathione status (synthesis and catabolism), export into plasma and ultrastructural changes in chronically TAA-induced liver injury.

Material and Methods

Female Wistar rats were pretreated with TAA as described earlier (Kretzschmar et al 1989). Blood was taken for determination of plasma GSH concentration according to Adams and Lauterburg (1983). Liver samples were processed as described below. Tissue GSH level and the activity of GSH synthesizing enzymes were measured (Kretzschmar et al 1989). Hepatic GGT activity was determined as described previously (Kretzschmar and Klinger 1989). Tissue GSSG concentration was detected according to Hissin and Hilf (1976), enzyme

Table 1. Biochemical parameters in thioacetamide (TAA)-induced micro- (3 months) or macronodular (6 months treatment) experimental liver cirrhosis. Arithmetic means ± SEM are given (n = 6), significant differences are marked by * ($p \leq 0.05$, Student's t-test)

parameters	controls	TAA 3 months	controls	TAA 6 months
GSH (μmol/g)	8.59±0.08	8.65±0.14	6.44±0.26	7.66±0.62*
GSSG (μmol/g)	0.48±0.01	0.51±0.05	0.36±0.26	0.36±0.04
GSH/GSSG	17.8±1.0	17.2±1.0	17.7±1.2	21.6±1.8*
GCS (nmol/mg×min)	21.5±2.4	28.5±4.0	25.9±2.3	19.24±1.4*
GSHS (nmol/mg×min)	11.6±1.3	8.1±1.0	10.2±1.4	16.8±1.1*
GGT (nmol/mg×min)	0.31±0.04	1.81±0.18*	0.27±0.02	7.66±0.62*

histochemical visualization of tissue GGT according to Lojda et al (1976). Details of the technical processing of liver samples for ultrastructural investigations have been described elsewhere (Franke et al 1983).

Results

The results of the biochemical investigations are demonstrated in Table 1. A small to moderate hepatocellular GGT activity was observed histochemically after 3 months of TAA treatment. GGT activity was also demonstrable in epithelia of the proliferated bilary ducts of single enlarged portal tracts. After 6 months of TAA administration the hepatocellular GGT activity was moderate to strong. It was demonstrable both in parenchymal (preneoplastic) nodules and in cholangiocellular/cholangioductular proliferates. A GGT activity of mesenchymal cells was not demonstrable.

The electron microscopic studies revealed a series of characteristics structural changes, such as disorganization and total lack of microvilli border, appearance of basement membrane-like deposits within the narrowed space of Disse and disappearance of the highly porus endothelial cell lining.

Discussion

The GSH content in cirrhotic livers is found to be increased after 6 months of TAA indicating that the GSH/GSSG redox potential is shifted in favour of the reduced form. This agrees with the unchanged GSSG levels after TAA observed in this study. In correspondence with investigations on patients with cirrhosis (Burgunder and Lauterburg 1987) a decreased tGSH plasma level could be shown. Surprisingly, the GSH synthesis rate was normal in cirrhotic livers. It can be hypothesized that depression of plasma tGSH content could be caused by a disturbance of GSH export from liver into plasma. The electron microscopic studies support this view. Since GSH export takes place by a carrier transport localized at the microvilli border and that toxically induced lesions of the microvilli may be accompanied by alterations of the plasma membrane leading to changes in the function of membrane bound enzymes, the alterations of the microvilli border appear to be of greatest importance among all morphologic changes here described.

The increase in hepatic GGT activity after chronic TAA administration can be correlated with the process of development of preneoplastic nodules. TAA is a mild carcinogen (Neal and Halpert 1982). A connection between increased GGT activity and the process of fibroplasia is not probable. A complex change of hepatic GSH metabolism under stages of premalignency and malignency in the liver can be assumed. This mechanism leads to increasing amounts of GSH in the hepatocytes and an increasing detoxification potential. Both factors provide the altered hepatocyte a strong growth advantage in the face of toxic effects of carcinogenic agents.

References

Adams JD, Lauterburg BH, Mitchell JR (1983) Plasma glutathione and glutathione disulfide in the rat: Regulation and response to oxidative stress. J Pharmacol Exp Ther 227:749–754

Bannai S, Tataishi N (1986) Role of membrane transport in metabolism and function of glutathione in mammals. J Membrane Biol 89:1–8

Burgunder JM, Lauterburg BH (1987) Decreased production of glutathione in patients with cirrhosis. Eur J Clin Invest 17:408–414

Franke H, Zimmermann T, Dargel R (1983) Qualitative and quantitative changes in hepatic lipoprotein particles following acute injury of the rat liver induced by thioacetamide. Virchows Arch (Cell Pathol) 44:99–113

Hissin PJ, Hilf R (1976) A fluorometric method for determination of oxidized and reduced glutathione in tissues. Anal Biochem 74:214–226

Kaplowitz N, Aw TY, Ookhtens M (1985) The regulation of hepatic glutathione. Ann Rev Pharmacol Toxicol 25:715–744

Kretzschmar M, Klinger W (1989) Gamma-glutamyltranspeptidase in liver homogenates of rats of different ages: enzyme kinetics and age course of K_m and V_{max}. Z Versuchstierkd 32:41–47

Kretzschmar M, Franke H, Zimmermann T, Dargel R, Klinger W (1989) Glutathione synthesis and export in experimental liver cirrhosis induced by thioacetamide. Exp Pathol 36:113–122

Kretzschmar M, Klinger W (1990) The hepatic glutathione system – influences of xenobiotics. Exp Pathol 38:145–164

Lojda Z, Gossrau R, Schiebler TH (1976) Enzymhistochemische Methoden. Springer Verlag Berlin-Heidelberg-New York

Meister A, Anderson ME (1983) Glutathione. Ann Rev Biochem 52:711–760

Neal RA, Halpert J (1982) Toxicology of thiono-sulfur compounds. Ann Rev Pharmacol Toxicol 25:321–339

Zimmermann T, Müller A, Machnik G, Franke H, Schubert H, Dargel R (1987) Biochemical and morphological studies on production and regression of experimental liver cirrhosis induced by thioacetamide in Uje:WIST rats. Z Versuchstierkd 30:165–180

Recent Developments in Toxicology:
Trends, Methods and Problems
Arch. Toxicol., Suppl. 14, 71–75 (1991)
© Springer-Verlag 1991

Hitherto Unknown Additive Growth Effects of Fluorene and 2-Acetylaminofluorene on Bile Duct Epithelium and Hepatocytes in Rats

M. Danz, A. Hartmann, M. Otto, and H. Blaszyk

Institute of Pathology "Robert Rössle" of the Friedrich-Schiller University Jena, Ziegelmühlenweg 1

Introduction

The liver, more than other organs, may respond with a wide range of functional and structural lesions to xenobiotics. To identify putative tumorigenicity of a compound, its potency to cause cell proliferation is considered to be most reliable. However, most genotoxic carcinogens inhibit division of their target cells. This occurs even in the early phase after non-necrogenic doses. Contrarily, non-genotoxic carcinogens of different organotropism elevate cell proliferation in their target tissues primarily, either by restorative (post-necrotic) or by additive (chemical promoters) growth of normal and/or previously initiated cells (Clayson et al 1989). It should be noted that in man initial genotoxic events are thought to be inevitable and that cell proliferation is the main driving force in tumor development. Keeping this in mind the discovery of non-genotoxic proliferogenic factors is as important in risk assessment as that of genotoxic carcinogens (Clayson et al 1989; Butterworth 1989).

Recently the structurally related compounds 2-AAF, Fen and Fon were proved in a short-term PAT (Danz and Urban 1980). Unexpectedly, Fen and Fon yielded positive test results like complete carcinogens and tumor promoters. Therefore, the objectives of the present experiments were, first, to study influences of Fen and its two derivatives on cell growth in the liver and, second, to examine the presumed tumorigenicity of Fen and Fon in a medium-term I/P-assay according Hasegawa et al 1986.

Abbreviations: 2-AAF, 2-Acetylaminofluorene; BDE, Bile duct epithelium; DEN, Diethylnitrosamine; Fen, Fluorene; Fon, Fluorenone; Hcy, Hepatocytes; I/P-Assay, Initiation/Promotion-Assay; MI, Mitotic index; PAT, Promoting-Activity-Test.

Materials and Methods

Substances: 2-AAF was purchased from Fluka AG (Switzerland; purissimum). Fen and Fon were gifts from WB für organische Polymerenchemie of this Univ. (purity controled by melting point, chromatography and spectrography), DEN was a gift from the Centr. Inst. on Canc. Res. (Berlin-Buch; purissimum).

Animals and treatment: The proliferation in the livers was studied in about 70 days old male Uje:Wist rats kept under standardized conditions (Inst. of Pharmacol. and Toxicol., Jena) in groups of 6–7 per cage, with free access to food (basal diet, Versuchstierproduktion Berlin) and tap water and at natural light cycle. The PAT (Table 1) was reproduced repeatedly in male SD (Berlin-Buch) and Uje:Wist rats, both under natural and controled (from 6 a.m. to 6 p.m.) light periods. The substances, dissolved in dimethyl sulphoxide (2 ml/kg), were given under light ether anesthesia by stomach tube between 9 and 10 a.m. The I/P-assay was performed preliminarily only in three months old male SD rats (5 per experimental group). Initiation was two doses of 100 mg DEN/kg within one week. Beginning in the 4th week, the animals received three times weekly 2 ml solvent (control), 100 mg Fen or Fon per kg b.w. between 6 p.m. and 7 p.m. over a period of 11 weeks. Partial hepatectomy was carried out in the 6th experimental week.

Mitotic counts and statistical evaluation: The MI of the BDE (ducts and ductules) and Hcy was calculated conventionally in samples of 2,000 and 5,000 cells per animal, respectively. For statistical evaluation the U-test by Mann and Whitney was used.

Results and discussion

The genotoxic 2-AAF inhibits Hcy division at 48 h (results not given), but raises the MI of the BDE (Fig. 1). Following a 2nd dose during the mitotic decline at day 5 the MI is elevated again (Fig. 2). This is associated with the appearance of prominent periportal ductular structures lined by large gamma-GTP-positive cells. This mitotic burst in the experiment is independent of a choline-deficiency and could be the correlate of the strong promoter effect of 2-AAF in I/P-assays. However, a transition of BDE or periductal progenitor cells into normal or neoplastic Hcy is not yet evidenced (Marceau et al 1989).

In contrast to 2-AAF, Fen increases dose-dependently (35, 50 and 150 mg/kg) the liver weight by about 20% and the MI of Hcy 6-fold after 48 h. Quite similar results are obtained using Fon (details not given, Table 1). These effects are consistent with known chemical promoters. They produce, like complete carcinogens, a transient extrahepatic "side effect" as they act mitogenically on the zona fasciculata of the adrenals (PAT, Danz and Urban 1980). Because of the positive results in the PAT, Fen and Fon were suspected to be endowed with promoting activities. This was confirmed by the I/P-assay (Table 1).

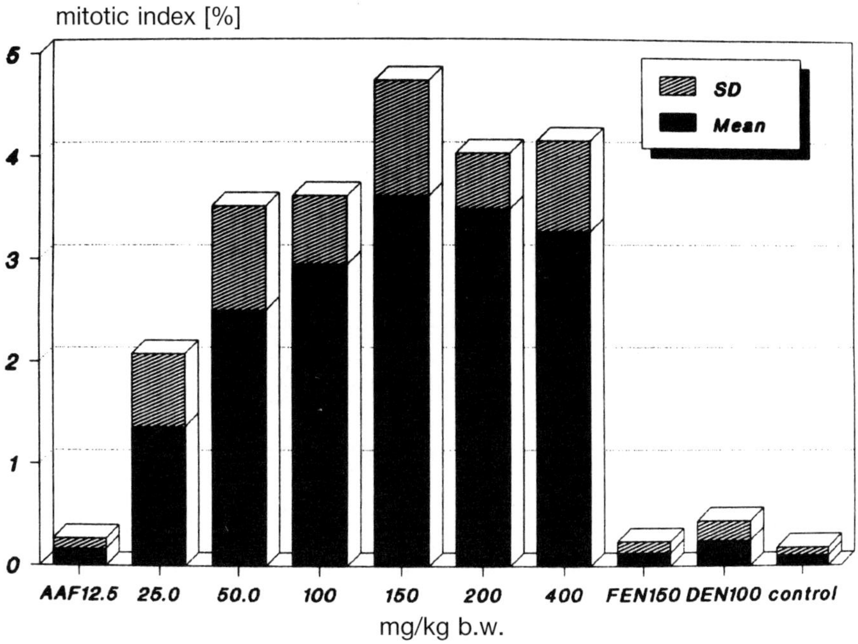

Fig. 1. Mitotic index (%) of the bile duct epithelium 48 h after single doses (mg/kg b.w.) of 2-AAF, DEN or Fen. The mitoses are stimulated significantly only by 2-AAF in doses larger than 12.5 mg (Mean and standard deviation)

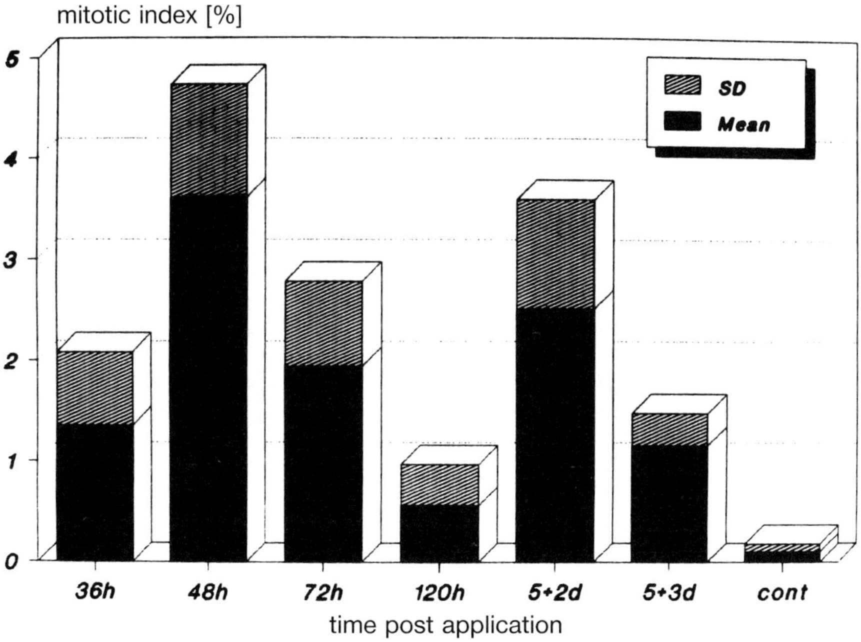

Fig. 2. Proliferation of bile duct epithelium is maximal 48 h after 2-AAF 150 mg/kg b.w. and declines thereafter. Two and three days following a 2nd dose of 150 mg given at day 5 cell division rises again, but to a lower extent

Table 1. Comparison of growth effects of fluorene (Fen), fluorenone (Fon) and 2-acetylaminofluorene (2-AAF) from literature compared with own results in a short-term promoting-activity-test (PAT) and in a medium-term initiation/promotion-assay (I/P-assay). The predicted tumorigenicity by the PAT of Fen and Fon is confirmed by the I/P-assay

Substance	from literature		own experimental results		
	carcinogen	promoter	liver cell mitoses	PAT	I/P-assay liver foci/cm^2 (% of control)
Fen	–	–	↑	↑	9.1 ± 5.1 (317)
Fon	–	–	↑	↑	11.4 ± 5.1 (393)
2-AAF	+	+	↓	↑	not tested

The following conclusions are derived. First, identical results by genotoxic and non-genotoxic carcinogens in the PAT may indicate similar early actions (endogenous mechanisms?) at the systemic level. Second, especially non-genotoxic compounds, not suspected to be tumorigenic in the past, will be identified, if they are examined in adequate short-term tests and/or in medium-term I/P-assays.

References

Butterworth BE (1989) Nongenotoxic carcinogens in the regulatory environment. Regulat Toxicol Pharmacol 9:244–256

Clayson DB, Nera EA, Lok E (1989) The potential for use of cell proliferation studies in carcinogen risk assessment. Regulat Toxicol Pharmacol 9:284–295

Danz M, Urban H (1980) Elevated mitotic number in the adrenal cortex as a reflection of early events in growth induction by carcinogens and promoters – a possible short-term assay. Arch Toxicol Suppl 4:19–21

Hasegawa R, Tsuda H, Shirai T, Kurata Y, Matsuda A, Ito N (1986) Effects of timing of partial hepatectomy on the induction of preneoplastic liver foci in rats given hepatocarcinogens. Cancer Lett 32:15–23

Marceau N, Blouin MJ, Germain L, Noel M (1989) Invited Review: Role of different epithelial cell types in liver ontogenesis, regeneration and neoplasia. In Vitro Cell Develop Biol 25:336–341

Recent Developments in Toxicology:
Trends, Methods and Problems
Arch. Toxicol., Suppl. 14, 75–79 (1991)
© Springer-Verlag 1991

Morphological Lesions of Pancreas and Bile Ducts in Rats Induced by Dibutyltin Dichloride

J. Merkord, L. Jonas*, and G. Hennighausen

Institute of Pharmacology and Toxicology, University of Rostock, Leninallee 70, 2500 Rostock and
* Institute of Pathology, University of Rostock, Strempelstr. 14, 2500 Rostock

Introduction

Dialkyltin compounds have been widely used as biocides, catalysts and plastic stabilizers (Kimmel et al 1977). DBTC is the final product of the stabilisation reaction of polyvinyl chloride resins and it is used as a catalyst for formation of polyurethane foam (Piver 1973).

In studies of the toxic effects of DBTC on bile ducts and of the biliary excretion of organotin in mice and rats (Barnes and Magee 1958, Hennighausen et al 1980, Merkord et al 1982) signs of acute pancreatitis were observed.

In the present paper the lesions of pancreas and bile ducts in rats after single oral and intravenous administration of DBTC are described.

Material and Methods

Male outbred rats (Rost: Wistar) weighing 80–100 g were used in this study. For intravenous administration (6 mg/kg body weight) DBTC (Schering AG, Berlin-West) was first dissolved in 100% ethanol (2 parts) and then mixed with glycerol (3 parts) and for oral administration (30 mg/kg b.w.) the organotin compound was dissolved in olive oil. The animals (6 per group) were killed 1 and 6 h, 1, 4 and 6 days after treatment.

Histology: The liver with bile duct and the pancreas were fixed in calcium formalin and processed for histopathological examination. Sections were stained with hematoxylin and eosin.

Bile duct diameter: The diameter of the bile duct was measured in the middle part of the ductus choledochus, using a microstereoprojector at a magnification of x 20.

Electron microscopy: The pancreas was quickly removed. Small cubes of 1 mm^3, cut with a razor-blade, were placed 1 h in 4% glutaraldehyde in 0,1 M phosphate buffer and then washed in the same buffer.

They were postfixed in 1% osmium tetroxide, dehydrated in a graded series of alcohol solutions and embeded in a Durcupan ACM. The tissue samples were thin-sectioned, stained with uranyl acetate and lead citrate and examined in the electron microscope (BS 540, Tesla).

Results

Signs of bile duct lesions (significant increase of serum alkaline phosphatase activity, enlargement of bile duct diameter, destruction of epithelial cells) were observed after i.v. and oral administration of DBTC. The morphological signs of bile duct injury was evident after 6 h, the maximal enlargement of bile duct was measured 6 d after treatment. Parts of bile duct wall have been destroyed by an acute inflammatory process and interstitial exudate spreads into the pancreas (Fig. 1).

In the liver, signs of acute inflammation of the portal tracts are present. The earliest stage of pancreas injury after i.v. or oral application was evident after 1 h and 6 h. The pancreas around the duct becomes oedematous and the duct

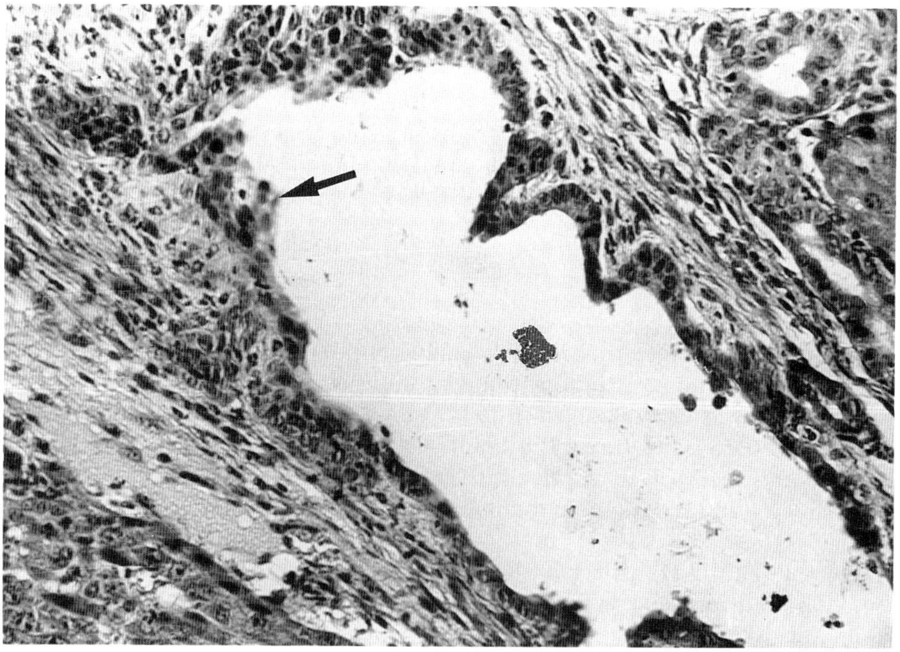

Fig. 1. Rat bile duct 4 days after i.v. administration of 6 mg/kg DBTC. Parts of the bile duct wall have been destroyed and the epithelium shows flattened cells with shrunken nuclei and loss of cytoplasm. HE, x 224

itself becomes dilated and contains serous fluid. Ultrastructural alterations in acinar cells and lysosomes were observed 1 h after i.v. and 6 h after oral administration of DBTC. Acinar cells showed multiple vacuoles with marked electron dense content. The Golgi apparatus is enlarged and filled with electron dense material. In the Golgi area dark autophagosomes with inclusions of endoplasmic reticulum, free ribosomes and membrane accumulations were visible. Myelin figures in big vacuoles occured in the subnuclear and basal cytoplasm (Fig. 2). Mitochondria damage was seen as myelinisations of cristae mitochondrales.

18 to 24 h after treatment an acute serous interstitial pancreatitis had developed. A severe interstitial pancreatitis with infiltrates of eosinophilic granulo-

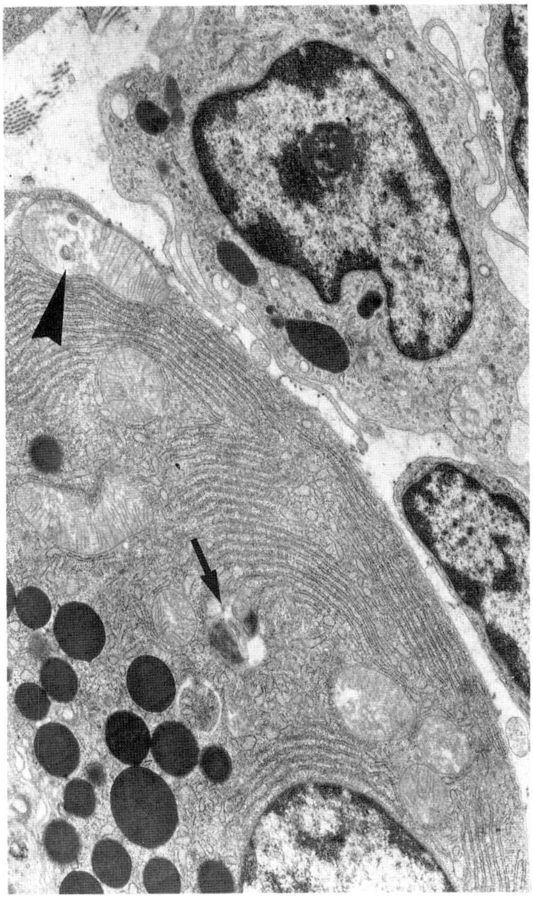

Fig. 2 a. Basal portion of an pancreatic acinar cell 6 h after i.v. administration of DBTC. Electron dense vacuole (↗) and a mitochondrion with myelin inclusions in the matrix (▲). Interstitice with macrophage like inflammatory cells in close contact to the basment membrane of the acinar cell. x 7300

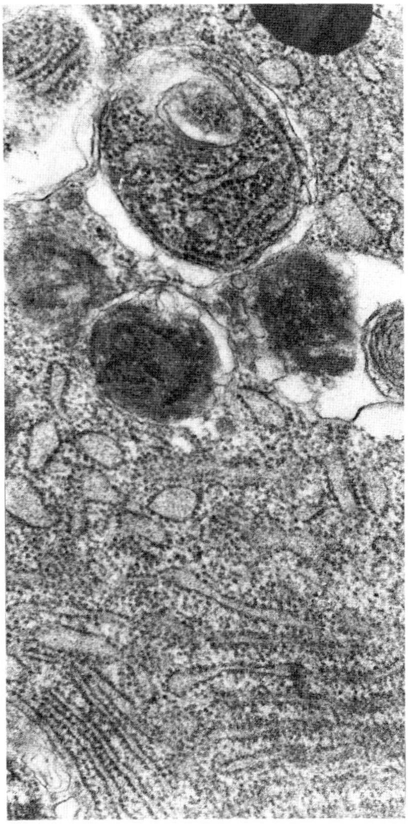

Fig. 2b. Autophagosomes and vacuoles with inclusions of elements of endoplasmic reticulum and ribosomes as well as of myelic accumulations of dark membranes. x 34000

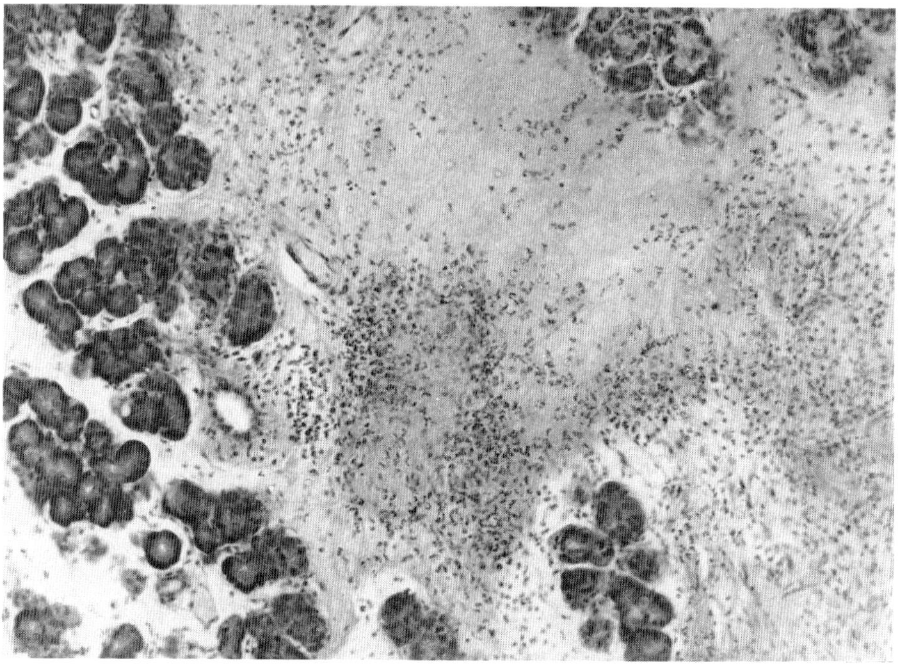

Fig. 3. Rat pancreas 24 h after i.v. administration of 6 mg/kg DBTC. Acute serous interstitial pancreatitis with infiltrates of eosinophilic granulocytes, histiocytes, lymphocytes and plasma cells. HE x 88

cytes, histiocytes, lymphocytes and plasma cells could be observed 4 to 6 days after treatment with DBTC (Fig. 3).

Discussion

In previous studies bile duct lesions in rats and mice after treatment with DBTC and some other dialkyltin compounds have been described (Barnes and Magee 1958, Hennighausen et al 1980, Merkord et al 1989). It has been suggested that the excretion of high organotin concentration in the bile may lead to a destruction of bile duct epithelial cells (Merkord et al 1982).

In the present study the development of acute pancreatitis could be demonstrated after i.v. and oral administration of DBTC. Destroyed epithelial cells may obstruct the ductus choledochus and disturb the flow of bile pancreatic fluid. Toxic concentration of DBTC in pancreatic fluid produced progressive ultrastructural alterations in the pancreas of rats. The changes were limited to acinar cells and consist of cytoplasmic vacuoles with zymogen like material and increased autophagocytosis. The described acute interstitial pancreatitis could be a result of this obstruction in combination with a toxic acinar cell damage. These findings at the electron microscopie level have not been previously reported. The possible use of DBTC for the development of a simple model of experimental serous pancreatitis are under further investigation.

Acknowledgement. The autors wish to thank Mrs. Ingeborg Neumann for her expert technical assistance.

References

Barnes JM, Magee PN (1958) The biliary and hepatic lesion produced experimentally by dibutyltin salt. J Pathol Bacteriol 75:267–279

Kimmel EC, Fisch RH, Casida JE (1977) Bioorganotin chemistry. Metabolism of organotin compounds in microsomal monooxygenase system and in mammals. Agric Food Chem 25:1–3

Hennighausen G, Lange P, Merkord J (1980) The relationship between the length of the alkyl chain of dialkyltin compounds and their effects on thymus and bile ducts in mice. Arch Toxicol Suppl 4:175–178

Merkord J, Hennighausen G, Karnstedt U, Lange P (1982) Die biliäre Exkretion von Organozinn bei der narkotisierten Ratte nach Gabe von Di-n-alkylzinndichloriden mit unterschiedlicher Länge der Alkylketten. Pharmazie 37:146

Merkord J, Hennighausen G (1989) Acute pancreatitis and bile duct lesions in rat induced by dibutyltin dichloride. Exp Pathol 36:59–62

Piver WT (1973) Organotin compounds: Industrial application biological investigation. Environm Health Perspect 61–73

Recent Developments in Toxicology:
Trends, Methods and Problems
Arch. Toxicol., Suppl. 14, 80–82 (1991)
© Springer-Verlag 1991

Toxic Action of Platinum Coordination Complexes on the Endocrine Pancreas of the Rat

H. G. Hillesheim[1], K. D. Kohnert[2], F. Fält[3], W. Besch[2], H. P. Schröer[1], and H. Hoffmann[1]

[1] Institute of Microbiology and Experimental Therapy, FRG O-6900 Jena;
[2] Institute of Diabetes, FRG O-2201 Karlsburg;
[3] Department of Pathology, University of Lund, General Hospital, S-214 01 Malmö

Introduction

Reports on toxic effects of platinum coordination complexes on the endocrine pancreas are missing with exception of cis-DDP itself (Goldstein et al 1983). Recently, oxoplatin (trans-dihydroxy-cis-dichlorodiammine-platinum [IV]) an oxygenated derivative of cis-DDP has been described by Presnov and Konovalova (1988), Fig. 1. As indicated by the maximally tolerated dose levels, oxoplatin has been found to be about 3 times less toxic than cis-DDP when administered daily to rats for 5 consecutive days (Härtl et al 1989).

Starting from preclinical screening of new antineoplastically active Pt(II)- and Pt(IV)-complexes the toxic action of trans-dihydroxy-cis-dichlorodiamine-platinum (IV) (oxoplatin) compared with that of cis-dichlorodiamine-platinum (II) (cis-DDP) on pancreatic islet B-cell function was investigated in the rat.

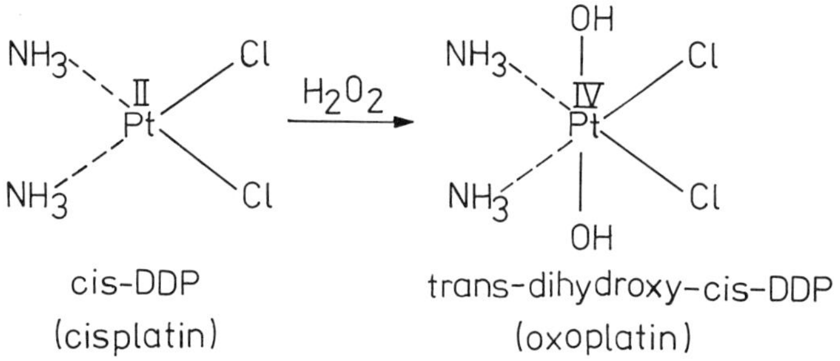

Fig. 1. Chemical structures of cis-DDP and oxoplatin

Material and Methods

Corresponding to this treatment schedule male Wistar rats (Jelei-Wist, Leipzig) weighing 100–150 g received 5 subsequent single doses of oxoplatin (6 mg/kg), cis-DDP (2 mg/kg) or saline (10 ml/kg) i.v. via the tail vein daily between 8 and 10 a.m. Both drugs were synthetized by Schröer, Institute of Microbiology and Experimental Therapy Jena, FRG. Intravenous glucose tolerance tests were performed either one day after drug pretreatment or 3 weeks later in animals under pentobarbital anaesthesia (Nembutal^R, 40 mg/kg i.p.) fasted for 20 hrs. Commercial glucose solution 40% (w/v) was injected into the exposed femoral vein (0.4 g glucose per kg b.wt.) and 2, 10, 20 or 40 min thereafter blood samples were collected by puncture of the vena cava abdominalis using heparin. Plasma glucose and plasma insulin (IRI) were measured by the glucose oxidase method and commercial insulin kits (Isocommerz, GDR), respectively. Finally, for the measurement of IRI content (Ziegler et al 1985) pancreatic tissue was prepared followed by an extraction procedure from half of the longitudinally divided pancreas. The other half was fixed in Bouin's fluid for immunocytochemistry (Sternberger 1979) and morphometric analysis (Kohnert et al 1989).

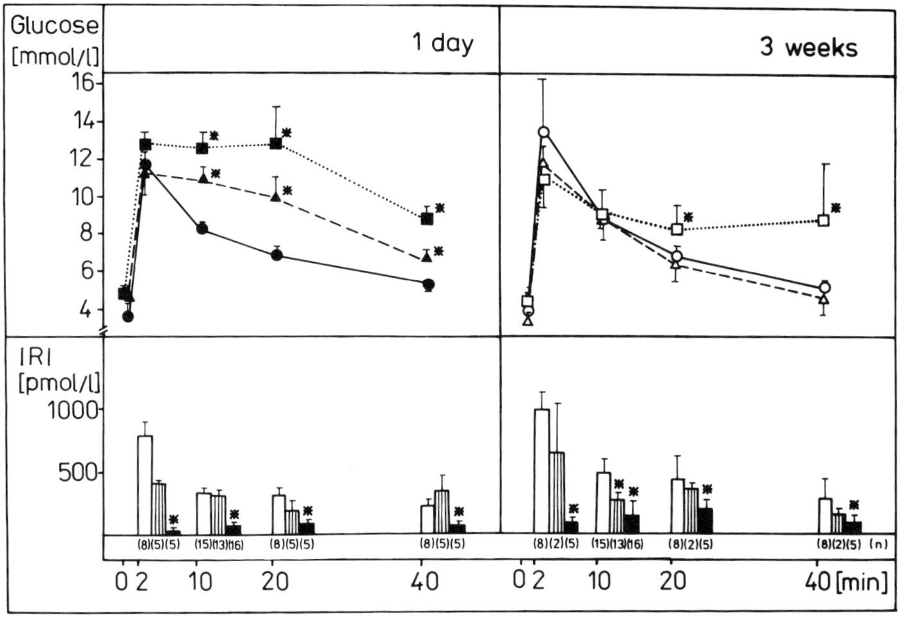

Fig. 2. Glucose tolerance in rats treated with 5 subsequent i.v. injections of 2 mg cis-DDP (▲----▲, △----△) 6 mg oxoplatin (■·····■, □·····□) per kg b.wt. or saline (●——●, ○——○) at one-day intervals. Glucose loadings were performed either 1 day or 3 weeks after the last treatment. Furthermore, the insulin response following cis-DDP ▥, oxoplatin ■, or saline □ treatment is shown. Mean ± SD; n = number; * $p < 0.01$ versus control (Mann-Whitney U-test)

Results and Discussion

One day after final drug treatment, oxoplatin and cis-DDP significantly impaired the glucose tolerance (Fig. 2). This was associated with a significant reduction of the glucose induced insulin response by oxoplatin only. Concomitant with this functional impairment the pancreatic insulin content (pmol/mg wet wt.) at 10 min after glucose loading was found to be decreased following treatment with oxoplatin (5.2 ± 0.8) or with cis-DDP (6.7 ± 0.8) compared with saline treated rats (11.9 ± 0.9), means ± SEM, n = 10 in each group. 3 weeks later the glucose intolerance continued only with oxoplatin as demonstrated by a persistent hyperglycemic state in connection with significantly decreased plasma IRI concentration (Fig. 2). In pancreas sections histopathological changes such as pyknotic nuclei in islets, degranulation and irreversible destruction of pancreas B-cells only were seen after administration of oxoplatin at day 1 and 3 weeks after drug administration. Morphometric analysis revealed a 42% ($p<0.01$) decrease in the volume density of insulin-immunoreactive cells following treatment with oxoplatin, whereas cisplatin had no significant effect on these cells.

Oxoplatin has been reported to be less toxic than cis-DDP (Presnov and Konovalova 1988; Härtl et al 1989), but this data provide evidence for a specific toxicity pattern following oxoplatin. The results focus attention on the pancreatic B-cells as an additional target of platinum coordination complexes in the rat. This demonstrates the possibility of some modifications in the toxicological spectrum of these compounds by changing the oxidation state of the molecule.

References

Goldstein RS, Mayor GH, Gingerich RI, Hook JB, Rosenbaum RW, Bond JT (1983) The effects of cisplatin and other divalent platinum compounds on glucose metabolism and pancreatic endocrine function. Toxicol Appl Pharmacol 69:432–441

Härtl A, Güttner J, Horn U, Jelinek F, Stöckel U, Schröer HP, Hoffmann H (1989) Akute und subakute Toxizität antineoplastisch wirksamer Pt(II)- und Pt(IV)-Koordinationsverbindungen bei Labornagern. Arch Geschwulstforsch 59:239–244

Kohnert KD, Ziegler B, Fält K, Ziegler M (1989) Augmentation of streptozotocin-induced hyperglycemia in mice by prior treatment with complete Freund's adjuvant. Internat Journ Pancreatol 4:321–333

Presnov MA, Konovalova AL (1988) Cycloplatam and oxoplatin – the new antitumor platinum compounds of the second generation. Arch Geschwulstforsch 58:43–49

Sternberger LR (1979) Immunocytochemistry, 2nd ed., Wiley and Sons, New York, pp 104–169

Ziegler B, Hahn HJ, Ziegler M (1985) Insulin recovery in pancreas and host organs of islets grafts. Exp Clin Endocrinol 85:53–60

Recent Developments in Toxicology:
Trends, Methods and Problems
Arch. Toxicol., Suppl. 14, 83–88 (1991)
© Springer-Verlag 1991

Investigations of Myelotoxic Effects in Rats

T. A. Moeller

Institute of Toxicology and Ecotoxicology, Stahnsdorfer Damm 81, Kleinmachnow – 1532

Possible targets of toxic substances in the bone marrow are the specific surface structures and enzymes of the cells, the specific regulation mechanisms and the very high cellular turnover rate. Nevertheless bone marrow is only seldom mentioned in toxicological publications. One reason may be uncertainties of the methods in the examination of this organ. The present study will attempt to answer the questions, in which cases and with which methods the bone marrow should be investigated in toxicological tests.

Materials and Methods

Myelogram of the rat: counting of 2000 cells, age of rats: 8 weeks, mean body weight (mbw) male (M): 308 g, s: 27 g, n: 50, mbw female (F): 214 g, s: 21 g, n: 50

Determination of minimum bone marrow cell number required to differentiate in smears for statistically acceptable results: $n > (\frac{\delta}{\varepsilon} \times z_{\alpha/2})^2$, same smears as for myelogram were used

Investigations of stress effects on bone marrow cell numbers: immobilisation in plastic tubes for 4 h, 24 h after begin of immobilisation, dissection and differential counting of cells in blood (100 cells) and bone marrow smears (500 cells) were carried out, age of rats (M): 8 weeks, n: 10 immobilised + 10 controls;

Investigation of hematopoiesis (erythrocytopoiesis) in spleen: hematopoiesis was stimulated by the pesticide Fenazox (96% Azoxybenzen), produced by VEB Fahlberg List Magdeburg. One oral dose of 5, 50, and 500 mg/kg body weight, suspended in 0.5% agar, control: agar.

Dissection 5 days after application: morphometric investigation of spleen according to the stereological principle of Weibel (1979), measurement with point counting method of 3 slides per organ, age of rats (M): 8 weeks, n: 10 per dose group; rats in all experiments: specific pathogenfree Shoe:WIST rats from VEB Versuchstierproduktion Schönwalde, maintained under barrier conditions. Bone marrow smears were made from the right femur, a suspension of

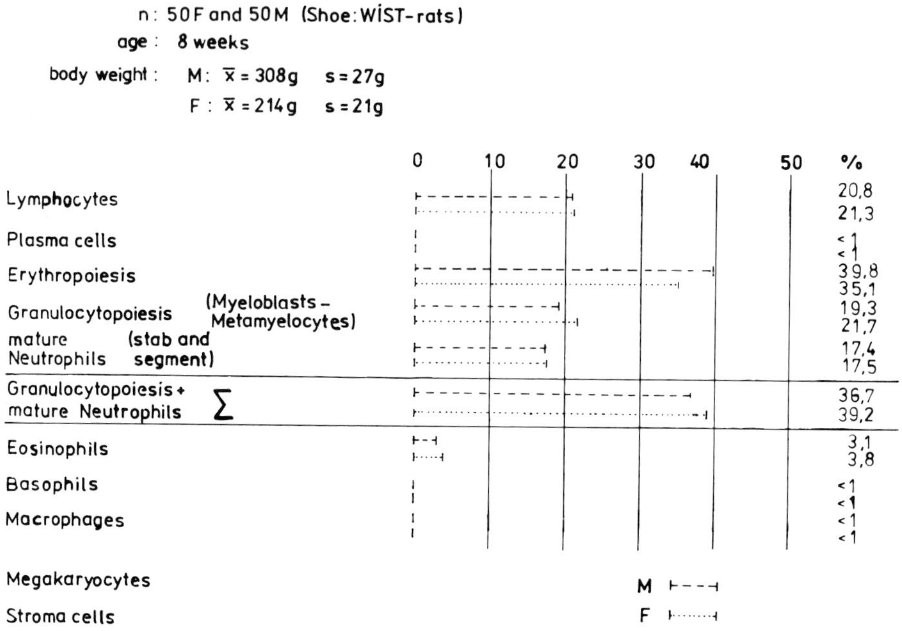

Fig. 1. Values for differential count of rat bone marrow cells

about 50% bone marrow and 50% rat serum was mixed, smeared, dried under air and stained with Pappenheim stain. Spleen: Bone marrow cells were distributed normally, but megacaryocytes were concentrated at the end of smears. Helley fixation, Giemsa staining; for a methodical comparison slides were prepared from the left femur of the same rats, mild preparation according Remmele (1986).

Results

For comparable differential counting a color atlas of bone marrow was produced. This atlas was used as a base for the production of a rat myelogram (Fig. 1). The minimum number of bone marrow cells required to differentiate in a smear so that a statistically acceptable number for every cell type occurs depends on the distribution of the different cell types in a smear. This investigation shows, that for the counting of lymphocytes with a statistical accuracy of 95% approximately 500 bone marrow cells and for the counting of erythropoietic cells approximately 200 cells are required. The uneven distribution of granulopoietic cells (myeoblasts – metamyeolocyte) and mature neutophils (stab and segment) requires a different method of counting. Even 2000 differentiated bone marrow cells give an accuracy far below 80%. Only when granulopoietic cells and mature neutrophils are counted together. 500 differentiated bone marrow cells are sufficient for 90% accuracy.

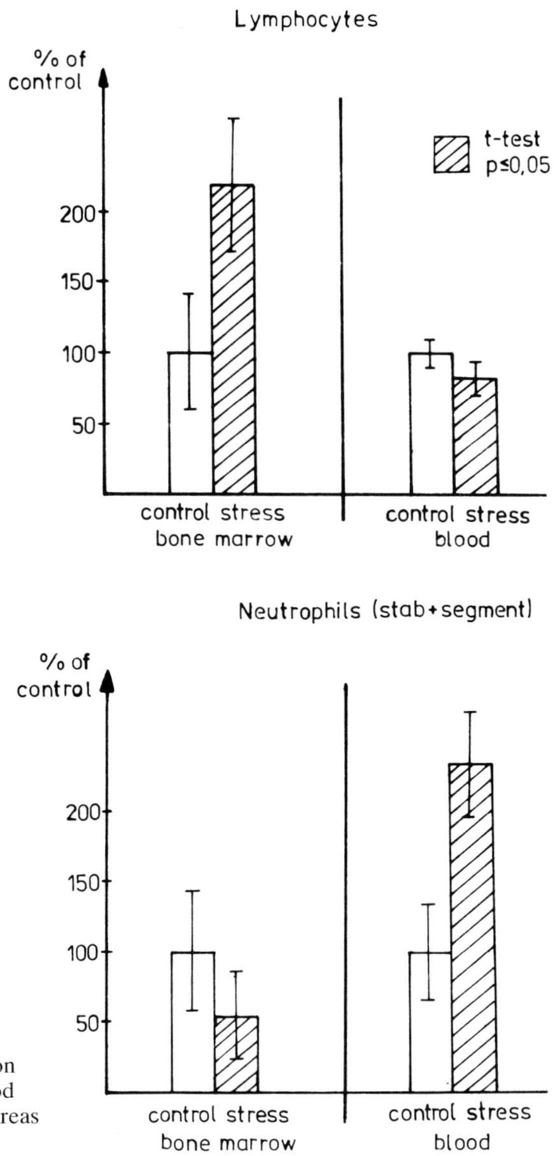

Fig. 2. The effect of immobilisation stress on rat bone marrow and blood cells (differential counts). Shaded areas indicate significance (t-test) $p \leq 0.05$

In the immobilisation test, the numbers of lymphocytes and neutrophils (stab and segment) were significantly altered (Fig. 2). The alterations in bone marrow and blood were reciprocal and were more than 100% of the control values. Other cell lines in blood and bone marrow did not show significant changes.

Fenazox, a pesticide which induces methemoglobinemia and hemolysis, caused a strong increased hematopoiesis (erythrocytopoiesis) in the spleen which was quantified by morphometric analysis (Fig. 3). Even in the lowest

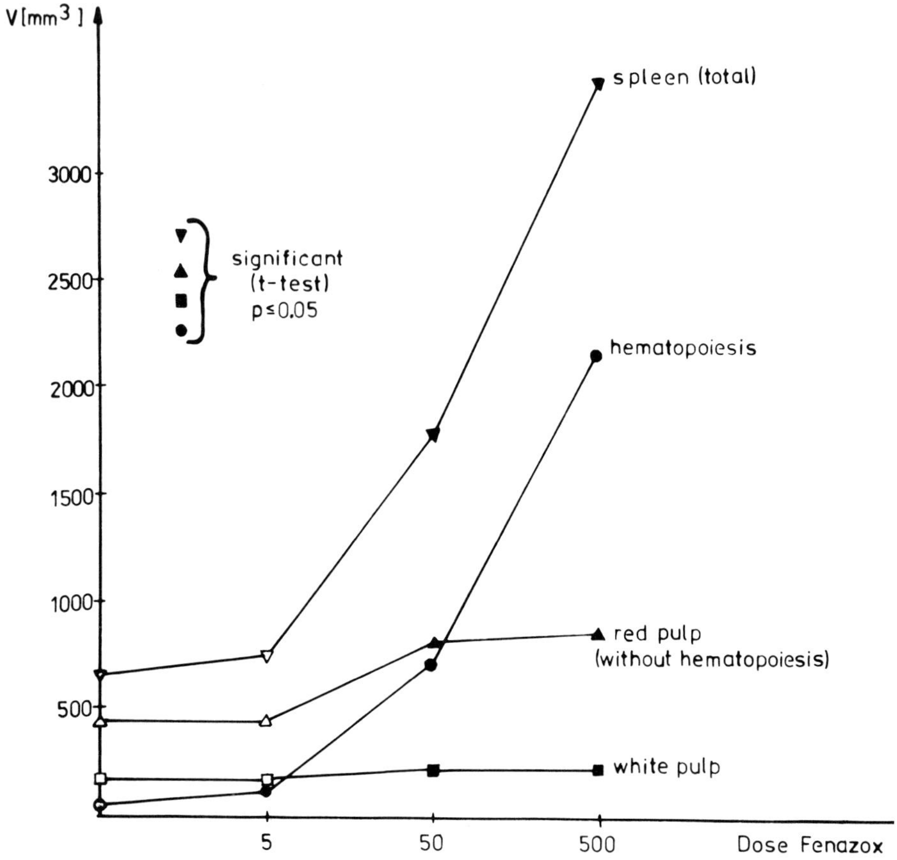

Fig. 3. The effect of a hemoglobin inducing substance (Fenazox) on component of the spleen. Dosage is in mg/kg bodyweight. Dark symbols indicate significance (t-test) $p \leq 0.05$

dose group (5 mg/kg KM) the increase of erythrocytopoesis in spleen was significant. In the bone marrow of the same animals the erythrocytopoiesis was only moderately increased with minimal significance.

Discussion

A comparison of methods for bone marrow investigation shows, that parallel investigation of slides and smears are optimal and supplement each other. The advantage of smears is the representation of cell plasma, which is especially important for the exact differentiation between lymphocytes and cells of erythrocytopoiesis. The advantage of slides is the possibility of investigation of stroma cells and the possibility of quantification of thrombocytopoiesis using morphometric methods. Megacaryocytes in smears are unevenly distributed. Other methods, such as nucleated cell counting are of subordinated importance because they give only a general information which may be, in cases of increased hematopoiesis, already observed in the spleen. The counted values of bone marrow cells in the presented myelogram strongly deviate from the findings of Cohrs et al (1958). They found 1.03% lymphocytes and about 47% cells of erythrocytopoiesis. The differences may not only be interpreted by different rat lines and maintenance, but suggest a difference in preparation or differentiation criteria. The previously determined minimum number of cells required to differentiate in smears shows, that in toxicological tests it is only useful to differentiate the larger cell lines such as granulocytopoiesis, and not the particular cell types. In this case with 500 counted cells the large cell lines can be determined with an accuracy of at least 90%. 500 cells is also the upper limit for routine counting. Previous toxicological investigations (Moeller 1990) indicated, that some parameters of toxicological routine tests are strong influenced by chemical stress. These results show, that also bone marrow cell numbers are strongly influenced by stress. In toxicological tests, such results could be mistakenly accepted as myelotoxic effects. An increased hematopoiesis can be noticed with the investigations of spleen (organ weight, histology), which are obligatory in the majoratory of toxicological tests. It is suggested that it is unnecessary in these cases to investigate the bone marrow because it supplies no further information. A screening programme to find effects on hematopoiesis can include investigations of blood and spleen in the first stage. Only in the case of a decreased blood cell number without the reaction of the spleen, should the bone marrow be investigated.

References

Cohrs P, Jaffe R, Meesen H (1958) Pathologie der Laboratoriumstiere. Springer-Verlag, Berlin

Moeller T (1990) Funktionelle Testung der Immuntoxizität der Phenocyalkansäure Mecoprop an Ratten und Ausschaltung immuntoxischer Effekte durch Adrenalektomie. Z gesamte Hyg 6:314–316

Remmele W (1984) Pathologie. Bd. 1. Springer-Verlag, Berlin, 435–436

Weibel ER (1979) Stereological methods. Academic press, London, 101–159

Recent Developments in Toxicology:
Trends, Methods and Problems
Arch. Toxicol., Suppl. 14, 88–91 (1991)
© Springer-Verlag 1991

The Effect of Aluminium Loading on Bones of Mice

G. Sahin[1], A. Isimer[2], A. Temizer[3], A. Sayal[2], R. Sunal[4], and S. Duru[1]

[1] Pharmaceutical Toxicology College of Pharmacy, Hacettepe University, 06100 Ankara, Turkey
[2] Department of Pharmacology and Toxicology, Gülhane Military Academy Ankara, Turkey
[3] Department of Analytical Chemistry College of Pharmacy, Hacettepe University, 06100 Ankara, Turkey
[4] Department of Pharmacology, College of Pharmacy, Hacettepe University, 06100 Ankara, Turkey

Introduction

A number of studies have been carried out on the toxic effect of aluminium on bone mineralization and its relation to calcium and parathroid-hormone metabolism. Exposure to aluminium by the means of either diasylate or ingestion of aluminium hydroxide or both leads to its deposition in tissues, particularly in bone. This deposition blocks incorporation of calcium into osteoids, leading to the osteomalacia characteristic of aluminium-induced bone disease (Drucke 1980; Alfrey 1980; Cournot-Witmer et al 1981; Walker et al 1982; Mayor et al 1980; Cann et al 1979; Cannata et al 1983; Boyce et al 1982). Therefore, the effects of aluminium hydroxide loading on bone of mice were examined. Besides aluminium, the calcium level of the same bone samples were determined and aluminium/calcium ratios were evaluated in this study. The aluminium and calcium measurements have been performed by graphite furnace and flame atomic absorption spectrometry respectively.

Methods

Instrumentation: Determination of aluminium was performed on a Perkin-Elmer 3030 Spectrometer, an HGA 400 electrothermal furnace and a Perkin-Elmer pr-100 recorder. A Perkin-Elmer aluminium hollow cathode lamp was operated at 25 mA. The atomic absorption of aluminium was monitored at 309.3 nm, the spectral band width was 0.7 nm and background correction was performed using a deuterium arc. The electrothermal furnace was purged with pre-purified argon gas during operation. The external argon flow rate was 900 ml/min and internal argon flow rate was 310 ml/min. Gas "stop condition" was used only during the atomization cycle. A Perkin-Elmer AS-40 autosampler was used to deliver 20 μl aliquots of sample solution. Pyrolytically coated graphite tubes (i.d. 5.8 mm- o.d. 8.0 mm and length 28.0 mm) were used.

The measurements of calcium in the same bone samples were performed by a Perkin-Elmer 403 Flame absorption spectrometer. In this method an air-acetylene mixture was employed.

Reagents and Solutions: Throughout the study triple distilled water was used for washing and standard preparation. Suprapure grade nitric acid (Merck) was used to acid leach containers and for the pre-treatment procedure. A dilute solution of triton X-100 (Sigma) was used together with magnesium nitrate (extra pure, Merck) was matrix modifier. In order to prepare the matrix modifier 2.0 g magnesium nitrate ($Mg(NO_3)_2$ $6H_2O$) was mixed with 2 ml concentrated triton-X 100 and the volume was adjusted to 1000 ml with triple distilled water. It was then transferred to polypropylene bottles.

A 1000 ppm certified atomic absorption aluminium reference solution (Fisher Scientific) was used to prepare secondary standards. Dilute aluminium solutions were prepared daily.

Experimental Design: In this study, 12 male Swiss albino mice (20 to 23.3 g) were used and fed with a commercial standard diet, and they were provided with distilled, deionized water or distilled, deionized water plus an aluminium hydroxide for 105 days. The mice were divided into 2 groups: Group 1 (n=5) received a commercial standard diet and distilled, deionized water: Group 2 (n=7) received a commercial diet and drank distilled, deionized water supplemented with aluminium hydroxide equivalent to 4.0 ± 1.7 μg/ml aluminium. Each animal had 5.0 ± 1.8 g commercial standard diet and drank 5.0 ± 1.2 ml distilled, deionized water daily.

For statistical analysis, Mann-Whitney U Test was used.

Sample Preparation: The whole femur was used as the bone sample. They were weighed and collected in polypropylene tubes which were decontaminated by soaking for at least two days in dilute nitric acid, rinsed three times with distilled water and dried in a laminar flow. The bone samples were stored and frozen at –20°C until they were analyzed.

Each bone sample was digested in 3 ml suprapure nitric acid (Merck) in an acid digestion bomb (Parr 4547) at 120°C for 2 hours. In 6 separate polypropylene tubes, 0.050 ml of digested bone samples were mixed with 0.100 ml of matrix modifier, then from 200 ppb secondary stock Al solution 0.025, 0.050, 0.075, 0.100, 0.150 ml were added in order. All tubes were filled up to 0.500 ml with triple distilled water. The mixtures were transferred to the automatic sampler of the graphite furnace and the results were taken from the printer after starting the furnace program.

Results and Discussion

Table 1 shows the furnace program which is obtained by careful optimization studies for measurements of Al. The control group had lower Al levels in the

Table 1. The Furnace Setting for aluminium measurement

Phases	Values
Drying temp. 1 (°C)	: 90
Time, ramph/hold (s)	: 15/10
Drying temp. 2 (°C)	: 120
Time, ramp/hold (s)	: 15/10
Ashing temp. (°C)	: 1500
Time, ramp/hold (s)	: 25/15
Atomization temp. (°C)	: 2650
Time, ramp/hold (s)	: 0/5
Cleaning temp. (°C)	: 2700
Time, ramp/hold (s)	: 1/3

bone samples (0.191 ± 0.059 mg/g) than those of the group treated with Al (0.235 ± 0.139 mg/g). In the control group, the average bone Ca level was (45.818 ± 15.188 mg/g) while bone Ca levels in the treated group averaged 28.580 ± 14.478 mg/g).

In the controls, the bone Al/Ca ratio was found to be 4.665.10 ± 1.839 and the average value in the group treated with Al was 8.638 ± 3.544.

The ratio was found to be significantly increased compared with the control group ($p<0.05$).

The results for aluminium agree with others (Susan et al 1983; Slanina et al 1984; Smeyer-Verbetke et al 1985). It is known that, for evaluation of accumulation and effect of aluminium, the bone sample is the best and the most logic at choice. However there may be individual differences among mineral contents of bones, for this reason, to consider the Al/Ca ratio could be an important and useful criterion in Al intoxication.

References

Alfrey AC, Hogg A, Crasswell P (1980) Metabolism and toxicity of aluminium in renal failure. Am J Clin Nutr 33:1509–1516

Boyce BF, Fell GS, Elder HY, Elliot HL, Fogelman I, Fell GS, Junor BJR, Beastall G, Boyle IT (1982) Hypercalcaemic osteomalacia due to aluminium toxicity. Lancet ii:1009–1013

Cann CE, Prussian SG, Gordon GS (1979) Aluminium uptake by the parathyroid glands, JClin Endocr Metab 49:543–575

Cannata JB, Briggs JD, Junor BJR, Fell GS, Beastall G (1983) Effect of acute aluminium overload on calcium and parathroid-hormone metabolism. Lancet March 5:501–503

Cournot-Witmer G, Zingraff J, Plachot JJ, Escalg F, Lefévre R, Boumato P, Bourdeau AS, Garabedian M, Galle P, Bourdon R, Drüeke T, Balsan S (1981) Aluminium localization in bone from hemodialysed patients; relationships to matrix mineralization. Kidney Int 20:375–385

Drüeke T (1980) Dialysis osteomalasia and intoxication. Nephron 26:207–210

Mayor MG, Sprague SM, Hourani MP, Sanchez TV (1980) Parathyroid hormone mediated aluminium deposition and regress in rat. Kidney Int 17:40–44

Ott MS, Maloney MA, Kline GL, Alfrey AC, Ament ME, Coburn JW, Sherrard DJ (1983) Aluminium is associated with low bone formation in patients receiving chronic parenteral nutrition. Ann Int Med 98:910–914

Slanina P, Falkenborn Y, Frech W, Cedergren A (1984) Aluminium concentrations in the brain and bone of rats fed citric acid, aluminium citrate or aluminium hydroxide. Fd Chem Toxic 22 (5):391–397

Smeyera-Verbeke J, Verbeelen D (1985) Determination of aluminium in bone by atomic absorption. Spectroscopy, Clin Chem 31 (7):1172–1174

Walker GS, Aaron JE, Peacock M, Robinson PJA, Davison AM (1982) Dialysate aluminium concentration and renal bone disease. Kidney Int 21:411–415

Recent Developments in Toxicology:
Trends, Methods and Problems
Arch. Toxicol., Suppl. 14, 92–95 (1991)
© Springer-Verlag 1991

Cobalt Particles, Identified by Energy-Dispersive X-Ray Microanalysis, in Diamond Polisher's Lung

D. Dinsdale, E. K. Verbeken[1], M. Demedts[2], and B. Nemery[3]

MRC Toxicology Unit, Woodmansterne Road. Carshalton, Surrey SM5 4EF, UK, and Departments of [1] Pathology, [2] Pneumology & [3] Lung Toxicology, Katholieke Universiteit Leuven, B-3000, Belgium

Introduction

Fibrosing alveolitis and bronchial asthma have long been recognised as hazards of occupational exposure to "hard metals".

The toxicity of these alloys has been shown to result from the presence of cobalt and attributed to the solubility of the metal within the body (Schepers 1955).

The development of fibrosing alveolitis, in humans exposed to cobalt, not in association with tungsten carbide, was first reported by Demedts et al (1984). High-speed diamond-cobalt grinding tools release fine particles of pure cobalt when they are used to polish hard materials, particularly diamonds. Polishing discs are made from half-hard steel and their cutting surface consists of micro-diamonds cemented into a layer of finely-divided cobalt.

The presence of particles of cobalt, which did not contain detectable levels of any other metal, has recently been demonstrated (Nemery et al 1990) in autopsy samples taken from a patient who died from cobalt-lung. This work also suggested that the disease may involve tissue injury by oxygen free-radicals.

The present study was undertaken to establish if, despite its reported solubility, particles of cobalt could be localised and identified within archive samples of clinical material which had been processed for paraffin histology.

Materials and Methods

Pulmonary biopsies were selected from 13 patients with various forms of interstitial lung disease. These samples had been fixed in Bouins solution and embedded in paraffin wax for routine histology. Sections (5–7 μm thick) were mounted on coded glass coverslips for transit.

The coverslips were de-waxed in xylene, rinsed in absolute ethanol and embedded, section uppermost, in Araldite resin using microscope-slide-shaped moulds (75 × 25 × 2 mm). The polymerised blocks were examined by light microscopy and areas rich in particulate material were selected for ultramicrotomy. Each coverslip was removed by immersion in hydrofluoric acid (40%), for 2 minutes, and the region of interest was mounted for sectioning. Sections (0.2–1.0 μm thick) were mounted on titanium grids and examined in a JEOL 100-CX electron microscope, using a customised titanium holder. Energy Dispersive X-ray Spectrometry (EDS) was conducted at 80 kV using a Model 860 X-ray spectrometer (Link Systems, U.K.).

Results

Particles were found in twelve of the lung biopsies. The morphology (Figs. 1 a–b) and composition of these particles was very diverse but the most common elements were aluminium and silicon (Figs. 1a and 2a). Cobalt was found in 2 of the 13 samples, it was present in elongate particles, 250–500 nm long, which were often aggregated into dense clusters up to 2 μm in diameter (Fig. 1b). Analysis of the particles indicated that the cobalt was not alloyed to any other metals (Fig. 2b), the only other element detected being sulphur. Carbon cannot be detected by this EDS system. No particles containing tungsten, or other metals with known toxicity, were detected in these samples although significant

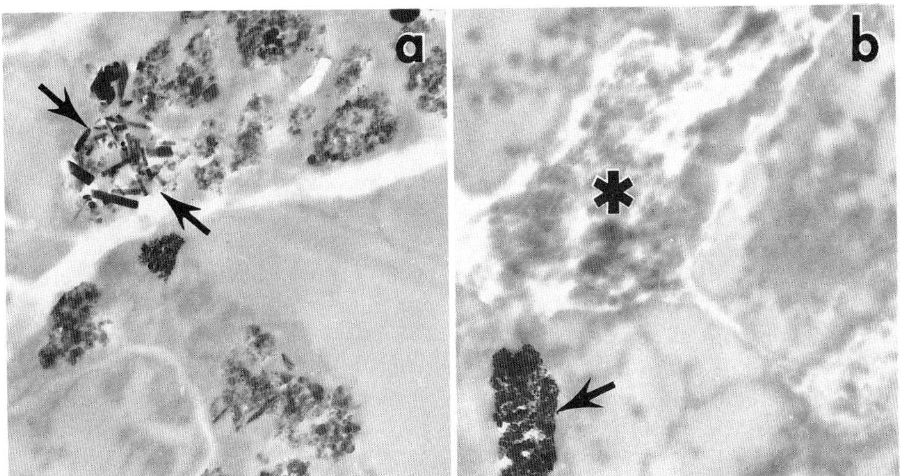

Fig. 1 a, b. Lung biopsies from patients with **(a)** eosinophilic pneumonia, showing the various sizes and forms of particle present, and **(b)** "cobalt-lung", showing a clump of dense particles (arrow) in the lung parenchyma. Fibrotic material is evident (*) in the alveolar wall. (Bars = 2 μm)

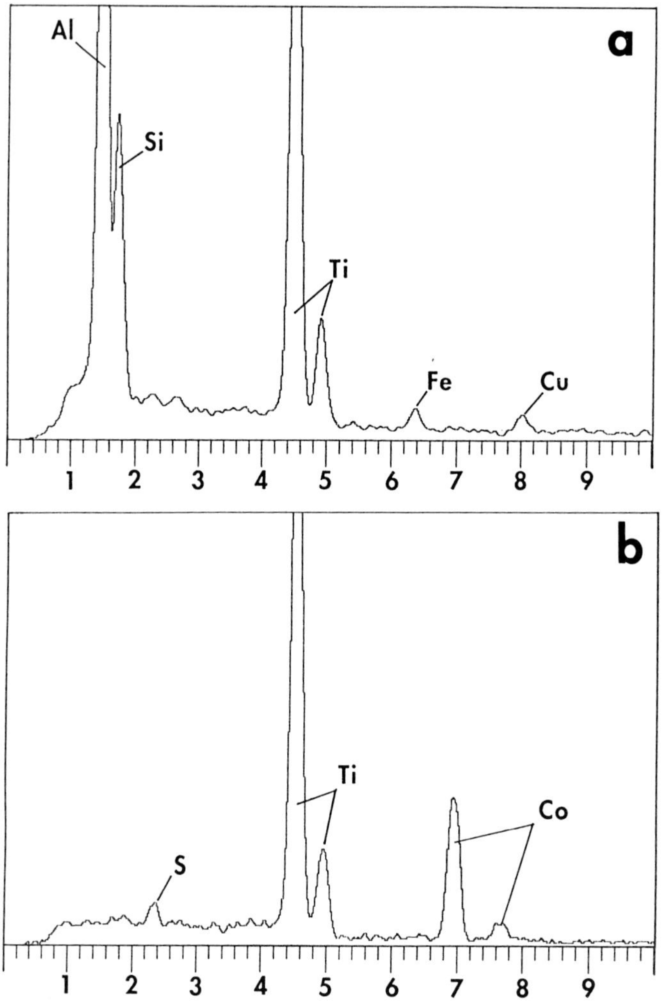

Fig. 2a, b. EDS Spectra, 0–10 keV, vertical scale = 511 counts, showing **(a)** the predominance of aluminium and silicon in the group of particles indicated in Fig. 1a and **(b)** the presence of cobalt in the clump of particles indicated in Fig. 1b

levels of iron and traces of copper were found in some of the larger, heterogeneous aggreates of particles. Trace levels of these metals were also detected in some groups of non-cobaltous particles (Fig. 2a).

Identification of the biopsies indicated that both samples containing cobalt had originated from diamond polishers. Independent diagnoses, based upon case histories and routine pathology, were consistent with the "cobalt-lung syndrome".

Cobalt was not found in one other sample from a suspected case of cobalt lung. The remaining biopsies had originated from patients with eosinophilic pneumonia (5), asbestosis (2), extrinsic allergic alveolitis (2) or sarcoidosis (1). Four of these samples showed the presence of significant numbers of particles containing heavy metals but no trace of cobalt was detected. The sarcoid granuloma was the only sample in which particles were not found.

Discussion

Inhaled particles are often sequestered and retained in human lung, without resulting in any clinical effects, but the toxicity of cobalt has been related to its solubility in protein-containing fluids (Harding 1950). Very fine particles are particularly susceptible to dissolution and so the retention of finely-divided cobalt, within archive material, was unexpected. The signal for sulphur, which was invariably associated with that for cobalt, may reflect the composition of the original material but this element is usually found during the analysis of lysosomes in many organs. The particles may thus have been sequestered within the acidic contents of phago-lysosomes but this could not be confirmed owing to the poor preservation of the tissue. The solubility of the metal is known to be much lower in saline than in plasma and so the retention of particles may depend upon minimising any delay before sampling. Such a delay may account for our failure to detect cobalt particles in the third case of suspected cobalt-lung. The sampling procedure, where regions for analysis are selected according to the presence of non-cobaltous particles, visible by light microscopy, must also contribute to the incidence of false-negative results.

This work has shown that cobalt particles are retained during the processing of samples for routine histology and electron microscopy. The microanalysis of these samples successfully eliminated samples from patients exposed to other dusts, including other heavy metals. Retrospective studies may thus be undertaken to examine other archive material where exposure to cobalt may have resulted in lung disease.

The results of this work are also compatible with the hypothesis that fibrosing alveolitis, in diamond polishers, is specifically associated with exposure to cobalt.

Acknowledgements. We are grateful to Stanley Preston for his valuable assistance.

References

Demedts M, Gheysens B, Nagels J, Verbeken E, Lauweryns J, Van den Eeckhout A, Lahaye D, Gyselen A (1984) Cobalt Lung in Diamond Polishers. Am Rev Respir Dis 130:130–135

Harding HE (1950) Notes on the Toxicology of Cobalt Metal. Brit J Ind Med 7:76–78

Nemery B, Nagels J, Verbeken E, Dinsdale D, Demedts M (1990) Rapidly Fatal Progression of Cobalt-Lung in a Diamond Polisher. Am Rev Respir Dis 141:1373–1378

Schepers GWH (1955) The Biological Action of Particulate Cobalt Metal. Am Arch Ind Health 12:127–133

Recent Developments in Toxicology:
Trends, Methods and Problems
Arch. Toxicol., Suppl. 14, 96–99 (1991)
© Springer-Verlag 1991

Density of Gastroendocrine Cells in Cynomolgus Monkey and Other Species

K. Tuch, D. Ockert, U. Hübel, and D. Hauschke[1]

Byk Gulden Pharmaceuticals, Institute of Pathology and Toxicologty,
Hamburg and Department of Biometry[1], Konstanz, FRG

Introduction

Potent acid suppression in the stomach induces hypergastrinemia and an increase of fundic enterochromaffin like cells ((ECL) Håkanson et al 1986). It is known that rats normally have a high content of ECL cells in the gastric mucosa in comparison to humans. This seems to be the reason that rats develop carcinoids under conditions of unsurmountable acid inhibition over 2 years (Ekman et al 1985; Koop et al 1987; Betton et al 1988). In man, however, the risk of inducing carcinoids by antiulcerative drugs is negligable (Lambert et al 1988). The basic number of ECL cells in the gastric fundus of 6 species were investigated.

Material and Methods

Untreated animals of 6 different species and different rat strains were studied. The number of animals, the ages and the body weights are listed in Table 1. Stomachs were fixed in Bouin's solution, embedded in paraplast and stained with silver solution according to Grimelius (1968). The number of Grimelius positive cells (GPC) were counted. GPC, i.e. argyrophilic cells, comprise ECL-cells and A-like cells, but the main portion is represented by ECL-cells.

The GPC were counted with a light microscope at 16 times magnification in 5 visual fields over the entire height of the fundic mucosa. The area was made up by the basic line of the visual fields and the mucosal height. The number of GPC was converted into 1 mm^2 mucosal area (GPC density).

Statistics: The mean and the standard error as well as the median with minimal and maximal values are displayed for each species (Table 1). The differences between the rat strains were evaluated by the Kruskal-Wallis test.

Table 1. Number of Grimelius positive cells (GPC) per mm^2 fundic mucosa

Species Strain	n = Sex	Age	Mean SP	Median min	 max
Rat					
Sprague Dawley	10 F	65 w	766 289	797 275	 1224
Wistar	9 F	65 w	482 218	509 186	 912
Fischer 344	10 F	65 w	646 107	642 446	 784
Mouse					
B6C3F1	15 F	10 w	165 25	161 128	 232
Guinea pig					
hairless, Pirbright	45 F	7 w	118 32	113 77	 262[1)]
	42 F	7 w	117 23	119 66	 176[2)]
Rabbit					
Himalaya Russian	35 F	8 m	96 29	95 46	 154
Dog					
Beagle	9 M	1 y	284 43	277 229	 355
	10 F	1 y	294 64	288 198	 392
Cynomolgus					
Macaca fascicularis	5 M	*	129 40	139 85	 173
	6 F	*	117 36	115 73	 160

* body weight M: 5–6.2 kg
F: 3–3.7 kg

1) Fundus I
2) Fundus II

Results

The density of GPC was the highest in rats. The Sprague Dawley strain has normally more GPC and shows more deviation than the two other rat strains (Fig. 1). The dog shows the second highest GPC density. As one can see in guinea pigs for example, a negative correlation exists between the height of the mucosa and the number of GPC. The low coefficient indicates a poor dependency between both parameters (Fig. 2).

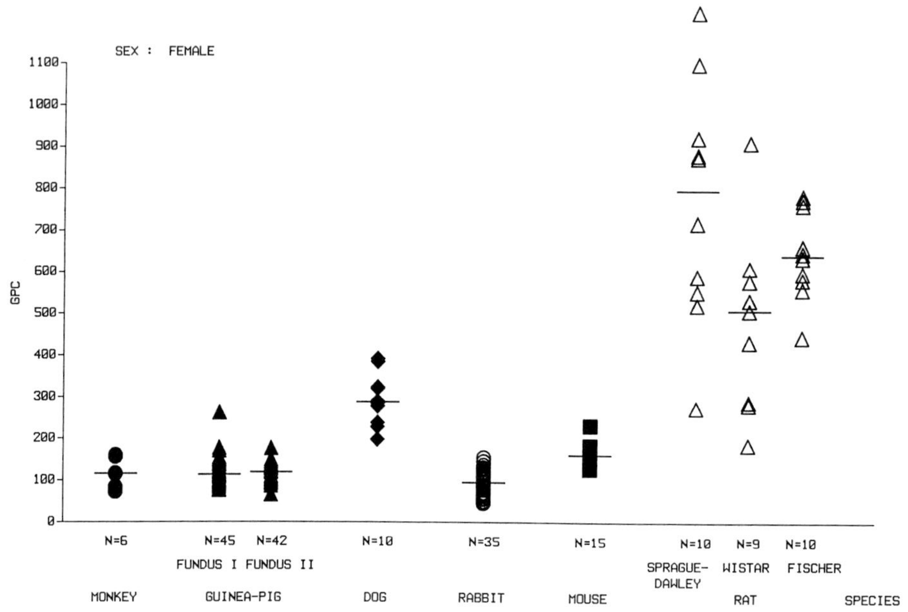

Fig. 1. GPC-density in the fundic mucosa in female animals of different species. In the guinea pig two localizations were evaluated

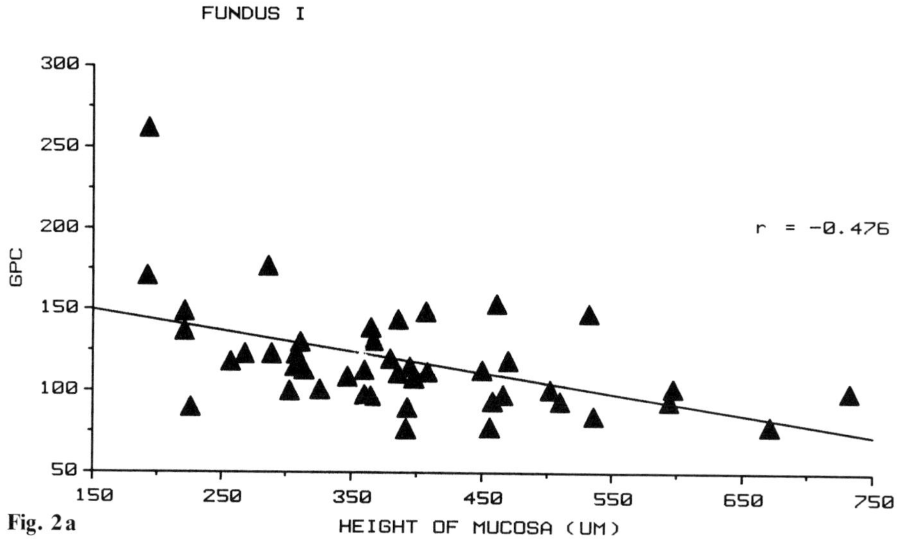

Fig. 2 a

Conclusion

Although it is possible to induce hypergastrinemia in different species due to acid suppression, it is possible that only the rat tends to develop carcinoids because of the normally high number of GPC in the fundic mucosa.

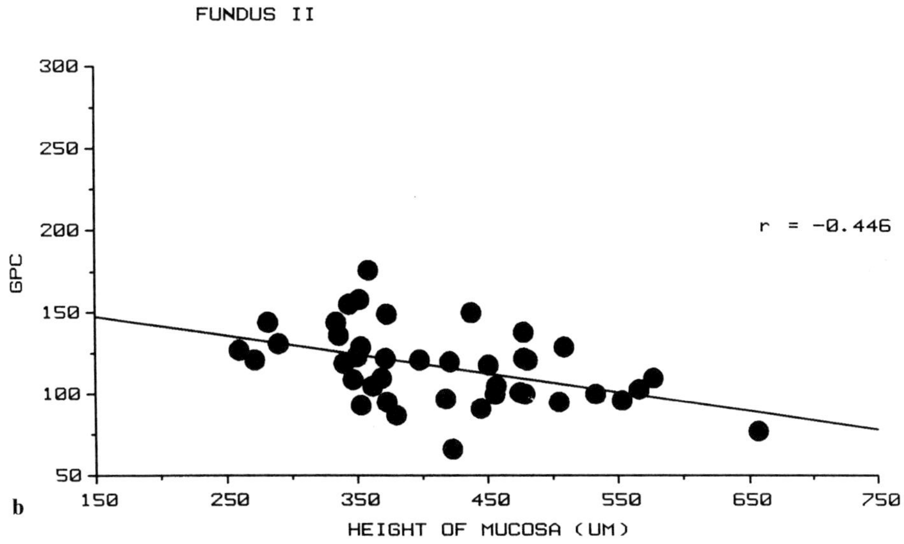

b

Fig. 2. Weak negative correlations between height of the mucosa and GPC in the guinea pig

References

Betton GR, Dormer CS, Wells T, Pert P, Price CA, Buckely P (1988) Gastric ECL-cell hyperplasia and carcinoids in rodents following chronic administration of H_2-antagonists SK&F 93479 and oxmetidine and omeprazole. Toxicol Pathol 16:288–298

Ekman L, Hansson E, Havu N, Carlsson E, Lundberg C (1985) Toxicological studies on omeprazole. Scand J Gastroenterol 20, Suppl 108:53–69

Grimelius L (1968) A silver nitrate stain for A_2-cells in human pancreatic islets. Acta Soc Med Ups 73:243–270

Håkanson R, Böttcher G, Sundler F, Vallgren S (1986) Activation and hyperplasia of gastrin and enterochromaffin-like cells in the stomach. Digestion 35, Suppl 1:23–41

Koop H, Willemer S, Steinbach F, Eissele R, Tuch K, Arnold R (1987) Influence of chronic drug-induced achlorhydria by substitued benzimidazoles on the endocrine stomach in rats. Gastroenterology 92:406–413

Lambert R, Creutzfeld W, Stöckmann F, Jacubaschke U, Maas S, Brunner G (1988) Long-term omeprazole treatment in man: Effect on gastric endocrine cell populations. Digestion 39:126–135

Recent Developments in Toxicology:
Trends, Methods and Problems
Arch. Toxicol., Suppl. 14, 100–103 (1991)
© Springer-Verlag 1991

Computerised Analysis of Pathological Findings in Longterm Trials with Phenylmercuric Acetate in Rats

R. Solecki, L. Hothorn*, M. Holzweissig*, and V. Heinrich

Institute of Toxicology and Ecotoxicology, Biological Research Centre Berlin, Stahnsdorfer Damm 81, Kleinmachnow, O-1532
* Produktentoxikologie Graupa GmbH, Bonnewitzer Str. 34, O-8304 Graupa

Introduction

Recording, analyzing and printing the massive amount of pathological data from animal experiments requires computerized pathology data systems with a systematized nomenclature of lesions.

This paper describes a specialized application of the software system CTPAT (ChronTox-Pathology) as a tool for the pathologist in the examination of a long-term bioassay with phenylmercuric acetate (PMA).

Experimental conditions and application of CTPAT

20 male Ico:WIST rats of each dose group, obtained from VTP Schönwalde, were treated for 103 weeks with PMA (VEB Fahlberg-List Magdeburg, purity 65%) in drinking water (dw) in doses of 0 (C), 5 (B1) and 50 (B2) mg PMA/l dw, respectively. The tested charge of PMA was identical with the basic compound of the the "Falisan"-preparations as "Falisan-Saatgut-Naßbeize", "Falisan-Universal-Flüssigbeize" and "Falisan-Universal-Trockenbeize 69" (PSM-Verzeichnis 1989/90), which was not characterized in detail by the producer. Hematological, serum and urine analysis as well as organ weights were recorded according to Solecki et al (1990). The software systems were developed by the authors. Numerical data were processed with the CHRONTOX III software system. The CTPAT system has the possibility to access and to use CHRONTOX III data. The software packages are supported on AT-PC (personal computer) under MS-DOS 3.x and prepared for use in local networks. All files are compatible with dBase III-tables. The outputs are possible as screen, printer or ASCII-files. The CTPAT system was used for evaluation, processing and reporting of mortality and pathological findings in this long-term bioassay. The macroscopic and microscopic examination was carried out on the basis of the systematized CAPLA-nomenclature for non-neoplastic and

neoplastic lesions in rats, which was developed in this Institute (ITÖ Kleinmachnow). All findings were graduated semiquantitatively. Topographical and descriptive qualifiers, also generated by the glossary, were added to each finding. Free-text comments were used to complete the description of pathological lesions, which were insufficiently characterized by the glossary terms. The statistical procedures were carried out according to Hothorn (1989) using the PASSING/84-procedure.

Results

The significantly changed parameters following chronic treatement with PMA (Fig. 1), estimated with the CHRONTOX III software were a dose-dependent increase of leucocytes and adrenal weight together with a decrease of hemoglobin, erythrocytes, hematocrit, and body weight, respectively.

Different tables were created by CTPAT. The summary table (Table 1) shows a significant increase of intercurrent mortality, animals with tumors and total amount of tumors in the B2 group. The incidence table for neoplastic lesions demonstrates the spontaneous tumor rate in untreated old Ico:WIST rats. In the kidneys a significant increase of renal cell adenomas was observed in the dose of 50 mg PMA/l dw (Table 1). The incidence table for non-neoplastic lesions (Table 2) demonstrates an increase in incidence of the fol-

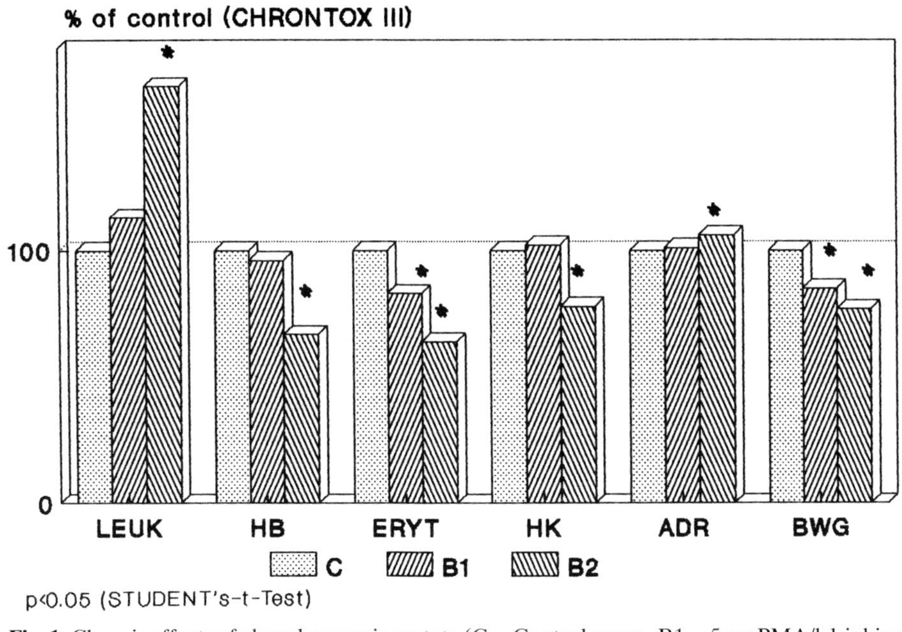

Fig. 1. Chronic effects of phenylmercuric acetate (C = Control group, B1 = 5 mg PMA/l drinking water, B2 = 50 mg PMA/l drinking water) on rats; LEUK = leucocytes, HB = hemoglobin, ERYT = erythrocytes, HK = hematocrit, ADR = adrenal weight, BWG = body weight

Table 1. Incidence table of intercurrent mortality and significant changed neoplastic lesions created by the "CTPAT" system

Groups:	:	C	B1	B2
No. animals:	:	20	20	20
Intercurrent mortality:	:	7	11	14 #
Animals with tumors:	:	9	8	16 #
Sum of tumors:	:	10	13	21 #
Kidney:	:	18	20	20
– oncocytoma	:	0	0 –	1 –
– renal cell adenoma	:	0	5 –	10 *

CHI-square test: # = P<0.05
Passing/84 : – = P>0.05; * = P<0.05
Treatment : C = Control; B1 = 5 mg PMA/l DW; B2 = 50 mg PMA/l DW

Table 2. Incidence table of significant changed non-neoplastic lesions created by the "CTPAT" system

Groups:	:	C	B1	B2
No. animals:	:	20	20	20
Cecum	:	15	15	17
– ulcerative cecitis	:	0	1 –	13 *
Kidney	:	18	20	20
– pigment deposits	:	5	18 *	14 *
– chronic renal disease	:	19	20 –	19 –
– chronic renal disease > grade 2	:	7	19 *	14 *
*grade 1	:	2	0	4
*grade 2	:	10	1	1
*grade 3	:	3	2	4
*grade 4	:	2	3	6
*grade 5	:	2	14	4

Passing/84 : – = P>0.05; * = P<0.05
Treatment : C = Control; B1 = 5 mg PMA/l DW; B2 = 50 mg PMA/l DW

lowing lesions: ulcerative cecitis in the large intestine in the B2 group and pigment deposits in the kidney cortex in the B1 and B2 groups. Significantly more animals in both dose groups showed a grading of chronic nephrosis >2 after splitting this age-associated lesion with the reporting table. The single animal sheets containing all gross lesions and the microscopic findings with all free-text comments were printed for better interpretation of the totality of all individual pathological findings. They showed a correlation between high degrees of pathological lesions in the stomach and intestine.

Discussion

In the present study, the application of CTPAT has shown a decreased time expense and an increased security of the pathological examination. The subjective error of the pathological investigation will be markedly reduced by the use of the systematized nomenclature and the grading of each finding. However, the total elimination of this error will not be possible. The operations of CTPAT are in compliance with the Good Laboratory Practise regulations. In previous studies it was demonstrated, that PMA can cause after an acute administration a high degree of epithelial necrosis in the kidney cortex (Solecki et al 1985), followed by an induction of adaptive reactions after subacute and subchronic treatment with PMA (Solecki et al 1989). The chronic treatment with PMA increased the development of age-associated chronic nephrosis in connection with the enhanced incidence of renal cell adenomas. Besides the nephrotoxicity, the chronic PMA treatment damaged the general health status, the digestive system, as well as the red and white blood cells. The incidences of other age-associated neoplastic and non-neoplastic lesions in untreated rats were reported to be in the same range for a larger control group Ico:WIST rats of the same stock (Solecki et al 1990). The results of this paper lead to the suggestion that the tested charge of PMA, which was used in agricultural practice (PSM-Verzeichnis 1989/90), is probably a potential conditional carcinogen of the promoting class on the basis of the nephrotoxic degenerative action of this compound, but this assumption must be validated in a carcinogenicity study. The DABACC has considered PMA a substance with carcinogenic action after both intravaginal application and implantation in the urinary bladder (Teichmann et al 1979).

References

Hothorn L (1989) A simple statistical procedure for testing tumour rates in animal carcinogenicity experiments. Arch Toxicol, Suppl 13:265–286

PSM-Verzeichnis (1989/90) Pflanzenschutzmittelverzeichnis der DDR 1989/1990. Ed. by Akademie der Landwirtschaftswissenschaften der DDR, VEB Deutscher Landwirtschaftsverlag 1988, 41–42

Solecki R, Gottschalk M, Mahro U (1985) Nephrotoxische Effekte von Phenylquecksilberazetat: Akute Wirkungen bei der Ratte. Ergebn Exper Med 46:459–461

Solecki R, Mahro U, Heinrich V (1989) Beeinflussung toxischer Wirkungen von Phenylquecksilber durch adaptive Reaktionen der Rattenniere. Arch exper Vetmed, Leipzig 43:541–550

Solecki R, Moeller T, Mahro U, Peetz U (1990) Influences of Age-Associated Changes in Specific-Pathogen-Free Ico:WIST – Rats on Long-terms Bioassays. Beiträge aus toxikol. Forschung i.d. WB der MLU 19:8–17

Teichmann B, Schramm T, Teichmann M (1979) Substanzen mit kanzerogener Wirkung, ZIK-DABACC, Berlin-Buch, 7

Cardiovascular Toxicology

Recent Developments in Toxicology:
Trends, Methods and Problems
Arch. Toxicol., Suppl. 14, 107–117 (1991)
© Springer-Verlag 1991

Preclinical Assessment of the Cardiotoxic Potential of Anthracycline Antibiotics: N-L-Leucyl-Doxorubicin

G. Zbinden*, D. DeCampeenere**, and R. Baurain**

* Institute of Toxicology, Swiss Federal Institute of Technology and University of Zurich, CH-8603 Schwerzenbach, Switzerland, and
** Medgenix Group, Brussels, Belgium

The Concept of Toxicological Screening

The primary task of industrial toxicologists is to provide a comprehensive assessment of the adverse effects of selected chemical substances in several animal species and over a broad range of doses. Increasingly, their collaboration is also required when a compound has caused serious toxicity in humans or animals, and when it is decided to modify the chemical structure, in an effort to reduce or to eliminate the undesirable properties. In such programs it is often necessary to assess a considerable number of newly synthesized substances. Moreover, the results of the toxicity tests must be available within a short period of time, and they must be quantitative, so as to provide a rational basis for the synthesis of new derivatives. It is evident that the standard toxicological procedures cannot be used for these purposes, and that a toxicological screening approach has to be developed which differs fundamentally from that on which the conventional safety assessment programs are based (Zbinden et al 1984a). The main differences between conventional safety testing programs and toxicological screening are summarized in Table 1.

The variables measured in a toxicological screening test are often selected to imitate, as closely as possible, the characteristic features of the toxicological syndrome seen in man. An example is a short-term *in vivo* model in rats, designed to detect nephrotoxic properties of chemicals, in which urinary excretion of cells (red and white blood cells, epithelia), proteins, enzymes, electrolytes, glucose and amino acids is used to assess the degree of kidney damage (Fent et al 1988; Zbinden et al 1988). On the other hand, very specific biological properties of chemicals can also be used as screening parameters, if it is

---

Abbreviations: BP: blood pressure; sec: second(s); DOX: doxorubicin; h: hour(s); ECG: electrocardiogramm; min: minute(s); AUC: area under the curve; LEU-DOX: N-L-Leucyl-doxorubicin; i.p.: intraperitoneal; i.v.: intravenous

Table 1. Characteristic features of conventional safety studies and toxicological screening

Conventional safety studies	Toxicological screening
Comprehensive assessment of the whole toxicological spectrum	Limited to selected target organ responses
Use of standard protocols	Flexible selection of test models
Mainly interested in no-effect levels of selected toxic effects	Interested in dose-effect relationships
Short-term and long-term studies	Only short-term studies are practical
Reference substances rarely tested	Reference substances always tested
In vitro models used only exceptionally	*In vitro* models frequently used
GLP procedures mandatory	GLP procedures not required
Acceptable for regulatory purposes	Not intended for regulatory purposes

established that the measured effect plays an important role in the causation of the toxic response. Examples are the determination of binding of benzodiazepines to human serum proteins and their cytotoxicity against *S. typhimurium* under the influence of ultraviolet light, as indicators of hazard for phototoxicity (Beijersbergen van Henegouven 1988), and the determination of lipid peroxidation in rat kidney slices as an assay for nephrotoxicity of cephalosporins (Cojocel et al 1988). Finally, it should be mentioned that toxicity of chemical substances can also be reduced if the formation of toxic metabolites can be prevented by appropriate structural modifications, or if one can reduce the uptake into sensitive target tissues. The development of N-L-leucyldoxorubicin (LEU-DOX) described in this paper is an example for this latter approach.

Screening Procedures for Anthracycline Cardiotoxicity

Anthracycline antibiotics are highly potent, broad-spectrum antitumor agents used for the treatment of many solid tumors and leukemias. Unfortunately, congestive heart failure due to toxic cardiomyopathy is a serious side-effect that can only be avoided by limitation of the total dose. Thus, many patients are deprived of the therapeutic benefits once the critical dose has been given. Efforts to reduce the cardiotoxic hazard of anthracyclines included the synthesis of derivatives with higher chemotherapeutic potency, the search for anthracyclines with lower intrinsic cardiotoxicity or with reduced uptake into the myocardium, and the development of antagonists against the cardiotoxic effects, e.g. ubiquinone.

The success of all attempts to reduce anthracycline antibiotics is critically dependant on the availability of reliable biological assay systems for cardiotoxicity. The models that have been proposed range from chronic treatment of rabbits and dogs with pathomorphological evaluation of the characteristic myocardial lesions, to acute toxicity studies in mice, and *in vitro* assays with cultured cardiocytes.

When it was found that cardiotoxic anthracycline antibiotics induced characteristic changes of the electrocardiogram (ECG) that preceded the appearance of pathomorphologic lesions of the heart (Zbinden and Brändle 1975), the rat *in vivo* assay that is based on repeated analysis of the ECG was adopted by many laboratories (Jensen et al 1984). Later, it was found that cardiotoxic anthracyclines also caused lowering of systolic blood pressure (BP) in rats, and serial determinations of this effect were also included in the screening system. In contrast to the findings in rabbits, the histopathological lesions of the cardiocytes that are characteristic for anthracyclines, i.e. dilatation of the sarcoplasmatic reticulum and loss of myofibrillar mass, are not as well developed in rats. Thus, the electron-microscopic analysis of the heart muscle does not contribute much to the early recognition of anthracycline cardiotoxicity in the rat model. In addition, morphologic lesions of the heart show focal distribution and are, therefore, difficult to quantify. For derivatives that are expected to have a lower propensity to accumulate in the heart muscle, determination of myocardial tissue concentrations and clearance from the heart are an integral part of the toxicological screening procedure (Zbinden et al 1984b).

Finally, it should be mentioned that anthracycline antibiotics cause nephropathy and sensory neuropathy in rats. Although these toxic responses have not as yet been a problem in humans, they are monitored in the rat screening model, together with the bone marrow depressant effects that serve as a general indicator of toxicity related to the intended pharmacodynamic effects.

Cardiotoxicity Screening of LEU-DOX in Rats

Rationale for Compound Selection

The synthesis of N-amino acids derivatives of anthracyclines was undertaken in the hope that they would act as pro-drugs and act more selectively in tumors with high aminopeptidase activity (Masquelier et al 1980). Based on the finding of high antitumor activity in animal models (Magis et al 1989), LEU-DOX (Fig. 1) was selected for further preclinical evaluation.

Materials and Methods

Female rats, IVA: SIV 50 strain (Ivanovas GmbH, Kisslegg, BRD) starting weight approximately 190 g, were housed 4 to 5 to a cage in Macrolone® cages Type IV with wood shavings as bedding. Food (Kliba 343 pellets) and water were freely available. Environmental conditions were: Temperature 22 ± 1°C, relative humidity 55 ± 10%, 12 h light-dark cycle. LEU-DOX and docorubicin (DOX) were obtained from IRE-CELLTARG S.A. LEU-DOX was dissolved in distilled water and DOX in 0.9% NaCl, and injected intraperitoneally (i.p.) on Mondays and Thursdays into groups of 10 rats. Solutions were freshly prepared immediately before each administration. The doses were as follows: LEU-DOX: 2.4, 4.8, and 9.6 mg/kg (2.0, 4.0, and 8.0 mg/kg of DOX equiva-

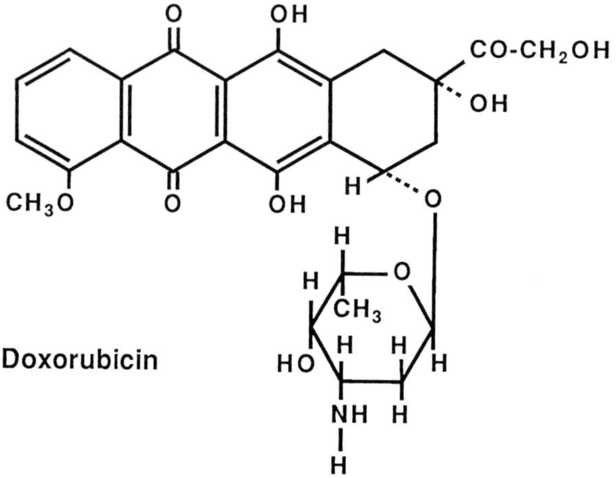

Doxorubicin

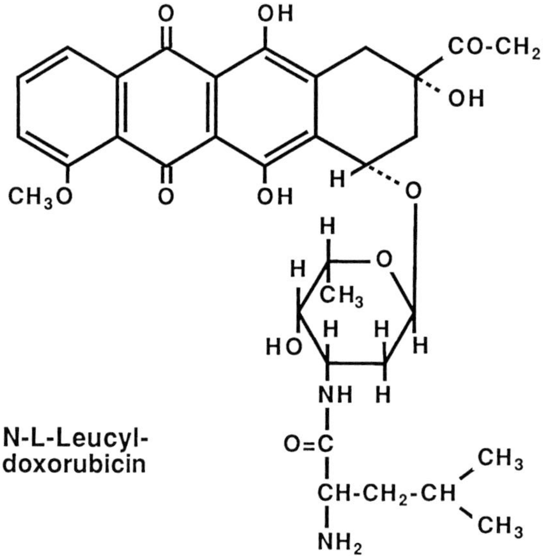

N-L-Leucyl-doxorubicin

Fig. 1. Structural Formula of LEU-DOX and DOX

lent, respectively). DOX: 2 mg/kg. A total of 12 doses were given to rats of the high dose LEU-DOX and the DOX group, the others received 15 doses. Controls were injected with equal volumes (2 ml/kg) of 0.9% NaCl i.p.

ECG (Zbinden 1981) and systolic BP (semi-automatic tail-cuff method with a piezoelectric sensor, Sandoz AG, Basel) were monitored without anesthesia once weekly before and during administration of the test substances. A neurobehavioral examination (Alder and Zbinden 1983) was performed weekly, a quantitative urine analysis (Zbinden et al 1988) after the 5th and the 10th dose, and blood cell counts every 2 weeks. Plasma biochemistry was performed with blood taken prior to sacrifice from the retroorbital sinus and by cardiac puncture.

All animals were killed by i.p. Nembutal® overdose 3 to 6 days after the last dose. They were autopsied, heart, spleen, kidney, adrenals, thymus and liver were weighed, and 20 organs, including heart, kidney and sciatic nerve, were prepared for light-microscopic examination. At least 6 small samples of both ventricles were taken from 3 rats per group and prepared for transmission electron microscopy (Zbinden et al 1978).

Results

All rats survived the experiment. All treatments caused a statistically significant reduction in body weight gain, but there was no statistically significant difference between the group treated with the high dose of LEU-DOX and that receiving 2 mg/kg DOX. Small nodular indurations were seen at the injection sites, particularly in rats receiving the high dose of LEU-DOX and DOX. Neurobehavioral disturbances were minimal and limited to the groups receiving 9.6 mg/kg LEU-DOX and 2 mg/kg DOX.

A precipitous drop in systolic BP was observed in rats treated with the high dose of LEU-DOX and 2 mg/kg DOX after administration of 4 doses. It should be noted that the equipment does not permit measurement of blood pressures

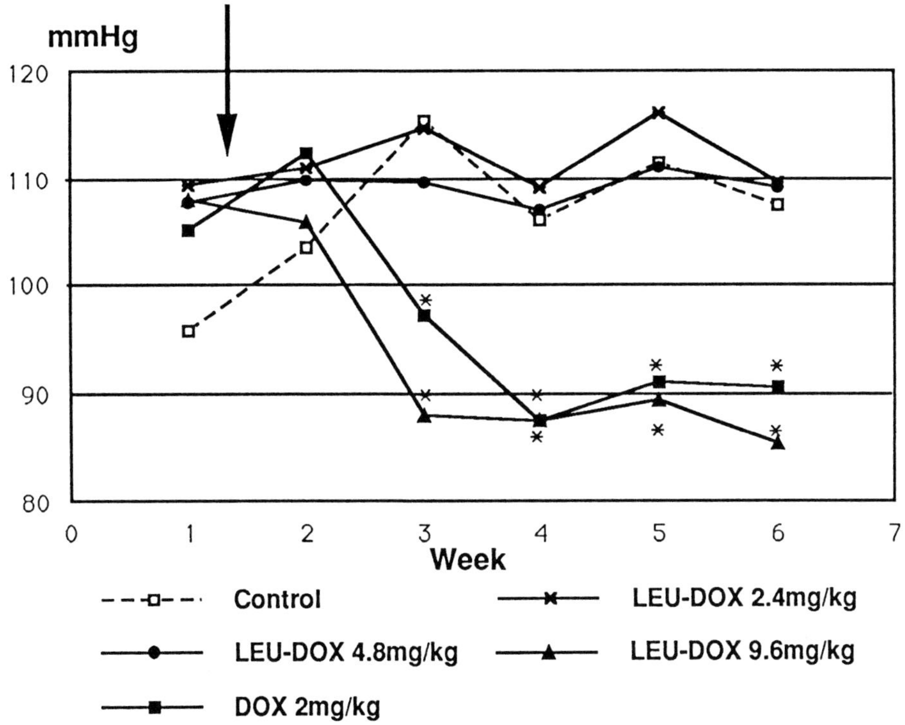

Fig. 2. Systolic blood pressure in groups of rats treated with LEU-DOX and DOX twice per week, i.p. * statistically significant difference in comparison with controls $p<0.05$ (U-test of Mann and Whitney, 2-tailed)

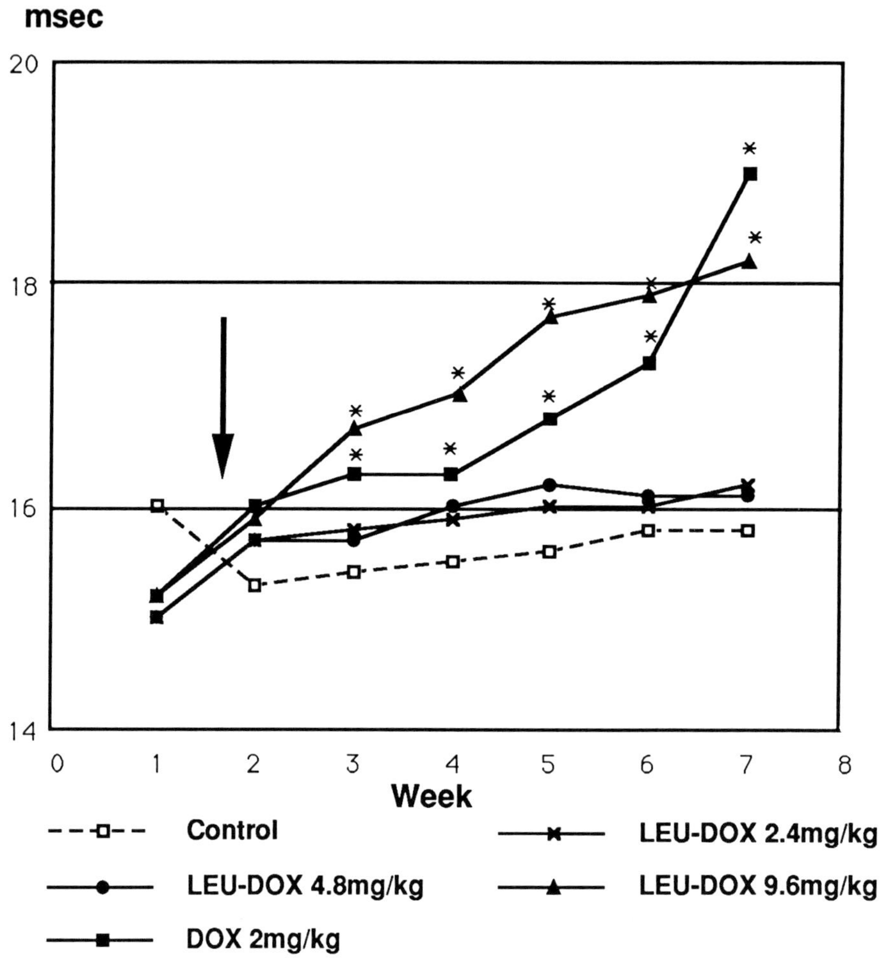

Fig. 3. QRS intervals in groups of rats treated with LEU-DOX and DOX twice per week, i.p. * statistically significant difference in comparison with controls $p<0.05$ (U-test of Mann and Whitney, 2-tailed)

lower than 80 mmHg. In all rats in which no blood pressure signal was obtained, a value of 80 mmHg was arbitrarily taken. No change in BP was seen in rats of the low and intermediate dose groups (Fig. 2).

The analysis of the ECG showed a statistically significant prolongation of the QRS interval after the 4th injection of the high dose of LEU-DOX and 2 mg/kg DOX (Fig. 3). With LEU-DOX the QαT interval was also prolonged at these time points, whereas in DOX-treated rats prolongation of QαT became statistically significant only after the 8th dose (Fig. 4). There were no changes of the PQ intervals and the P-, Q-, R-, S- and T-wave voltages (data not shown). With the lower doses of LEU-DOX no ECG changes were observed. A small number of arrhythmias occurred: single premature ventricular contractions (PVC) were seen in a rat on high dose LEU-DOX on 2 occasions and in 2 rats

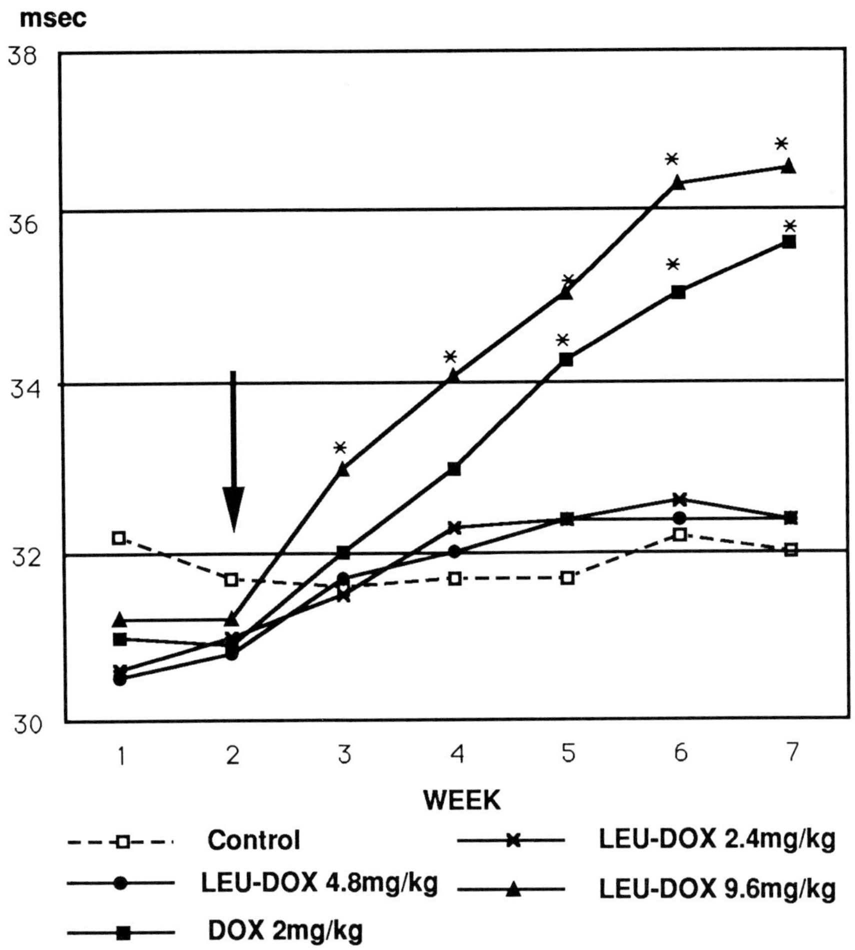

Fig. 4. QαT intervals in groups of rats treated with LEU-DOX and DOX twice per week, i.p. * statistically significant difference in comparison with controls p<0.05 (U-test of Mann and Whitney, 2-tailed)

of the intermediate LEU-DOX group on one occasion. A burst of interpolated PVC was seen in a DOX-treated rat, and an animal on the intermediate LEU-DOX dose had a single event of sinus arrest.

Leukopenia developed in all treated groups in a dose-and time-related fashion, but the red blood cells were not much affected. The white blood cell (WBC) counts did not differ significantly between the rats treated with the high dose of LEU-DOX and the animals receiving 2 mg/kg DOX. The urine analysis showed a dose-dependent increase in excretion of WBC and epithelia in the LEU-DOX-treated rats. The effect seen with the high dose was comparable to that measured in rats receiving 2 mg/kg DOX. Protein and enzyme (malate dehydrogenase and γ-glutamyltranspeptidase) excretion were not altered. The only statistically significant changes compared with controls seen

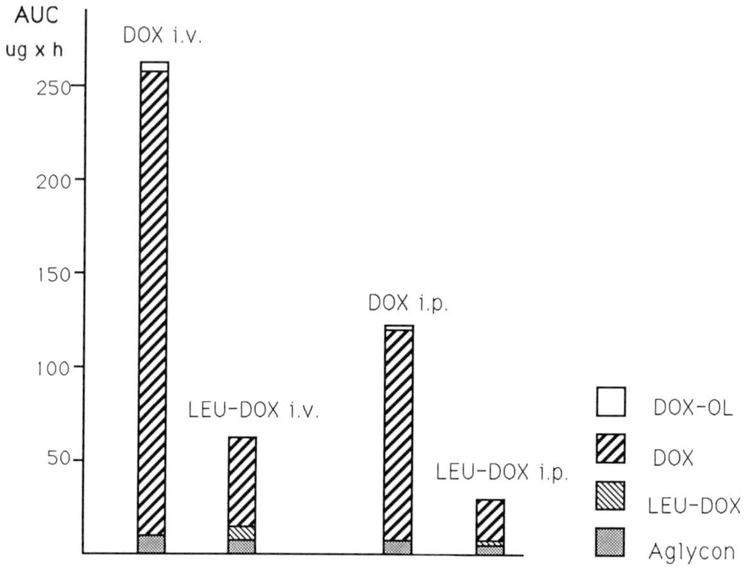

Fig. 5. Heart exposure after i.v. and i.p. administration of LEU-DOX and DOX to rats at equimolar doses, as measured by AUC ($\mu g \times h$)

in plasma biochemistry were increases in lactate dehydrogenase, aspartate aminotransferase, and creatine kinase in rats receiving 9.6 mg/kg LEU-DOX, and an increase in total cholesterol in DOX-treated animals, but there were no statistically significant differences between these 2 groups for any of these variables.

At autopsy ascites and peritoneal adhesions were seen in all DOX-treated rats and in a small number of animals receiving injections of LEU-DOX. Absolute spleen weight was decreased in intermediate and high dose LEU-DOX groups ($p<0.02$), relative kidney weight was increased in high dose LEU-DOX and DOX groups ($p<0.01$), absolute adrenal weight was decreased in the high dose LEU-DOX group ($p<0.05$), absolute and relative thymus weights were decreased in the low and intermediate LEU-DOX dose groups ($p<0.01$–0.05), and absolute heart weight was decreased in the DOX-treated rats ($p<0.05$). Light microscopic changes seen in the heart were slight myocardial atrophy (high dose LEU-DOX and DOX), and occasional focal necrosis and myocyte vacuolation (DOX). Atrophy of the lymphatic tissue and hemosiderosis were seen in rats receiving the high dose LEU-DOX and DOX. Modest evidence of proteinuria was observed only in 3 DOX-treated animals. Evidence of minimal to slight, focal neuropathy, i.e. ballooning and disintegration of myelin sheath, was obtained in 1, 4, and 7 rats treated with low intermediate and high dose LEU-DOX, respectively, and in 6 DOX-treated animals.

The electron microscopic examination of the hearts showed occasional and slight thinning of myofibrillar mass and minimal dispersion of the Z-line material in rats receiving the low and intermediate LEU-DOX doses. Occasionally slight crystolysis in mitochondria was present. In rats treated with the high dose of LEU-DOX focal mycocyte degeneration, slight dilatation of the sarcoplasmatic reticulum, moderate thinning of the myofibrillar mass and moderate crystolysis of the mitochondria were noted. In animals exposed to DOX moderate thinning of myofibrillar mass, slight dispersion of Z-line material and moderate crystolysis of mitochondria were observed.

Discussion

The results described in this paper show that a comprehensive assessment of the toxicological spectrum of anthracyclines is possible with the help of a relatively simple experiment in rats. Many of the toxic end-points can be measured quantitatively, i.e. the indices of general toxicity (survival, body weight gain), hematotoxicity (leukopenia), and the signs of cardiotoxicity (ECG changes, hypotension). The morphological lesions in the heart seen after short-term treatment with non-lethal doses are slight and show focal distribution. Therefore, they cannot be used for quantitative evaluation of cardiotoxicity. The release of marker enzymes is an equally unreliable indicator of myocardial damage caused by anthracyclines in rats.

The nephrotoxicity of anthracyclines can best be quantified by determining urinary protein and cell excretion (Hu et al 1983), but in the present study, these effects were too slight to be used as a screening parameter. For the evaluation of the weak neurotoxic effects of anthracyclines, the neurobehavioral check list approach that is useful for other neurotoxic agents (Alder and Zbinden 1983), is not sufficiently sensitive. With histopathological examination of the peripheral nerves it is possible to detect the lesions, but since the damage is focal and affects only small numbers of axons, quantitative conclusions cannot be drawn.

The evaluation of LEU-DOX in the rat screening model has shown that its spectrum of toxicity is identical to that of DOX. However, a dose that is four-fold higher than that of DOX had to be administered to induce the same degree of cardiotoxicity, i.e. hypotension and prolongation of the QRS and QαT intervals. The high dose of LEU-DOX and the four-fold lower dose of DOX also caused comparable signs of general toxicity (i.e. decrease in body weight gain, peritoneal irritation), and the bone marrow depressant effect als judge from the WBC counts was also the same. As mentioned before, signs of nephrotoxicity were slight with both drugs, and the degree of neuropathy observed in sections of the sciatic nerves appeared to be comparable.

The development of amino acid derivatives of anthracycline antibiotics was initiated in the hope that they might accumulate less in non-tumor tissues, particularly the heart.

Biopharmaceutical studies proved that this is indeed the case. Fig. 5 shows the exposure of the heart muscle, expressed as area under the curve (AUC), of rats after single intravenous (i.v.) and i.p. administrations of equimolar doses of DOX and LEU-DOX. It has been proved that LEU-DOX is converted to a large extent to DOX. With both drugs the total exposure of the heart after i.p.

administration is about 50% of that measured after i.v. injection. A comparison of the AUC of parent drugs and metabolites taken together indicates that the exposure of the heart with LEU-DOX is about 25% of that measured for DOX. If one considers only DOX exposure resulting from injections of LEU-DOX, the AUC is about 20% of that measured after DOX administration. These findings are in excellent agreement with the results of the toxicological screening in rats that showed identical cardiotoxic effects of LEU-DOX and DOX at a dose ratio of approximately 4:1.

Pharmacokinetic studies have uncovered significant species differences with regard to the uptake of LEU-DOX into the heart. For example, in rabbits the AUC of LEU-DOX after i.v. injection measured in terms of relative DOX accumulation was about half of that obtained with equimolar doses of DOX. Nevertheless, in a 11 and 16-week study in rabbits, biweekly i.v. injections of LEU-DOX induced significantly less severe cardiotoxicity than DOX given at a dose that was three-fold lower (Jaenke et al 1980). The chemotherapeutic activity of LEU-DOX against L1210 murine leukemia (subcutaneous and i.p.) and various human tumor xenografts is superior to that of DOX when administered at equitoxic doses (Magis et al 1989). If these observations and the preclinical findings of lower drug exposure of the heart and reduced cardiotoxicity are also relevant for man, it is expected that LEU-DOX will permit a tumor chemotherapy, that is more aggressive and can be maintained for longer periods of time than that possible with currently available anthracyclines.

References

Alder S, Zbinden G (1983) Neurobehavioral tests in single and repeated-dose toxicity studies in small rodents. Arch Toxicol 54:1–23

Beijersbergen van Henegouven GMJ (1988) In vitro and in vivo research on phototoxic xenobiotics: structure-activity relationships. Arch Toxicol Suppl 12:3–9

Cojocel C, Göttsche U, Tölle KL, Baumann K (1988) Nephrotoxic potential of first-, second-, and third-generation cephalosporins. Arch Toxicol 62:458–464

Deprez-DeCampeneere D, Baurain R, Masquelier M, Trouet A (1980) Toxic and chemotherapeutic effects of L-leucyl-daunorubicin and L-leucyl-doxorubicin after intravenous administration in mice. Current Chemotherapy and Infectious Diseases. Proc. 11th ICC and 19th ICAAC. American Society of Microbiology, pp. 1692–1694

Fent K, Mayer E, Zbinden G (1988) Nephrotoxicity screening in rats. A validation study. Arch Toxicol 61:349–358

Hu ST, Brändle E, Zbinden G (1983) Inhibition of cardiotoxic, nephrotoxic and neurotoxic effects of doxorubicin by ICRF-159, Pharmacology 26:210–220

Jaenke RS, Deprez-De Campeneere D, Trouet A (1980) Cardiotoxicity and comparative pharmacokinetics of six anthracyclines. Cancer Res 40:3530–3536

Jensen RA, Acton EM, Peters JH (1984) Doxorubicin cardiotoxicity in the rat: comparison of electrocardiogram, transmembrane potential, and structural effects. J Cardiovasc Pharmacol 6:186–200

Magis AJ, Atassi G, Baurain R, DeCampeneere D, Trouet A, Zbinden G (1989) Preclinical development of N-L-leucyl-doxorubicin. Proc. 3rd ARTAC (Association pour la Recherche Thérapeutique Anticancéreuse) Workshop. Paris October 19–20

Masquelier M, Baurain R, Trouet A (1980) Amino acid and dipeptide derivatives of daunorubicin. 1. Synthesis, physicochemical properties, and lysosomal digestion. J Med Chem 23:1166–1170

Zbinden G (1981) Assessment of cardiotoxic effects in subacute and chronic rat toxicity studies. In: Cardiac Toxicology, Ed. T. Balazs, Vol. III, CRP Press Inc. Boca Raton, Fla, pp. 7–32

Zbinden G, Fent K, Thouin M (1988) Nephrotoxicity screening in rats: General approach and establishment of test criteria. Arch Toxicol 61:344–348

Zbinden G, Elsner J, Boelsterli UA (1984a) Toxicological screening. Regulatory Toxicol Pharmacol 4:275–286

Zbinden G, Beilstein AK, Bachmann E (1984b) Toxicological studies with a novel synthetic anthracycline, the bis-hydrazone bridge analog of 4-demethoxydaunorubicin (SC-33428). Arzneimittelforsch 34:1298–1301

Zbinden G, Bachmann E, Holderegger Ch (1978) Model system for cardiotoxic effects of anthracyclines. Antibiotics and Chemother 23:255–270

Recent Developments in Toxicology:
Trends, Methods and Problems
Arch. Toxicol., Suppl. 14, 118–124 (1991)
© Springer-Verlag 1991

Hypertension is not exacerbated by Chronic Carbon Monoxide Exposure, with or without Added Salt, in the Borderline Hypertensive Rat

D. G. PENNEY, and R. M. SKIKUN

Department of Physiology, Wayne State University School of Medicine, Detroit, Michigan USA

Introduction

Hypertension is a major health problem in the western world, with approximately 25% of the adult population showing borderline to frank high blood pressure (BP). Regardless of the enormous human exposure to carbon monoxide, only a few studies have examined its possible role in the development of hypertension from sources such as environmental pollution (Hernberg et al 1976) and cigarette smoke (Reid et al 1966). Borderline hypertensives, for example, are usually younger and tend to have elevated cardiac output as compared with frank hypertensives (Eich et al 1966). Possibly such individuals are more sensitive to CO. Malignant hypertension, for example, has been found to be far more common among hypertensive smokers (Isles et al 1979).

Analysis of the chronic effects of CO on BP is difficult in humans, due to factors such as stress, diet, and the lack of adequate controls. The borderline hypertensive rat (BHR), the F1 progeny of a spontaneously hypertensive rat (SHR) female and a normotensive Wistar Kyoto (WKY) male, displays a high normal to modestly hypertensive resting BP, and is reported to be sensitive to various environmental stimuli such as shock, cold, restraint, and dietary salt (Lawler et al 1988), rapidly developing arterial hypertension compared with unstressed littermates. It was reasoned that the BHR would be an excellent animal model with which to further evaluate the effect of chronic CO exposure on BP, and to explore the possible interacting effects of CO and a pre-existing stress factor such as high Na intake. The questions addressed by this study are as follows:

1. Is the BHR sensitive to Na in terms of BP?
2. Does age at the time of salt treatment affect the BP response in the BHR?
3. Does chronic CO exposure lower BP in the BHR?
4. Is the CO-induced polycythemic response similar or different in the BHR with or without salt treatment?

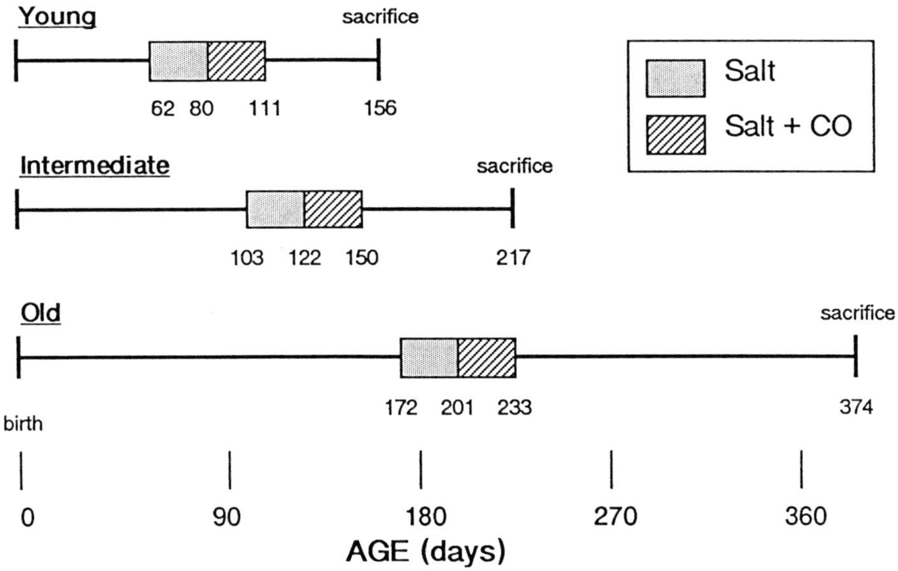

Fig. 1. Design of the study examining the responses of Young, Intermediate age, and Old borderline hypertensive rats to chronic 500 ppm carbon monoxide exposure, a high salt (8.8%) diet, and a combination of chronic 500 ppm carbon monoxide exposure and a high salt (8.8%) diet

5. Does salt treatment alter the BP and polycythemic responses of the BHR to CO?
6. Is BP, hematocrit and heart weight altered in the BHR after CO exposure is discontinued?
7. Does salt treatment alter BP, hematocrit and heart weight in BHR after termination of CO and salt treatments?

Methods

Three groups of male BHR were obtained, the progeny of SHR females and WKY males (Taconic Farms Inc., Germantown, NY). The study design is shown in Fig. 1. Standard rat chow and water were available *ad libitum*. Systolic BP and heart rate measurements were begun at 62, 83 and 115 days of age, using the tail-cuff method (Kanten et al 1983). These three study groups are designated as the "Young", "Intermediate", and "Old" BHR. Beginning at 62, 103, and 172 days of age the rats in each study were randomly assigned to one of two groups, one group thereafter receiving powdered chow containing 8.8% NaCl, the other standard chow (0.8% NaCl). At 80, 122, and 202 days of age, each of the high Na groups and the low Na groups were again randomly assigned, one-half for exposure to CO, the other one-half to remain in room air (AIR). Thereafter, each study contained four groups:

1. no salt/air,
2. no salt/CO,
3. salt/air,
4. salt/CO.

Each group in each study contained 6–8 rats. CO exposure and the high Na diet were terminated 28–31 days later. The rats remained in room air until sacrifice.

The rats, housed in plastic shoe-box cages, were continuously exposed to 500 ppm CO in large transparent plastic bags inflated by an air-CO mixture (Penney 1984). The flow rate through each bag was approximately 25 l/min to ensure rapid removal of waste gases. CO concentration was monitored by a Beckman infrared gas analyzer. Hematocrit was determined in triplicate by the microhematocrit method. The rats were sacrificed by decapitation, the hearts removed, dissected, blotted, and the left and right ventricles weighed individually. To obtain dry weight and water content, tissue was dried at 60°C for 48 hrs.

Results

Ingestion of the high Na diet increased water consumption by 50–80% in the Old study group prior to CO exposure. Related to this, the rate of body weight gain was slowed in the Intermediate and Old BHR, while weight gain of the two No salt groups increased steadily. This indicates that salt ingestion at 11-fold the normal amount constituted a considerable stress on the animals. CO exposure had no effect on body weight gain.

Hematocrit was increased from approximately 52–53% to 68–72% in all three study groups by exposure to 500 ppm CO for 4 weeks. With the exception of the Young group, the presence of salt did not modify the polycythemic response to CO. Hematocrit had returned to or slightly below the control values by the time of sacrifice.

Systolic BP increased modestly with age in both the Young and Intermediate groups, but tended to fall in the Old group. The addition of 8% NaCl to the diet had no significant effect on BP in any of the three study groups. Figure 2 shows the response in the Young group. Exposure to CO resulted in significant decreases in BP in all three study groups. The changes were of about the same magnitude regardless of age (ie. 10–16 mmHg), and were not modified by the presence of salt. BP recovered to control levels following termination of CO exposure. With one exception, chronic exposure to CO resulted in no longterm alteration of BP post-CO, with or without salt.

Neither salt nor CO, separately or in combination, had a significant effect on heart rate in any of the three study groups, either during or following CO and salt treatment.

At sacrifice, relative weight of the left ventricle plus the interventricular septum (LV+S) remained 3–9% greater in the CO-exposed BHR, in the three study groups (Fig. 3). This was particularly the case in the Young study group, possibly related to the shorter regression period. Evidence of persistent cardio-

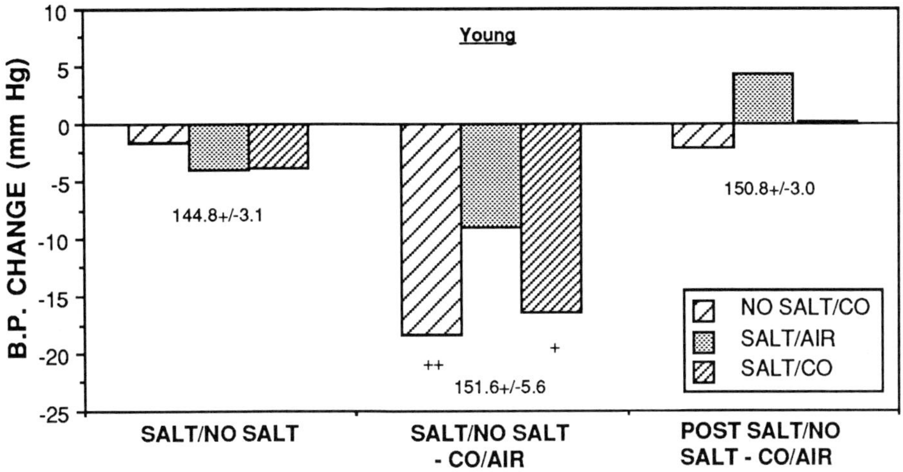

Fig. 2. Systolic blood pressure changes (relative to an untreated control group) of Young adult borderline hypertensive rats treated with a high salt (8.8%) diet, exposed to 500 ppm carbon monoxide for 31 days, and allowed to recover in room air while ingesting a low salt (0.8%) diet. Numbers below the vertical bars are control values +/– the standard error of the mean. Compared with the control value: +, $P<0.05$; ++, $P<0.01$

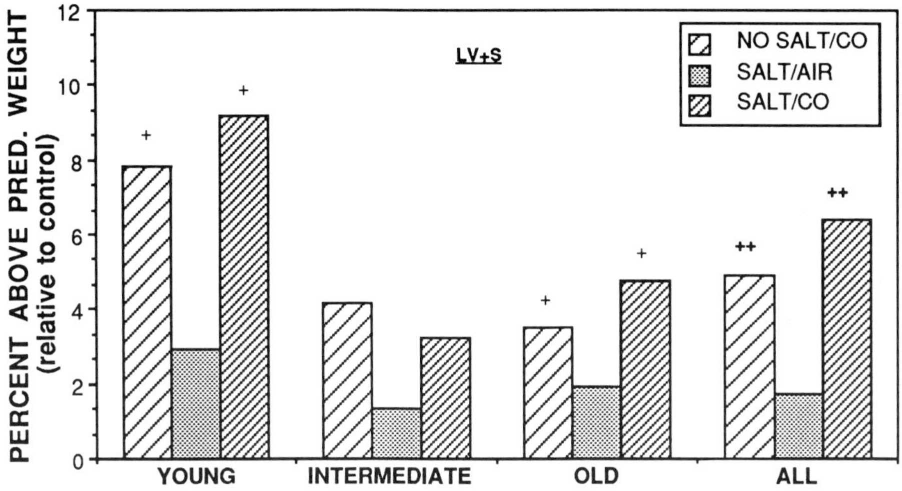

Fig. 3. Weight of the left ventricle plus interventricular septum (relative to an untreated control group) of adult borderline hypertensive rats of various ages treated with a high salt (8.8%) diet, exposed to 500 ppm carbon monoxide for 28–31 days, and allowed to recover in room air while ingesting a low salt (0.8%) diet. Compared with the control value: +, $P<0.05$; ++, $P<0.01$

megaly became more highly significant when all the data were combined. Salt treatment did not modify this apparent lack of complete regression of CO-induced cardiac hypertrophy. The right ventricle (RV) of CO-exposed rats also showed incomplete regression. Percent dry weight of the LV, S, and RV was unaltered by previous salt, CO, and salt and CO treatments.

Regression analysis of relative LV+S weight against BP post-CO showed the weight of the CO-exposed LV+S to be unrelated to BP, while that of the rats not exposed to CO was significantly ($r = 0.52$, $P<0.001$) related to BP.

Discussion

The failure to observe an elevation in BP in rats ingesting the high Na diet is inexplicable, since the parents were procured from the same breeding unit, and the BHR tested were of the appropriate age and received exactly the same Na supplement as in previous studies reporting increased BP (Lawler et al 1987; Sanders et al 1988). The high Na was clearly stressful as indicated by the increased water uptake and reduction in body weight gain.

All three groups of BHRs showed decreases in BP during CO exposure. Furthermore, the degree of hypotension was similar to that seen in previous studies with the Sprague-Dawley rat (Penney 1984) and the SHR (Penney and Robinson 1990).

Although the development of cardiac hypertrophy was not verified immediately post-CO in this study, it is a well known and consistent response to chronic CO exposure in the rat (Penney 1988; Penney and Robinson 1990), as is polycythemia. Thus, cardiac hypertrophy is assumed to have occurred in the BHR. Polycythemia of course was observed. Regression of cardiac hypertrophy following chronic exposure to CO requires 25–35 days in adult rats (Styka and Penney 1978). Under certain conditions the process may be incomplete, resulting in a persistent cardiomegaly. This apparently was the case here.

Contrasting sharply with the present study, Shiotsuka et al (1984) reported that 500 ppm CO exposure for 62–63 days enhanced development of systemic hypertension in the Dahl salt-sensitive rat fed a diet containing 8% NaCl. They suggested that this was due to the CO raising cardiac output at the same time that high dietary Na was elevating peripheral resistance. This explanation, however, fails to recognize that CO usually also lowers peripheral vascular resistance (Kanten et al 1983). It is possible that there are major physiological differences between the Dahl rat and the BHR.

In summary, the following points may be drawn from this study:

1. The BHR strain is not sensitive to salt in terms of BP elevation.
2. The age at the time of salt treatment does not alter the BP response in the BHR.
3. CO exposure lowers BP in the BHR.
4. The polycythemic response to CO is the same in the BHR with or without Na treatment.
5. High Na does not alter the BP or polycythemic responses of the BHR to CO.
6. Heart weight remains elevated after termination of CO in the BHR, but BP and hematocrit do not.

7. High Na does not alter BP, hematology, or heart weight in the BHR after termination of CO and salt treatments.

In conclusion, there was no evidence to suggest a provocatory role for CO in the development of immediate or persistent hypertension in this animal model, although the BHR is reputed to have increased sensitivity to environmental stimuli, both in the presence and in the absence of high Na. Instead CO produced marked hypotension. This contrasts with the reported elevation of BP in the Dahl salt-sensitive rat when exposed to CO. The results also contradict the reputed salt sensitivity of the BHR. Some degree of cardiac hypertrophy, however, was found to persist for 1.5–5 times the normal period required for the regression process. Finally, in respect to human hypertension, this animal study fails to support the notion that longterm CO exposure can trigger a BP increase, and/or exacerbate the elevation of BP induced by increased Na intake.

Acknowledgements. This study was supported by a grant from St. John Hospital and Medical Center, Detroit.

References

Eich RH, Cuddy RP, Smulyan H, Lyons RH (1966) Hemodynamics in labile hypertension: A follow-up study. Circulation, 34:299–307

Hernberg S, Karava R, Koskela RS, Luoma K (1976). Angina pectoris, ECG findings and blood pressure of foundry workers in relation to carbon monoxide exposure. Scand J Work, Environ Health 2 (Suppl 1):54–63

Isles C, Brown JJ, Cumming AMM, Lever AF, McAreavey D, Robertson JIS, Hawthorne VM, Stewart GM, Robertson JWK, Wapshaw J (1979) Excess smoking in malignant-phase hypertension. Brit Med J 1:579–581

Kanten WE, Penney DG, Francisco K, Thill JE (1983) Hemodynamic responses to acute carboxyhemoglobinemia in the rat. Am J Physiol 244:H320–H327

Lawler JE, Cox RH, Sanders BJ, Mitchell VP (1988) The borderline hypertensive rat: A model for studying the mechanisms of environmentally induced hypertension. Health Psychology 7:137–147

Lawler JE, Sanders BJ, Chen YF, Nagehama S, Oparil S (1987) Hypertension produced by a high sodium diet in the borderline hypertensive rat (BHR). Clin Exper Hypertension – Theory Practice 9 (Suppl A):1713–1731

Penney DG (1984) Carbon monoxide-induced cardiac hypertrophy. Growth of the Heart in Health and Disease. New York, Raven Press

Penney DG (1988) A Review: Hemodynamic response to carbon monoxide. Environ Health Perspect 77:121–130

Penney DG, Robinson MC (1990) Effects of verapamil and carbon monoxide on blood pressure and heart masss in the spontaneously hypertensive rat. Europ J Pharmacol 182:29–36

Reid DD, Holland WW, Humerfelt S, Rose G (1966) A cardiovascular survey of British postal workers. Lancet 1:614–618

Sanders BJ, Cox RH, Lawler JE (1988) Cardiovascular and renal responses to stress in borderline hypertensive rats. Am J Physiol 255:R431–R438

Shiotsuka RN, Drew RT, Wehner RW (1984) Carbon monoxide enhances development of hypertension in Dahl rats. Toxicol Appl Pharmacol 76:225–233

Styka PE, Penney DG (1978) Regression of carbon monoxide-induced cardiomegaly. Am J Physiol 235:H516–H522

Recent Developments in Toxicology:
Trends, Methods and Problems
Arch. Toxicol., Suppl. 14, 124–127 (1991)
© Springer-Verlag 1991

Cardiovascular System and Kidney as Specific Targets of Chronic Exposure to Vanadate in the Rat: Functional and Morphological Findings

M. Carmignani[1,2], P. Boscolo[3], A. R. Volpe[1,2], G. Togna[2], L. Masciocco[1], and P. Preziosi[2]

[1] Department of Cell Biology and Physiology, University of L'Aquila, I-67100 L'Aquila, Italy
[2] Institute of Pharmacology, Catholic University School of Medicine, I-00618 Rome, Italy
[3] Occupational Medicine, University of Chieti, I-66100 Chieti, Italy

Introduction

The physiological significance of vanadium (Va) and its mechanisms of cardiovascular (CV) and renal toxicity (following chronic exposure) remain contradictory (Nechay 1984). Pentavalent vanadate (Va^{+5}, as analogue of phosphate) and tetravalent vanadyl (resembling the size of Mg^{+2}) have many potential sites of action. However, it is yet to be established whether CV and renal effects of Va involve inhibition and/or stimulation of various enzymatic activities (for instance, Na^+, K^+, ATPase and adenylate cyclase, respectively) and are influenced by other ions (like Na^+, K^+, Mg^{+2} and Mn^{+2}) and/or by the redox and metabolic states of the involved cells. Contradictory results also concern the possibility that Va is implicated in certain types of arterial hypertension (Carmignani et al 1990).

This research was designed to help answer the question of whether and how long-term exposure to Va^{+5} may affect CV homeostasis and renal function.

Methods

Male Sprague-Dawley rats (n = 10 in each group) were given, for seven months starting from weaning, deionized drinking water alone (control group) or containing 100 μg/ml of Va (as $NaVO_3$).

At the end of the treatment aortic blood pressure (BP), maximum rate of rise of the left ventricular pressure (dP/dt, an index of cardiac inotropism) and heart rate (HR) were monitored polygraphically, as previously described (Carmignani et al 1983). The protocol included: a) the i.v. injection of veratrine

(VE; 0.10 μg/kg; Sigma; a hypotensive drug increasing neurogenic discharge from the baroreceptors areas), phenylephrine (10 μg/kg; Merck; an α_1-adrenoreceptor agonist), acetylcholine (ACH; 2.50 μg/kg; Merck), isoprenaline (ISO; 0.25 μg/kg; Merck; a $\beta_{1,2}$-adrenoreceptor agonist), angiotensin I and II (0.50 μg/kg; Ciba), dibutyryl-cAMP (dcAMP; 5000 μg/kg; Sigma; an analogue of cAMP able to cross plasma membrane) and tyramine (TYR; 250 μg/kg; Merck; a drug releasing stored catecholamines from the post-ganglionic adrenergic terminals); b) the bilateral vagotomy below the nodose ganglion; c) the i.v. injection of ouabain (OUA; 20 μg/kg; Merck; a drug binding Na^+, K^+ ATPase from the extracellular side), verapamil (VER; 100 μg/kg; Knoll AG; a drug blocking the receptor-operated calcium channels, mainly in the myocardial cells), hexamethonium (HEX; 625 μg/kg; Sigma; a ganglioplegic drug) and clonidine (CLO; 10 μg/kg; Boehringer Ingelheim; an α_2-adrenoreceptor agonist). This protocol was performed according to Carmignani et al (1983).

Tissue samples were stained for observation by light microscopy; ultrathin sections of kidney and heart were observed in a Philips EM 4000, as previously described (Boscolo et al 1988).

The levels of Va in various tissues were determined by instrumental neutron activation analysis (Carmignani et al 1990). Standard techniques were used to determine, in 24-hour urine samples, osmolarity and levels of total nitrogen, proteins, Na^+, K^+, and Ca^{+2}.

Data were expressed as means ± SE and compared by analysis of variance; a p value below 0.05 was considered to be significant.

Results

Systolic and diastolic BP (mmHg) and HR (beats/min) were 144 ± 4, 115 ± 5 and 288 ± 6, respectively, in the Va^{+5}-treated rats, and 122 ± 3, 95 ± 3 and 239 ± 4 in the control group ($p<0.05$). dP/dt and CV responses to phenylephrine, acetylcholine, angiotensin I and angiotensin II were unaffected by Va^{+5}, which potentiated CV effects of ISO and reduced those of VE.Va^{+5} increased BP responses to dcAMP, TYR, VER, HEX and CLO, reduced the effects of vagotomy on BP and HR and opposed those of OUA and CLO on dP/dt (Table 1).

In the Va^{+5}-exposed rats, the highest levels (μg/g wet weight) of the metal were found in the femur (10750 vs 20 in the control group) and in the kidney (4950 vs 3); the aorta (1763 vs < 20) contained much more Va than the heart (151 vs < 5).

Light (LM) and transmission electron microscopy (EM) did not show cardiac or vascular alterations. In the kidney, the lumen of the proximal tubule cells was narrowed and contained protein amorphous material; Va was accumulated within spheroidal electrondense structures in the cytoplasm of these cells, in which many swollen mitochondria were observed (Fig. 1). No other renal alterations were found by LM and EM.

Table 1. Maximal cardiovascular changes after i.v. administration of several drugs or after vagotomy in control and metavanadate-exposed male rats[a]

Drug	Dose (μg/kg)	Mean blood pressure (mmHg) control	Mean blood pressure (mmHg) exposed	Heart rate (beats/min) control	Heart rate (beats/min) exposed	dP/dt[b] (mmHg/sec) control	dP/dt[b] (mmHg/sec) exposed
Veratrine	0.10	−16±5	−6±1*	−20±4	+6±2*	−1605±171	−805±127*
Isoprenaline	0.25	−12±1	−22±3	+17±5	+39±6*	+5619±430	+10940±984*
Dibutyryl-cyclic AMP	5000	−18±1	−29±2*	+31±5	+23±7	+5885±535	+5136±327
Tyramine	250	+20±6	+37±5*	+34±1	+29±6	−3479±270	−3745±431
Vagotomy		+18±3	+9±1*	+28±4	+5±1*	+1070±711	+1284±215
Ouabain	20	+14±2	+18±4	−13±2	−10±4	+5457±320	+2311±684*
Verapamil	100	−15±4	−35±5*	−40±4	−35±8	−4280±203	−4260± 68
Hexa-methonium	625	−21±3	−39±3*	−12±3	−15±4	−2944±269	−2568±322
Clonidine	10	+15±2	+29±3*	−33±5	−35±3	−5350±580	−3424±107*

[a] All values are means ± SE (n = 10 in both groups)
[b] Maximum rate of rise of the left ventricular pressure
* Significantly different from the control ($p<0.05$)

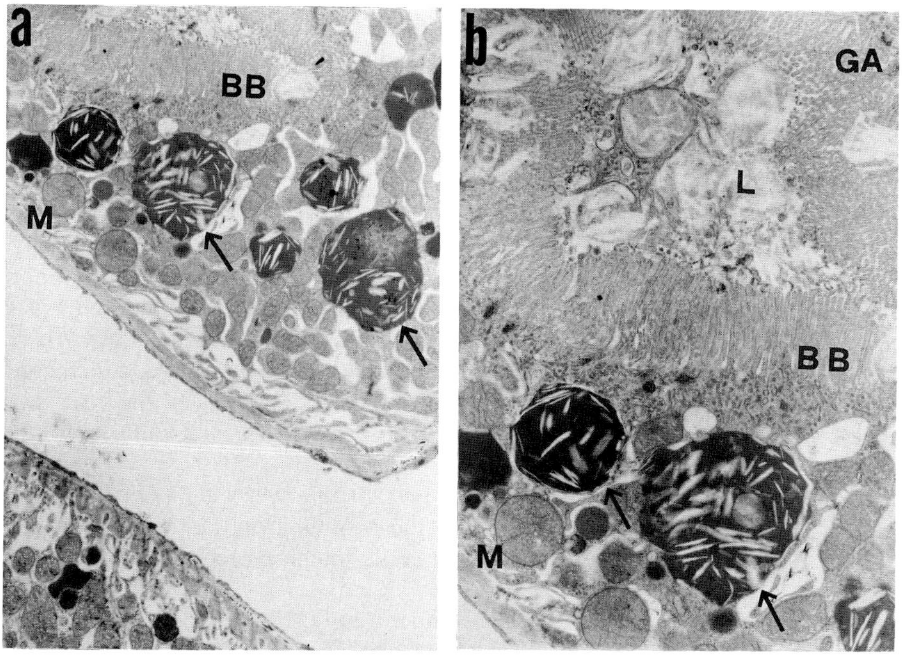

Fig. 1a, b. Electron micrographs (**a**) x12000; **b**) x28000) showing proximal tubule cells from the kidney of a metavanadate-exposed rat. Vanadium is accumulated within electrondense cytoplasmatic structures (arrows). Many mitochondria are swollen (M). Tubular lumen (L) is reduced in size and contains protein amorphous material. B: brush border. GA: junctional apparatus

Urinary osmolarity and urinary levels of total nitrogen, creatinine (creat), protein and Ca^{+2} were unchanged in the Va^{+5}-exposed rats. However, Va^{+5} strongly increased the levels (mEq/g creat) of urinary Na^+ (8.6 ± 3.9 vs 4.7 ± 0.8 in the control group; $p<0.05$) and K^+ (23.7 ± 3.6 vs 7.4 ± 1.2; $p<0.05$).

Discussion

This study indicates that chronic exposure to Va^{+5} induces arterial hypertension, depending on increase of both HR and peripheral vascular resistance and associated to selective ultrastructural alterations of the proximal tubule cells in the kidney. Such hypertension is also associated to strong increase of the urinary excretion of Na^+ and K^+, with specific accumulation of Va in the proximal tubule cells, and to accumulation of the metal in arterial vessels much more high than in cardiac muscle.

The data show that Va^{+5} reduces baroafferent sensitivity, decreases vagal parasympathetic activity and enhances central sympathetic tone. Moreover, Va^{+5} stimulates the release of noradrenaline from the peripheral sympathetic endings and increases the α_2-, β_1- and β_2-adrenoreceptor-dependent CV responsiveness (also through effects on both Ca^{+2} fluxes and cAMP-dependent intracellular availability of Ca^{+2}, mainly in vascular myocells) (Carmignani et al 1990). It is likely, considering the effects of Va^{+5} on the inotropic response to OUA, that the *in vivo* effects of this metal on myocardium cannot be explained only by the binding of Na^+, K^+ ATPase (Nechay 1984).

It is suggested that CV homeostasis and renal function may be primary targets of both physiological and toxic effects of Va. These effects appear to depend on multiple levels and sites of action.

Acknowledgements. This work was supported by a grant of Italian CNR, No. 70874.

References

Boscolo P, Carmignani M, Sacchettoni G, Ranelletti FO, Artese L, Preziosi P (1988) Ultrastructure of the testis in rats with blood hypertension induced by long-term lead exposure. Toxicol Lett 41:129–137

Carmignani M, Finelli VN, Boscolo P (1983) Mechanisms of cardiovascular regulation following chronic exposure of male rats to inorganic mercury. Toxicol Appl Pharmacol 69:442–450

Carmignani M, Sabbioni E, Boscolo P, Pietra R, Ripanti G (1990) Cardiovascular effects of long-term exposure to vanadate in rats. In: Anke M, Baumann W (eds) 6th International trace element symposium 1989 (Mo, V), vol 1, Karl-Marx-Universität, Leipzig, pp 106–113

Nechay BR (1984) Mechanisms of action of vanadium. Ann Rev Pharmacol Toxicol 24:501–524

Recent Developments in Toxicology:
Trends, Methods and Problems
Arch. Toxicol., Suppl. 14, 128–131 (1991)
© Springer-Verlag 1991

Acute Effects of Dichloromethane on Arrhythmia Development During the Early Phase of Myocardial Ischemia and Reperfusion in the Rat

J. Scholz, M. Klapperstück, M. Weise, and P. Hoffmann

Institute of Industrial Toxicology, Martin Luther University Halle, Franzosenweg 1 a, FRG-4020 Halle

Introduction

Serious toxicity and death can occur with dichloromethane (DCM) overexposure. The initial coma in acute poisoning is assumed to be caused by the direct anesthetic effect of DCM (Hall and Rumack 1990), but it could also result from the sudden onset of a nonperfusing cardiac arrhythmia as suggested earlier (Aviado and Belej 1974). The purpose of the study was to evaluate the arrhythmogenic properties of DCM exposure. An ischemia-reperfusion model was chosen because DCM was shown to produce a transient decrease of coronary blood flow in dogs (Juhasz-Nagy and Aviado 1977).

Materials and Methods

The surgical technique was a modification of the method described recently (Hoffmann 1987). Urethane-anesthetized (0.9 g/kg) male Wistar rats bred in the institute were intubated and ventilated. Arterial blood pressure (femoral artery) and lead II ECG were monitored throughout the experiment. After finishing the open-chest preparation, infusion of 3.5 mmol DCM/kg/h (Riedel-de Haën 99.8%) dissolved in water containing polyethylene glycol 600 (PEG) (50% v/v, Serva, pract) into the jugular vein was started. There were two control groups of rats receiving the same volume of PEG (PEG-control) and 0.9% saline (saline-control) respectively. After a stabilization period of 30 min, a regional myocardial ischemia was produced by occlusion of the left descending coronary artery for 5 min followed by a reperfusion period of 10 min.

After ischemia and reperfusion, hearts were quickly excised by freeze-clamp technique and stored at –80°C until analyzed. Lactate was determined using lactate dehydrogenase according to Hohorst (1970). Glutathione (GSH) was extracted by homogenizing the tissue in a ten-fold volume of 0.2 mol/l Na-phosphate buffer solution, pH 8 containing 0.005 mol/l EDTA. To 3 ml of the homogenate 1 ml of 25% metaphosphoric acid was added. After centrifugation GSH was estimated in an aliquot of the clear supernatant by determining the nonprotein sulfhydryls using Ellman's reagent (1959). The arterial blood level of DCM was analyzed using a gas chromatograph with flame ionization detector. Data are expressed as mean ± SEM and statistically analyzed by the Chi^2-method or analysis of variance with subsequent multiple t-test. $p<0.05$ was used as a criterion of significance.

Results and Discussion

During the course of experiment the DCM-concentrations of 1.12 ± 0.06, 1.14 ± 0.07, and 1.04 ± 0.11 μmol/ml were measured 30, 45, and 60 min after commencing infusion. This concentration is within the range of the levels observed in intoxicated humans (Hall and Rumack 1990).

The vehicle PEG per se reduced reperfusion-induced tachyarrhythmias and mortality (= irreversible fibrillation). The incidence of atrioventricular block during the reperfusion phase was markedly increased by DCM-infusion (Fig. 1).

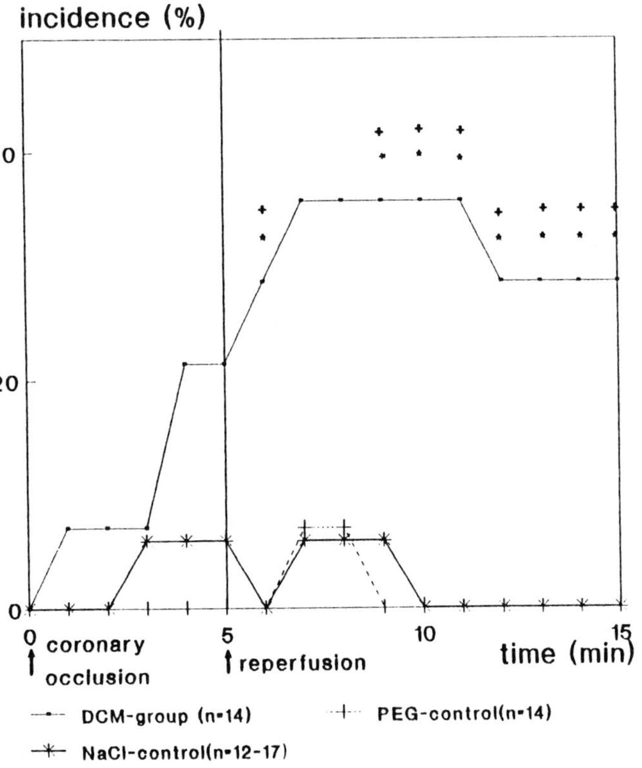

Fig. 1. Effect of DCM-PEG-infusion on the development of passive heterotopias (AV-blocks) during the course of the experiment. * $p<0.05$ compared to saline-control + $p<0.05$ compared to PEG-control

Table 1. Lactate and glutathione contents in rat heart tissue following ischemia and reperfusion

	5 min ischemia			5 min ischemia followed by 10 min reperfusion		
	Lactate content (μmol/g wet weight)					
	n	x	SEM	n	x	SEM
DCM-group	11	16.25	1.29	14	6.39	0.72
PEG-control	11	17.71	1.37	14	5.69	0.17
saline-control	10	16.06	1.43	12	8.82	2.05
	Glutathione content (nmol/g wet weight)					
	n	x	SEM	n	x	SEM
DCM-group	10	213.5	12.1	13	116.2	11.3
PEG-control	10	513.0	42.3	14	180.0	7.6
saline-control	9	183.4	25.7	12	117.4	13.2

The prolongation of the PR-interval of the ECG in rats of the DCM-group compared with the PEG-control ($p<0.05$, data not shown) confirmed the negative dromotropic DCM action.

Myocardial ischemia resulted in a high lactate content in all treatment groups compared with non-ischemic myocardium (4.14 ± 0.56 μmol/g, $p<0.05$). After 10 min of reperfusion myocardial lactate content tended to decrease more in the DCM-group and the PEG-control ($p>0.05$, Table 1). This effect seems to be related to the reduced incidence of the high energy consuming tachyarrhythmias in these groups.

Compared with the GSH content in nonischemic myocardium (543 ± 69 nmol GSH/g) the GSH content was decreased after 5 min ischemia in the DCM-group and in the saline-control ($p<0.05$), whereas it remained unaffected in the PEG-control. Reperfusion resulted in a decrease of the GSH content in all treatment groups, but the highest values were still measured in the PEG-control ($p<0.05$). The protection of the GSH content by PEG may be interpreted as a radical scavenging effect of this substance.

The results support the view that the initial coma in DCM poisoning could result not only from the anesthetic effect, but also from sudden onset of cardiac arrhythmias.

References

Aviado DM, Belej MA (1974) Toxicity of aerosol propellants on the respiratory and circulatory systems. I. Cardiac arrhythmia in the mouse. Toxicol 2:31–42

Ellman GL (1959) Tissue sulfhydrylgroups. Arch Biochem Biophys 82:70–77

Hall AH, Rumack BH (1990) Methylene chloride exposure in furniture-stripping shops: Ventilation and respirator use practice. J occupat Hlth 32:33–37

Hoffmann P (1987) Cardiotoxicity testing of organic solvents by coronary artery ligation in closed-chest rats. Arch Toxicol 61:79–82

Hohorst HJ (1970) α-(+)-Lactate: Bestimmung mit Lactatdehydrogenase und NAD. In: Bergmeyer HU (ed), Methoden der enzymatischen Analyse. Akademie-Verlag, Berlin 1970, pp. 1424–1429

Juhasz-Nagy A, Aviado DM (1977) Role of carboxyhemoglobin formation in the cardiac effects of methylene chloride. CFSC Progress Report No 11, CFSC-C-75-078, pp. 1–45

Recent Developments in Toxicology:
Trends, Methods and Problems
Arch. Toxicol., Suppl. 14, 132–135 (1991)
© Springer-Verlag 1991

Pharmacokinetics of Total and Free Digoxin and Fab Fragments in 5 Intoxicated Patients after Administration of Specific Anti-Digoxin Fab Fragments

M. Urtizberea[1], A. Sabouraud[1], M. Lachaise[1], O. Chappey[1], V. Cosson[1], F. J. Baud[2], and J. M. Scherrmann[1]

[1]INSERM U_{26}, [2]Service de Réanimation Toxicologique, Hôpital Fernand Widal, 200 rue du Faubourg Saint Denis 75475 Paris Cedex 10

Introduction

Life-threatening digitalis overdose can now be safely and rapidly reversed by administration of specific anti-digoxin Fab fragments (Antman et al 1990). The aim of the therapy is to lower the intravascular free drug concentration by changing the equilibrium of the free drug between the intravascular and intracellular body water compartments (Hursting et al 1987). Thus, a toxin concentration gradient from intracellular to intravascular space is created (Scherrmann et al 1989). Fab fragment administration has been shown to alter digoxin disposition (Schaumann et al 1986). To monitor the efficacy of Fab fragments, both free and total digoxin concentrations have to be determined. This paper reports the pharmacokinetics of free and total digoxin concentrations in five patients poisoned by digoxin and of Fab fragments for two of them.

Material and Methods

Patients: The five patients had taken digoxin doses ranging from 5 to 80 mg with suicidal intent (Table 1). On admission to intensive care unit, all patients presented life-threatening cardiac arrhythmias which were the direct reason for administration of Fab fragments.

Administration of Fab fragments: Anti-digoxin Fab fragments (Digidot®, Boehringer Mannheim GmbH, France) were infused at doses ranging from 320 to 800 mg over 30 minutes (Table 1).

Determination of digoxin and Fab fragment plasma levels: Digoxin was assayed by a specific radioimmunoassay (Gammacoat Digoxin, Baxter, France). Determination of the total digoxin concentration required denaturation of Fab frag-

Table 1. Pharmacokinetic parameters of total and free digoxin and Fab fragments in patients intoxicated by digoxin. Cl_T = total body clearance, Vd_β = apparent distribution volume, $t_{1/2\beta}$ = elimination half-life, t_{reapp} = reappearance time of free digoxin

Pat. no.	Sex	Age	Digoxin dose (mg)	Fab dose (mg)	TOTAL DIGOXIN			FREE DIGOXIN		Fab FRAGMENTS		
					$C_{adm.}$ (ng/ml)	C_{max} (ng/ml)	$t_{1/2\beta}$ (h)	C_{max} (ng/ml)	$t_{reapp.}$ (h)	$t_{1/2\beta}$ (h)	Vd_β (l)	Cl_T (ml/min)
1	f	72	20	800	31.5	180	75	4	32	12.6	12.9	11.8
2	f	90	80	400	24.8	244	22.9	2.25	20	–	–	–
3	f	73	5	400	8	95	38.7	2.05	30	–	–	–
4	m	75	10	320	8.8	71	19.5	0.6	17	–	–	–
5	f	14	10	800	17.5	95	20	–	–	10.7	12.5	13.4

ments by heating 4-fold water diluted plasma at 100°C for 1 hour. Ultrafiltration (Centrikon, Kontron, Paris, France, 3000 RPM, rotor A24-24, 10°C) preceded drug assay for the measurement of the free fraction using the Centrifree micropartition system (Amicon Co., Paris, France) with YMT membranes. Concentrations of specific Fab fragments were determined by laser nephelemetry (Nephelemeter, Behring, France) with rabbit anti-sheep IgG. The detection limit was 1 mg/l.

Pharmacokinetic parameters: Non-linear regression analysis was performed to determine pharmacokinetic parameters using the SIPHAR/pharmacokinetics program (Simed, France).

Results

All digoxin concentrations after patient admission were in the toxic range (C_{adm} 8 to 31.5 ng/ml). Infusion of the Fab fragments resulted in a dramatic increase in the total digoxin concentration (6 to 12-fold) (Fig. 1). Maximal concentrations (C_{max}) ranged from 71 to 244 ng/ml and were reached at the end of the Fab infusion. Table 1 summarizes the pharmacokinetic parameters. Free digoxin concentrations decreased to an undetectable level about 30 minutes after Fab administration. Free digoxin concentrations reappeared at a significant level 2 hours later and reached a maximum with values ranging from 0.6 to 4 ng/ml on 24 hours (Table 1 and Fig. 2a). 36 hours after Fab fragment infusion, the kinetics of total digoxin was described by a monoexponential decay and elimination half-lives were between 19.5 and 75 hours (Table 1). Fab fragment concentrations were determined for patients 1 and 5 (Fig. 2b). The disposition kinetics involved a distribution phase with $t_{1/2\alpha}$ = 1 and 1.5 h and an elimination phase with $t_{1/2\beta}$ = 12.6 and 10.7 hours, respectively, for patients 1 and 5. The volume of distribution (Vd_β) and the total body clearance (Cl_T) were in a similar range for the two patients (Table 1). After 48 hours, Fab fragments were not recovered in plasma.

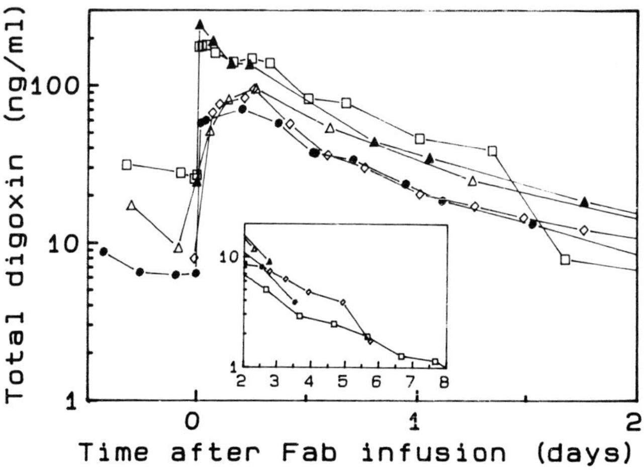

Fig. 1. Influence of specific digoxin Fab fragments infusion on total digoxin in patients 1 (□), 2 (▲), 3 (◇), 4 (●), 5 (△). (Fab administered at t = 0)

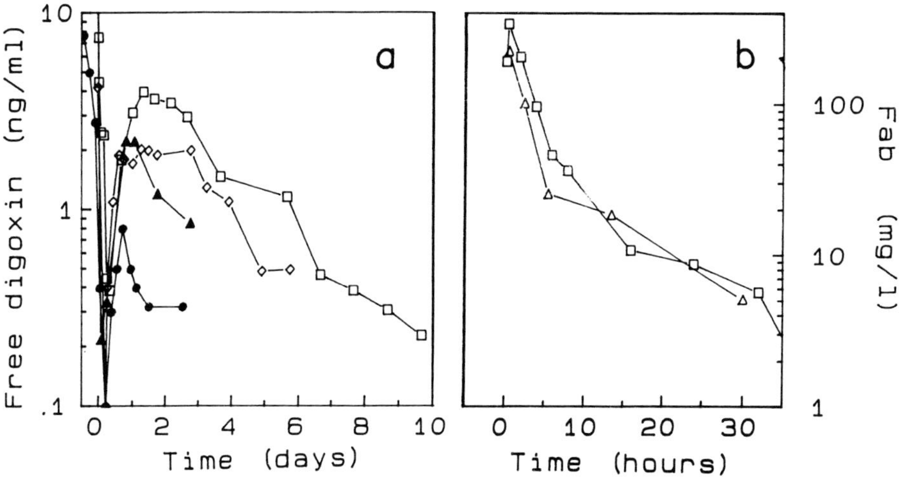

Fig. 2 a, b. **a)** Influence of specific digoxin Fab fragment infusion on free digoxin in patients 1 (□), 2 (▲), 3 (◇), 4 (●). (Fab administered at t = 0); **b)** Fab fragment disposition in patients 1 (□) and 5 (△)

Discussion

The use of specific antidigoxin Fab fragments to treat potentially life-threatening digoxin intoxication in humans is now well established (Antman et al 1990). The clinical outcome was good in all these patients and led to full recovery. The efficacy of the Fab fragment infusion can be evaluated by determination of total and free digoxin. These were respectively increased and decreased after Fab fragment infusion. The magnitude of the increase in total digoxin concentration indicated that substantial amounts of the drug were sequestrated in the vascular compartment after being removed from tissue action sites. The drop in the plasma digoxin free fraction is an indicator of the decrease of the bioactive drug levels. Presumably, extensive binding of digoxin by the Fab fragments is one of the most important mechanisms in reversing the toxic effects of digoxin. Because of the short elimination half-life of Fab fragments (11 hours) which is in the same range as that previously reported (14.3 hours) (Schaumann et al 1987), no detectable Fab fragments were present 48 hours after infusion. As a result there is no influence on digoxin elimination: total digoxin concentrations showed a monoexponential decay characterized by half-lives ranging from 19.5 to 75 hours as classically described in acute digoxin poisoning without treatment with Fab fragments (Smith et al 1984). As a consequence of this short elimination half-life and of the low Vd_β (0.4 l/kg) (Schaumann et al 1986) in comparison with that of digoxin (5–7.3 l/kg) (Mooradian 1988), free digoxin rapidly reappeared as early as 12 hours and bradycardia reoccurred in 3 of the 5 patients. This clinical drawback could perhaps be resolved by changing the dosage schedule of the Fab fragments. As suggested by Schaumann et al (1986) an infusion lasting 30 minutes seem to be too short. To prevent redistribution of digoxin and to optimize the immunoneutralization yield of the Fab fragment dose, Fab fragment could be infused over a period of several hours. New pharmacokinetic investigations are required to define the optimal conditions of Fab fragment administration in the treatment of digoxin poisoning.

Acknowledgement. This work was supported by a Grant-in-aid for scientific research from Inserm, CNAMTS and Fondation M. Merieux.

References

Antman EM, Wenger TL, Butler VP, Haber E, Smith TW (1990) Treatment of 150 cases of life-threatening digitalis intoxication with digoxin-specific Fab antibody fragments. Final report of a multicentric study. Circulation 8 (6):1744–1752

Hursting MJ, Raisys VA, Opheim KE, Bell JL, Trobaugh GB, Smith TW (1987) Determination of free digoxin concentration in serum for monitoring Fab treatment of digoxin overdose. Clin Chem 33:1652–1655

Mooradian AD (1988) An update of clinical pharmacokinetics, therapeutic monotoring techniques and treatment recommendations. Clin Pharmacokin 15:165–179

Schaumann W, Kaufmann B, Neubert P, Smolarz A (1986) Kinetics of Fab fragments of digoxin antibodies and of bound digoxin in patients with severe digoxin intoxication. Eur J Clin Pharmacol 30:527–533

Scherrmann JM, Terrien N, Urtizberea M, Pierson P, Denis H, Bourre JM (1989) Immunotoxicotherapy: Present status and future trends. Clin Toxicol 27:1–35

Smith TW, Antman EM, Friedman PL, Blatt CM, Marsh JD (1984) Digitalis glycosides: Mechanisms and manifestations of toxicity. Prog Clin Toxicol 26:413–540; 27:21–56

Recent Developments in Toxicology:
Trends, Methods and Problems
Arch. Toxicol., Suppl. 14, 136–138 (1991)
© Springer-Verlag 1991

Use of Spontaneously Beating Heart Muscle Cells Differentiating from Pluripotential Embryonic Stem Cells for Testing of Chronotropic Agents

G. Wallukat*, and A. M. Wobus

Institute of Genetics and Crop Plants Research, O-4325 Gatersleben
* Central Institute of Cardiac and Circulation Research, O-1115 Berlin-Buch

Introduction

Different methods for the cultivation of beating heart muscle cells were established to investigate the chronotropic activity of pharmaceutical agents (Chevalier et al 1987; Wollenberger 1984). Because permanent cells lines of beating heart muscle cells are not available, the possibility of using spontaneously beating heart muscle cells to differentiate from pluripotent embryonic stem cells of the mouse was examined. Pluripotent embryonic stem cell (ESC) lines were cultivated from inner cell mass cells of mouse blastocysts (Evans and Kaufman 1981; Martin 1981; Wobus et al 1984). Undifferentiated ES cells were able to differentiate into derivatives of the three primary germ layers. *In vitro*, ES cells can be specifically cultivated to differentiate into spontaneously beating heart muscle cells (Doetschmann et al 1985; Wobus unpublished).

The object of the present study was to investigate whether these cells react in comparable manner with neonatal heart myocytes with respect to the chronotropic activity of adrenergic and cholinergic agents (Karsten et al 1977; Wallukat and Wollenberger 1980).

Material and Methods

Embryonic stem cells of line D3 (Doetschmann et al 1985) were cultivated on feeder-layer of embryonic mouse fibroblasts (Wobus et al 1984) in cultivation medium described by Robertson (1987). The test procedure (Wobus and Wallukat, in preparation) is schematically shown in Fig. 1.

The following agents were investigated on differentiated cardiac myocytes of 20 individual embryoid body outgrowths in 2 independent experiments: (-)isoprenaline (Sigma, St. Louis), forskolin, IBMX (all Serva, Heidelberg),

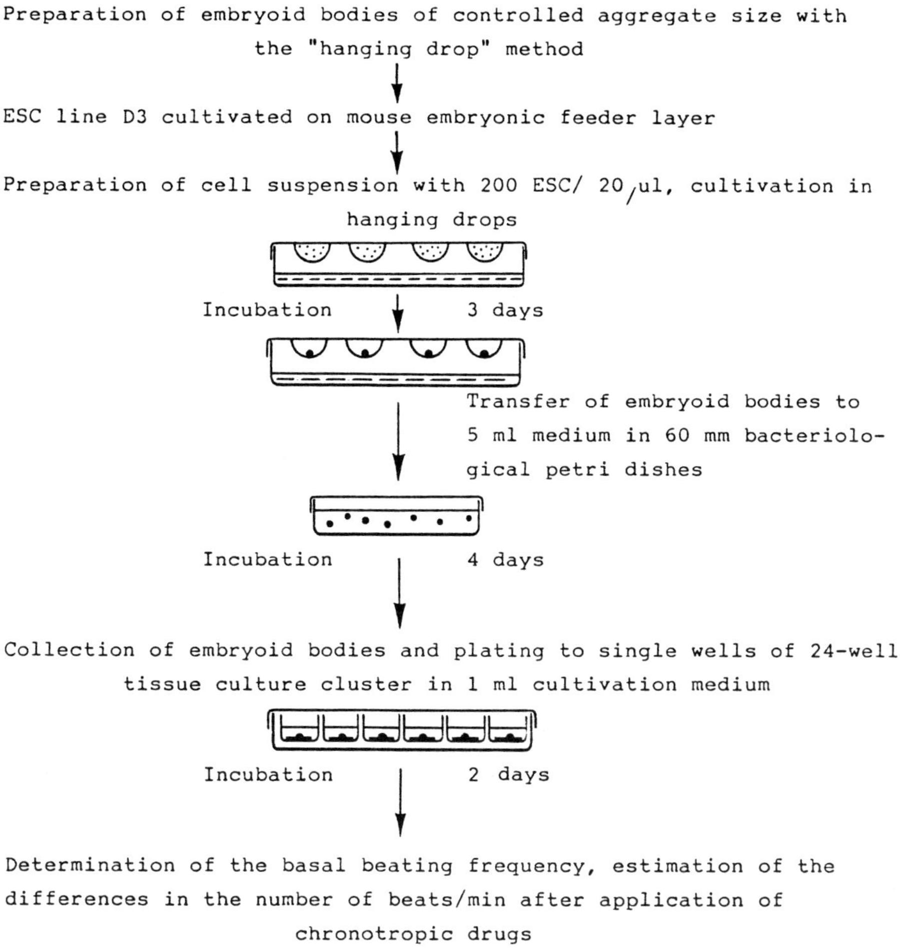

Fig. 1. Cultivation of spontaneously beating heart muscle cells differentiating from pluripotent mouse embryonic stem cells (ESC) for testing of chronotropic drugs

ICI 118.551 (Imperial Chemical Industries plc), bisoprolol (E. Merck, Darmstadt), prazosin (VEB Arzneimittelwerk Dresden).

Results and Discussion

All experiments were performed on myocytes which differentiated for 2 days after plating of 7 days (= 7+2d) old embryoid bodies. The basal level of the beating frequency was between 101 to 128 beats/min.

The chronotropic effects of five drugs and blocking agents were evaluated after 5 minutes incubation (Table 1).

The specific blocking agents, such as the β_1-adrenergic antagonist bisoprolol, the α_1-adrenergic antagonist prazosin, but not the β_2-adrenergic antagonist ICI

Table 1. Chronotropic effects of drugs on spontaneously beating heart muscle cells differentiated (7+2d) from pluripotent embryonic stem cells of line D3

Maximal effects (Molar/concentration)	Function	chronotropic reaction (beats/min)
(-)isoprenaline (10-5)	β-agonist	positive (+42,5)
IBMX (10-5)	inhibitor of phosphodiesterase	positive (+44,1)
forskolin (10-6)	activator of adenylate cyclase	positive (+38,5)
phenylephrine (10-4)	α_1-agonist	positive (+35,0)
carbachol (10-4)	cholinergic agonist	negative (−51,6)

118.551), completely abolished the positive chronotropic action of (-) isoprenaline (bisoprolol, ICI 118.551) and phenylephrine (prazosin), respectively. The inhibitor of cholinergic muscarinic receptor, atropine, specifically compensated the negative chronotropic effect of carbachol (Wobus and Wallukat in preparation). All the results are comparable to those observed in cultured neonatal rat heart myocytes (Karsten et al 1977; Wallukat and Wollenberger 1980). The results indicate the formation of adrenergic and cholinergic receptor-mediated signal transduction systems in cardiac myocytes which differentiated from pluripotent ES cells. Therefore, the system will be useful as a test procedure for investigating the action of chronotropic drugs thus avoiding the use of live animals for the preparation of primary cultures of neonatal cardiac myocytes.

Acknowledgements. The authors wish to thank Prof. R. Kemler, Freiburg/Br. for the ESC line D3.

References

Chevalier B, Mouas C, Mansier P, Aumont MC, Swynghedauw B (1987) Screening of inotropic drugs on isolated rat and guinea pig heart. J Pharmacol Methods 17:313–326

Doetschmann TC, Eistetter, HR, Katz M, Schmidt W, Kemler R (1985) The in vitro development of blastocyst-derived embryonic stem cell lines: formation of visceral yolk sac, blood islands and myocardium. J Embryol exp Morph 87:27–45

Karsten U, Wallukat G, Wollenberger A (1977) Alpha- and beta-adrenergic receptor stimulation in cultures of beating rat heart cells and its effects on automathicity and cyclic AMP level. Acta biol med germ 36:k63–k69

Robertson EJ (1987) Embryo-derived stem cell lines. In: Robertson EJ (ed) Teratocarcinomas and embryonic stem cells. A practical approach, IRL press Oxford, Washington DC, p. 71–112

Wallukat G, Wollenberger A (1980) Differential α- and β-adrenergic responsiveness of beating rat heart myocytes after stationary and non-stationary cultivation. Acta biol med germ 39:k7–k13

Wobus AM, Holzhausen H, Jäkel P, Schöneich J (1984) Characterization of a pluripotent stem cell line derived from a mouse embryo. Exp Cell Res 152:212–219

Wollenberger A (1984) The cultured myocardial cell as as model in heart research. In: Abe H, Ito Y, Tada M, Opic LH (eds) Regulation of cardiac function. Molecular, cellular and pathophysiological aspects. Japan Scientific Soc. Press, Tokyo, VNU Science Press, BV Utrecht, pp. 269–284

Recent Developments in Toxicology:
Trends, Methods and Problems
Arch. Toxicol., Suppl. 14, 140–143 (1991)
© Springer-Verlag 1991

Characterization of the Hypotensive Side Effect of Macrolide Antibiotics in Animals

A. Härtl, E. Kirchner, and H. Hoffmann

Institute of Microbiology and Experimental Therapy, Beutenbergstraße 11, O-6900 Jena, FRG

Introduction

Neuromuscular blocking activities have been reported for aminoglycosides, for antibiotics of the polymyxin group, and has also been discussed in relation to macrolides such as tylosin, erythromycin and turimycin (Singh and Marshall 1977; Viswanath and Jenkins 1978; Anderson et al 1966; Härtl et al 1979). Investigations have shown that turimycin (T) at doses ranging from 10–50 mg/kg i.v. induced a reversible dose-dependent hypotensive effect in cats and dogs. *In vitro* T successfully reduced both the tonus of the smooth muscle of the rat ileum and the amplitude of the guinea-pig atrium (Härtl et al 1979). The present study was performed to characterize this hypotensive response *in vivo* and *in vitro* experiments.

Methods and Materials

In vivo experiments: Mature cats of either sex, weighing 1.5–3.3 kg, were anaesthetized with choralose/urethane (55/300 mg/kg i.p.). The decapitation of anaesthetized cats was performed according to Sherrington (in Ther 1949). Mongrel dogs of either sex, weighing 12.2–16.5 kg, were anaesthetized with chloralose/urethane (30/800 mg/kg i.v.). In both species a femoral vein was cannulated for injection. The blood pressure was measured in the femoral artery using a polyethylene catheter (internal diameter 2 mm) which was connected via a pressure transducer (W 102, RFT) with BMT401 and 503 and recorded on a 3-NEK-1-recorder (VEB Meßgerätewerk Zwönitz, GDR). The blood pressure effect of graded doses of T given by bolus i.v. injection was investigated alone and in the presence of the antihistaminic substance hibernone. The carotid occlusion response caused by the sudden occlusion of the two common carotid arteries of dogs for 15 s was recorded 3 min after i.v. turimycin injection. The changes of the blood pressure were expressed as a percentage of the initial values. The smooth muscle contraction of the nictitating membrane of anaesthetized cats were recorded via a thread which was connected with nictitating membrane and a frontal writing lever on a kymograph. The pre- and postsynaptic fibers of the cervical sympathetic trunk were dissected free and the electrical pre- and postganglionic stimulation of the superior cervical ganglion was performed with rectangular pulses (0.5 ms impulse width, 10 cycles/s, 10 V, 10–15 s duration) via bipolar platinum electrodes.

In vitro experiments: The spasmolytic effect of T was measured using a strip of isolated guinea-pig ileum which was in an oxygenated bath containing Tyrode's solution kept at 37°C. The isotonic contractions of the strips after $BaCl_2$ ($1.2 \cdot 10^{-3}$ mol/l) were recorded on a Hellige recorder. Increasing concentrations of T (10^{-5} to 10^{-2} mol/l) or of the standard papaverine (10^{-8} to 10^{-4} mol/l) on the following barium spasm were studied. The response were expressed as percentage of the initial contractions.

α-Adrenergic activity of T was measured using the isolated vasa deferentia of rats which were suspended isotonically in oxygenated Tyrode's solution kept at 37 °C. The contractions induced by norepinephrine ($1.2 \cdot 10^{-5}$ mol/l) were recorded as described above. The effects of increasing concentrations of the standard tolazoline ($2.5 \cdot 10^{-5}$ to $1.25 \cdot 10^{-4}$ mol/l) or the test compound T ($1 \cdot 10^{-6}$ to $1 \cdot 10^{-3}$ mol/l) given prior to norepinephrine addition were investigated. The responses were expressed as percentage of the initial concentrations. The IC_{50} values were estimated graphically. The results were presented as mean ± S.E.M.

Drugs used: turimycin (Jenapharm), azamethonium (Ganlion® – Isis-Chemie, Zwickau), barium chloride, papaverine (Papaverinhydrochlorid – Jenapharm), tolazoline (Pridazol® – Polfa-Warsaw), epinephrine (Noradrenalinbitartrat – Jenapharm), hibernone (Hann, München).

Results and Discussion

The acute pharmacological action of T is a reversible dose-dependent decrease of blood pressure after i.v. injection of 10–50 mg/kg bw (Härtl et al 1979). This effect was identical in extent in intact cats as well as in decapitated cats. The pressure response of T was also not modified by pretreatment of the animals with an antihistaminic substance. If dogs were pretreated with T the hypertensive pressure response due to carotid occlusion was found to be reduced but not abolished completely (Fig. 1) whereas the hypertensive response to carotid occlusion after azamethonium was strongly inhibited. Likewise, T did not block the contraction of the nictitating membrane in response to electrical preganglionic stimulation of the superior cervical ganglion. Therefore, the hypotensive action of T seems not to be mediated by a central site by ganglion-blocking action or by a histamine-like effect. *In vitro* experiments (Table 1) using the isolated guinea-pig ileum, T showed a mild papaverine-like spasmolytic effect against barium induced spasm. T inhibited the contractile response of the vas deferens to norepinephrine and T was just as potent as tolazoline in antagoniz-

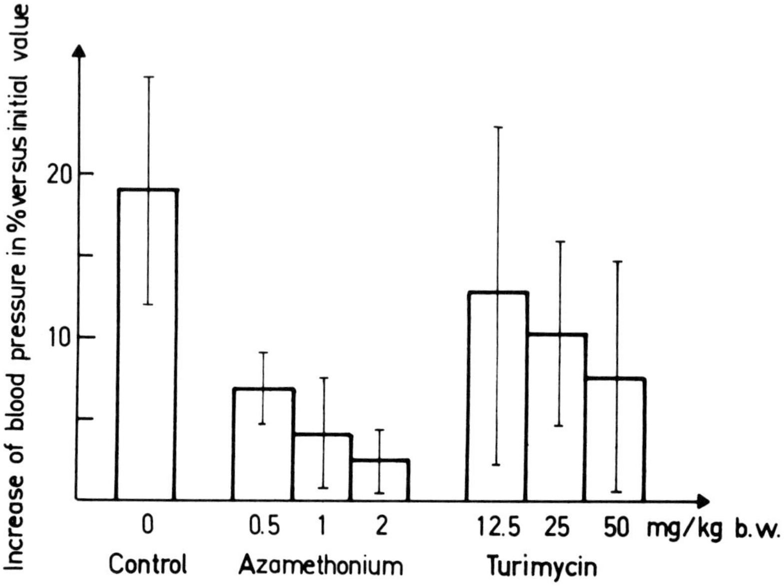

Fig. 1. Influence of turimycin and azamethonium on hypertensive response to carotid occlusion in anaesthetized dogs. Values are the means ± S.E.M. of 3 to 6 different experiments. Carotid occlusion for 15 s 3 min after i.v. drug administration

Table 1. Inhibitory effect of the macrolide antibiotic turimycin on isolated (guinea-pig) ileum and (rat) vas deferens. IC_{50} values were estimated graphically from each 6 experiments

Organ	Spasmogen mol/l	IC_{50} (mol/l)		
		Turimycin	Papaverine	Tolazoline
Ileum	$BaCl_2$ $1.2\cdot10^{-3}$	$1.3\cdot10^{-3}$	$3\cdot10^{-5}$	
Vas deferens	Norepinephrine $1.2\cdot10^{-5}$	$1\cdot10^{-4}$		$5.6\cdot10^{-5}$

ing the response to this spasmogen. A comparison of IC_{50} values for the spasmolytic and alpha-adrenolytic action of T shows a weaker spasmolytic effect, therefore, the spasmolytic activity of T on smooth muscles seems to be subordinated to the alpha-adrenolytic action. Taken together the findings demonstrate that the hypotensive effect of T may be caused by the alpha-adrenolytic potency of the drug resulting in reduced sympathetic tonus of the blood vessels of the peripheral circulatory system.

References

Anderson RC, Harris PN, Lee CC, Maze N, Small RN, Worth HM (1966) The Lilly Research Laboratories Product Registration
Härtl A, Kirchner E, Horn U, Stöckel U, Güttner J, Hoffmann H (1979) Pharmakologisch-toxikologische Untersuchung von Turimycin. Arch exper Vet med 33:111–119
Singh JN, Marshall AL (1977) The effect of amikacin on autonomic and neuromuscular transmission. J Pharmac Pharmacol Supp 29:50
Ther L (1949) Pharmakologische Methoden. Wissenschaftliche Verlagsgesellschaft MBH Stuttgart, 376–378
Viswanath DV, Jenkins HJ (1978) Neuromuscular block of the polymyxin group of antibiotics. J Pharmac Sci 67:1275–1280

Toxic Effects on Haemostasis

Recent Developments in Toxicology:
Trends, Methods and Problems
Arch. Toxicol., Suppl. 14, 147–152 (1991)
© Springer-Verlag 1991

Toxic Influences on Platelet Function

K. Schrör

Institut für Pharmakologie, Heinrich-Heine-Universität Düsseldorf, Moorenstraße 5, D-4000 Düsseldorf 1, Germany

Introduction

An acquired modification of platelet function is considered to be "toxic", when drugs or other chemicals interfere with the haemostatic potential of platelets but are not administered with the intention to treat disturbances in hemostasis. These unwanted platelet defects might either be inhibitory and clinically result in a prolonged bleeding time or stimulatory, causing a tendency for thrombosis. Because of the complex nature of platelet physiology, most agents affect more than one mechanism and in most cases it is not yet possible to classify these and other acquired disorders of platelet function by their molecular site of action (Vermylen and Blockmans 1989). Table 1 summarizes some classes of chemicals, including ethanol, that exert toxic effects on the platelet. The more interested reader is referred to several recent reviews dealing with acquired disturbances in platelet function in more detail (Barthel 1981; Carvalho and Rao 1987; Fass et al 1987; Vermylen and Blockmans 1989; Mikhailidis et al 1990).

Table 1. Selected drugs and chemicals with toxic influences on platelet function

Agents that interfere with platelet eicosanoid formation ("aspirin-type drugs", arachidonic acid)
Membrane stabilizing compounds (Tricyclic antidepressants, phenothiazines, local and general [volatile] anaesthetics)
Agents that reduce receptor-mediated platelet activation (β-lactam antibiotics, verapamil-type Ca^{++}-antagonists; ticlopidine)
Agents that enhance receptor-mediated platelet activation (streptokinase, cyclosporin A [?])
Agents with complex sites of action (heparin, ethanol, LDL/HDL-cholesterol)

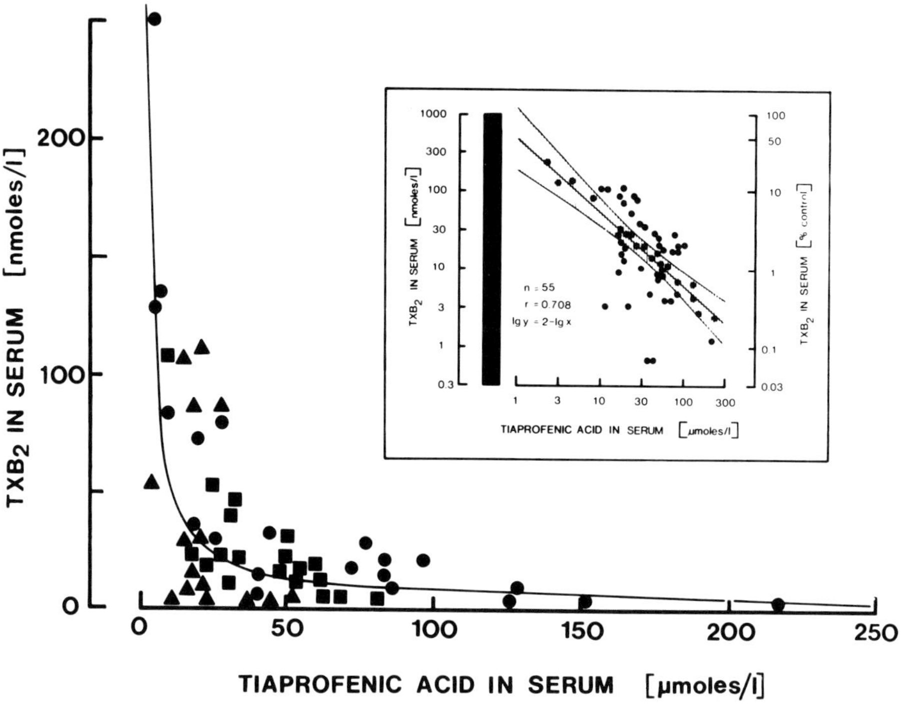

Fig. 1. Correlation between inhibition of serum thromboxane (TXB_2) levels and serum levels of tiaprofenic acid. TXB_2 levels in serum were determined after single oral dosing of 100 (▲), 200 (■) or 400 (●) mg tiaprofenic acid to healthy volunteers. Each point represents one determination. Note the exponential type of correlation which can be linearized in a double log plot (insert). (After data from Schrör et al 1985)

Non-steroidal antiinflammatory drugs (NSAID)

Human platelets are considered "all-purpose" inflammatory cells that are highly adapted for haemostasis and wound healing (Weksler 1988). Consequently, anti-inflammatory drugs are to be expected to inhibit platelet function. Indeed, inhibition of platelet function is a common feature of all non-steroidal anti-inflammatory drugs due to the specific inhibition of the platelet cyclo-oxygenase pathway. The opposite, namely a considerable increase in platelet activity, is observed after administration of arachidonic acid to man (Seyberth et al 1975). With the possible exception of acetylsalicylic acid, the inhibition of platelet thromboxane formation and the thromboxane-dependent secretion reaction by all of the other cyclooxygenase inhibitors probably has to be considered an unwanted side effect. Figure 1 demonstrates the close correlation between inhibition of platelet-derived thromboxane formation and the drug-serum level for one of the newer arylproprionic acid derivatives, tiaprofenic acid (Schrör et al 1985).

β-lactam antibiotics

A defect in platelet aggregation subsequent to administration of β-lactam antibiotics was first described for carbenicillin and subsequently detected in most of the other penicillins, including penicillin G, ticarcillin, methicillin, ampicillin, oxacillin and others but only a few cephalosporins, e.g. cephalotin (Fass et al 1987). Significant bleeding might occur at higher doses ($\geq$ 12 g/day) and requires continuous treatment. Thrombocytopenia, older age and additional chemotherapy are additive factors (Fass et al 1987). Azotemia and hypoprothrombinaemia as well as reduced antibiotic clearance, i.e. renal and hepatic failure will also enhance the bleeding diathesis (Carvalho and Rao 1987; Vermylen and Blockmans 1989). *In vitro* data suggest that penicillins disturb platelet membrane function by impairing the interaction of several platelet agonists with their specific receptor at the platelet membrane (Fass et al 1987).

Agents affecting the hemostatic system

All compounds influencing bone marrow function and megakaryocyte formation will also influence platelet count and, consequently platelet function. This involves also influence platelet count and, consequently platelet function. This involves several classes of cytostatic compounds, including vincristine, cyclophosphamide, methotrexate, thio-Tepa and others (see Carvalho and Rao 1987). High-molecular weight heparin may also cause platelet dysfunction which might be the basis of undesirable haemorrhagic side effects (Salzman et al 1980). Dextran, in addition to its effects on rheology and fibrin polymerization, may also "coat" platelet receptors, resulting in less responsiveness to physiological agonists and prolonged bleeding time (Vermylen and Blockmans 1989).

An enhanced tendency for haemorrhage by fibrinolytic agents, e.g. by dissolution of haemostatic fibrin, fibrinogenolysis or lysis of the fibrinogen bridges between aggregated platelets, is merely an amplification of the therapeutic effect that becomes unwanted and potentially toxic because of its generalization. More important is the enhanced platelet activation and excretion of platelet-derived thromboxane in patients suffering from myocardial infarction and treated with fibrinolytics (FitzGerald et al 1988). This might result in a partial resistance to anticoagulation by heparin and, less pronounced, to warfarin (Zagher et al 1990).

Membrane stabilizing agents

Membrane stabilizing agents (tricyclic antidepressants, phenothiazines and general and local anaesthetics) are a chemically distinct group of compounds that have the common feature to stabilize cell membranes including those of platelets. Although the precise mechanism of the antiplatelet effects is unknown (Carvalho and Rao 1987), it is obvious that the general depression of platelet function is associated with the inhibition of membrane signal transduction. This includes inhibition of phospholipase A_2 activation, release of endogenous arachidonic acid and eicosanoid formation while the conversion of exogenous arachidonic acid into eicosanoids is unchanged (Vanderhoek and Feinstein 1979). The clinical significance is probably low.

Cyclosporin A

The immunosuppressant cyclosporin A is now widely used to prevent transplant rejections and might cause an increased incidence of thrombotic complications, subsequent to enhanced platelet aggregation (Van Renterghem et al 1985). Platelets from both healthy volunteers and renal allograft recipients treated with cyclosporin were hyperreactive and exhibited increased thromboxane formation. This leads to the hypothesis that these events may be involved in cyclosporin nephrotoxicity (Grace et al 1987). However, platelet hyperreactivity may also result from endothelial injury (Vickers et al 1987/88).

Ethanol

Acute alcohol ingestion of medium doses (1 ml/kg body weight) by healthy volunteers causes inhibition of platelet function within 30 min (Mikhailidis et al 1987). The prolonged bleeding time is entirely due to a defect in primary haemostasis and there is no alteration of coagulation or fibrinolysis even at high doses: 2 ml/kg 40% ethanol, resulting in plasma levels of 19 mM. Interestingly, acute ingestion of 60 ml of ethanol potentiates acetylsalicylic acid-induced inhibition of platelet function and prolongation of bleeding time even in normal subjects (Deykin et al 1982). Moreover, a large and dose-dependent decrease in t-PA during several weeks of alcohol ingestion has been reported in healthy volunteers (Pikaar et al 1987).

The precise mechanism of ethanol action is unknown but appears to be multifactorial, including alterations in membrane lipid composition and reactivity (acidosis!) as well as reduced mediator release (PF 3, 5-HT, ADP) and reduced responsiveness to apparently all types of platelet agonists. At a cellular level, the 5-HT-induced phosphatidylinositol breakdown was impaired in alcoholics (Simonsson and Alling 1988). Neiman et al (1987) found a reduced percentage of arachidonic acid in the phosphatidylinositol fraction of platelets of alcoholics.

Another interesting aspect is the action of ethanol on lipid metabolism. Hennekens et al (1979) reported platelet hyporeactivity in alcoholics and suggested an increase in HDL as responsible factor. This action disappears in chronic alcoholics within a few days after alcohol withdrawal (Vermylen and Blockmanns 1989) and is followed by a prothrombotic withdrawal syndrome. Desai et al (1986) confirmed overall platelet hypoaggregability in alcoholics as well as a potent platelet inhibitory action of HDL_2 but in addition reported a hyperaggregability within 1–2 weeks of cessation of drinking with subsequent gradual return to normal levels. Platelet hyperreactivity was parallelled by an enhanced plasma LDL-cholesterol/HDL-cholesterol ratio which was increased by more than 50% in abstinence. This suggests that changes in plasma lipoproteins during acute alcohol withdrawal may partly account for haemostatic changes in these patients. Hillbom et al (1984) had previously reported increased platelet-derived thromboxane formation 1–2 weeks after alcohol withdrawal.

In conclusion inhibition of platelet function usually results in a prolonged bleeding time and a tendency for bruising. This is clinically relevant with β-lactam antibiotics and ethanol and may also occur with non-steroidal anti-inflammatory drugs. Enhanced platelet function may result from phospholipase A_2 or prothrombin activation, e.g. by several animal toxins and during ethanol withdrawal. There is also a platelet stimulating potential by several anticoagulants and fibrinolytics, including heparin and streptokinase. The molecular mechanisms of action are not completely understood but are probably multifactorial in nature.

References

Barthel W (1981) Toxische Beeinflussung der Plättchenfunktion. Erg Exp Med 38:11–42

Benigni A, Chiabrando C, Piccinelli A, Perico N, Gavinelli M, Furci L, Patino O, Abbate M, Bertani T, Remuzzi G (1988) Increased urinary excretion of thromboxane B_2 and 2,3-dinor-TXB_2 in cyclosporin A nephrotoxicity. Kidney Int 34:164–174

Carvalho ACA, Rao AK (1987) Aquired qualitative platelet defects. In: Colman RW, Hirsh J, Marder VJ, Salzman EW (eds) Lippincott Philadelphia (Hemostasis and Thrombosis, pp 750–772)

Desai K, Owen JS, Wilson DT, Hutton RA (1986) Platelet aggregation and plasma lipoproteins in alcoholics during alcohol withdrawal. Thromb Haemostasis 55:173–177

Deykin D, Janson P, McMahon L (1982) Ethanol potentiation of aspirin-induced prolongation of the bleeding time. New Engl J Med 306:852–854

Elmer O, Göransson G, Zoucas E (1984) Impairment of primary hemostasis and platelet function after alcohol ingestion in man. Haemostasis 14:223–228

Fass RJ, Copelan EA, Brandt JT, Moeschberger ML, Ashton JJ (1987) Platelet-mediated bleeding caused by broad-spectrum penicillins. J Infect Dis 155:1242–1248

FitzGerald DJ, Cattela F, Roy L, FitzGerald GA (1988) Marked platelet activation in vivo after intravenous streptokinase in patients with acute myocardial infarction. Circulation 77:142–150

Grace AA, Barradas MA, Mikhailidis DP et al (1987) Cyclosporin A enhances platelet aggregation. Kidney Intern 32:889–895

Hennekens CH, Willet W, Rosmer B, Cole DS, Mayrent SL (1979) Effects of beer, wine, and liquor in coronary death. J Am Med Ass 242:1973–1974

Hillbom ME, Kangasaho M, Hjelm-Jäger M (1984) Platelet aggregation and thromboxane B_2 formation after ethanol abuse: Is there a relationship to stroke? Act Neurol Scand 70:432–437

Mikhailidis DP, Barradas MA, Epemolu O, Dandona P (1987) Ethanol ingestion inhibits human whole blood impedance aggregation. Am J Clin Path 88:342–345

Mikhailidis DP, Barradas MA, Jeremy JY (1990) The effect of ethanol on platelet function and vascular prostanoids. Alcohol 7:171–180

Neiman J, Curstedt T, Cronholm T (1987) Composition of platelet phosphatidylinositol and phosphatidylcholine after ethanol withdrawal. Thromb Res 46:295–301

Pikaar NA, Wedel M, van der Beck EJ et al (1987) Effects of moderate alcohol consumption on platelet aggregation, fibrinolysis and blood lipids. Metabolism 36:538–543

Salzman EW, Rosenberg RD, Smith MH et al (1980) Effect of heparin and heparin fractions on platelet aggregation. J Clin Invest 65:64–73

Schrör K, Seidel H, Neuhaus V (1985) Potent and long-lasting inhibition of human serum thromboxane formation by tiaprofenic acid. New Trends Rheumatol 3:19–30

Seyberth H-W, Oelz O, Kennedy T (1975) Increased arachidonate in lipids after administration to man: Effects on prostaglandin biosynthesis. Clin Pharmacol Ther 18:521–529

Shattil SJ, Bennett JS, McDonough M, Turnbull J (1980) Carbenicillin and penicillin G inhibit platelet function in vitro by impairing the interaction of agonists with the platelet surface. J Clin Invest 65:329–337

Shulman NR, Jordan JV (1987) Platelet Immunology. In: Colman RW, Hirsh J, Marder VJ, Salzman EW (eds) Lippincott Philadelphia, (Hemostasis and Thrombosis, pp. 452–529)

Simonsson P, Alling C (1988) The 5-hydroxytryptamine stimulated formation of inositol phosphate is inhibited in platelets of alcoholics. Life Sciences 42:385–391

Vanderhoek JY, Feinstein MB (1979) Local anaesthetics, chlorpromazine and propranolol inhibit stimulus-activation of phospholipase A_2 in human platelets. Mol Pharmacol 16:171–180

Van Renterghem Y, Roels L, Lerut T, Gruwez J, Michielsen P, Gresele P, Deckmyn H, Colucci M, Arnout J, Vermylen J (1985) Thrombembolic complications and haemostatic changes in cyclosporin-treated cadaveric kidney allograft recipients. Lancet 1:999–1002

Vermylen J, Blockmans D (1989) Acquired disorders of platelet function. Bailliere's Clin Haematol 2:729–748

Vickers A, Guertler J, Spaans C, Tapparelli C, Donatsch P, Fischer V (1987/88) Cyclosporine A nephrotoxicity studied by the combined application of kidney cell lines, hepatocytes, and endothelial-platelet cocultures. Mol Toxicol 1:351–361

Weksler BB (1988) Platelets. In: Gallin JI, Goldstein IM, Snyderman R (eds) Raven Press, New York (Inflammation: Basic Principles and Clinical Correlates, pp. 543–557)

Zagher D, Maaravi Y, Matzner Y, Gilon D, Gotsman MS, Weiss AT (1990) Partial resistance to anticoagulation after streptokinase treatment for acute myocardial infarction. Am J Cardiol 66:28–30

Recent Developments in Toxicology:
Trends, Methods and Problems
Arch. Toxicol., Suppl. 14, 153–156 (1991)
© Springer-Verlag 1991

Hereditary and Acquired Disturbances of Haemostasis in Animals

J. Güttner

Institute of Microbiology and Experimental Therapy, Beutenbergstraße 11, D-6900 Jena

A great number of animal models for disturbances of the haemostasis is known (Dodds 1981; Gites et al 1981). They are found in farm animals, domestic (outpatients) and laboratory animals. As in man, lesions such as haemorrhagic diseases, thrombotic conditions and disseminated intravascular coagulation (DIC) can occur as hereditary or acquired diseases and, from the clinical point of view, as an isolated form or as consequence of other pathological conditions. In this short review, spontaneously occurring lesions are predominantly considered.

Haemorrhagic disorder models due to hereditary deficiency of isolated clotting factors are mostly available from some well-defined animal breeds (Dodds 1981, Moake 1988), predominantly of dogs and cats (Table 1). Breeding and maintenance of affected animals often requires special facilities and experienced personnel. Moreover, some species are normally lacking in clotting factor(s) as has been described for the Hageman factor in marine mammals, birds and reptiles (Robinson et al 1969). The combined deficiency of clotting factors can frequently be found in laboratory animals that suffer from infections or that are held on diets deficient in the certain ingredients (Table 2).

Isolated hereditary thrombopathies are mostly known from dogs, rats and mice, and other forms which ware associated with the Chediak-Higashi syndrome also develop in many other species (Dodds 1981; Rolovic et al 1985). The canine thrombasthenic thrombopathia is similar to Glanzmann's thrombasthenia of man but with bizarre giant platelets like those of Bernhard-Soulier syndrome. Thrombocytopenias arise predominantly from other, spontaneously occurring or experimentally induced, diseases (Table 3). Those forms which arise secondary to virus infections or tumour transplantation can easily be produced in routine laboratories. Platelet dysfunction has also been observed secondary to abnormal response to thromboxane A_2 in a minority of mongrels; the platelet rich plasma of which aggregates irreversibly when stirred with sodium arachidonate. Additionally, boxer breeds with familial hypersensitivity to warfarin, Sprague Dawley rat substrains with hereditary warfarin resistance

Table 1. Hereditary deficiency of single clotting factors

Factor	Lesion	Species/Strain/Breed
I	Hypofibrinogenaemia Dysfibrinogenaemia	Dogs: St. Bernard breed Dogs: Borzoi family
Ii	Hypoprothrombinaemia	Dogs: boxer (autosomal inheritance, newborns affected)
VII	Factor VII deficiency	Dogs: several beagle breeds (autosomal recessive trait)
VIII	Haemophilia A	Dogs: beagle, Irish setter, greyhound, Labrador (X-linked); Cats (X-linked)
VIII R:WF	Von Willebrand's disease	Many dog strains, swine, rabbits (autosomal incompletely dominant or autosomal recessive inheritance)
IX	Haemophilia B Double haemophilia AB	Cats, dogs (X-linked inheritance) Dogs (X-linked inheritance)
X	Stuart factor deficiency	Dogs: American cocker spaniel (autosomal recessive)
XI	PTA deficiency	Few breeds of dogs and cows (autosomal inheritance)
XII	Hageman factor deficiency	Cats (autosomal recessive inheritance)

Table 2. Multifactorial deficiency of clotting factors in laboratory animals

Factors	Aetiology	Lesion/Species
Hereditary		
II, V, VII, X, XII	Genetic defect	Hypolipaemic CFN rats
Acquired		
II, V, VII, VIII, X	Hepatotropic viruses	MHV-3 infection (mouse) Infectious ectromelia (mouse) Hepatitis contagiosa canis (dog) Herpes canis (dog)
	Hepatotropic bacteria	Leptospirosis (dog)
II, VII, IX, X	Deficiency of vitamin K	
VII, XI, XII	Deficiency of vitamin C	Scurvy (guinea pig)
VII, X	Iodine deficiency	Rat

Table 3. Platelet-related haemorrhagic disorders in animals

Disease	Species/Strain, Comment
Thrombasthenic thrombopathia (Glanzmann's disease)	Dog: Fox terrier, otterhound (autosomal recessive)
Familial thrombopathia	Cow (autosomal recessive) Dog: Basset hound (autosomal recessive) FH rats (autosomal dominant) C57BL/6J ep/ep mouse mutant (Hermansky-Pudlak syndrome, autosomal recessive)
Chediak-Higashi syndrome	Cow, mink, cat, bg/bg mouse mutant (autosomal recessive)
Hereditary thrombocytopenia	Dog (case reports)
Acquired thrombocytopenia	
Hepatitis contagiosa canis	Dog
Parvovirus infection	Dog
Graffi virus infection	XVII Mouse with leucaemia
RMuLV infection	Mouse with Rauscher's erythroleucaemia
Transplantation tumours	Dog, rat, mouse
Granulosa cell tumour	Dog
Graft-vs.-host reaction	Mouse

and animals with mild coagulation defects due to congenital C4-complement (C4D guinea-pigs) and C6-complement deficiency (Freiburg rabbits) have been described.

Thrombotic conditions and consumption coagulopathy are generally found in animals with disease complexes (Innes et al 1956; Wigton et al 1976). Spontaneous atrial thrombosis and consumption coagulopathy (deficiencies of the factors II, VII, VIII, and X; thrombocytopenia) occur in 70 to 90% of Syrian hamsters, with age-dependent increase in incidence. Females are more frequently afflicted than males and 90% of the animals have thrombosis in the left atrium. Atrial thrombosis has also been described in more than 60% of female breeder mice of a BALB/c substrain. Thrombosis develops also subsequent to acute, systemic infections (e.g. thrombosis of pulmonary veins of hamsters infected by S. enteritidis, thrombosis in the myocardium and CNS of parvovirus-infected dogs), in connection with canine hyperadrenocorticism (thrombosis of pulmonary arteries), and in dogs with renal amyloidosis (hyperfibrinogenaemia with thromboembolism). Disseminated intravascular coagulation (DIC) occurs sporadically in dogs, mink and chickens, but more often in Syrian hamsters with the above-mentioned consumption coagulopathy. DIC develops often secondary to many acute and chronic diseases in various species, such as systemic bacterial infections (rabbit dysentery by S. dysenteriae or S. flexneri, C. parvum infection of mice and rats), viral infections (hepatitis contagiosa canis, feline infectious peritonitis, MHV-3 infection of the susceptible mouse

Table 4. Peculiarities of spontaneously occurring animal models for pathologic haemostasis

Availability (special breeds in few laboratories only)
Special breeding and maintenance requirements (special facilities, blood products for treatment of bleeding episodes)
Mostly "large" animals
Cost
Short clinical course and difficult diagnosis (atrial thrombosis, thromboembolism)
Lack of experimental counterparts of hereditary deficiency of single clotting factors and of thrombasthenias
Many disorders of haemostasis develop secondary to other, rapidly progressing and lethal diseases
Models are models only and don't reflect the disease of the target species in all details

strains BALB/cJ, C57BL/6J, DBA), malignancies (promyelocytic leucaemia of rats), traumatic shock, heavy liver diseases and parasitoses (e.g. heart worm disease of dogs).

Although many models of disturbances of haemostasis are known from animals of various species, some peculiarities must be considered before theanimals under question will be used in research (Table 4). These facts may act as limiting factors for the unrestricted application of those models and induce the experimentors to search for further models.

References

Dodds WJ (1981) Second international registry of animal models of thrombosis and hemorrhagic diseases. ILAR News 24:R4–R50

Gites AR, Tinlin S, Hoggendoorn H, Greenwood P, Greenwood R (1984) Development of factor VIII: C antibodies in dogs with hemophilia A (Factor VIII:C deficiency). Blood 63:451–456

Innes JRM, Wilson C, Ross MA (1956) Epizootic Salmonella enteritidis infection causing septic pulmonary phlebothrombosis in hamsters. J Inf Dis 98:133–141

Moake J (1988) Von Willebrand factor and the pathophysiology of thrombotic thrombocytopenia: from human studies to a new animal model. Labor Invest 59:415–417

Robinson AJ, Kropatkin M, Aggeler PM (1969) Hageman factor (factor XII) deficiency in marine mammals. Science 166:1420–1422

Rolovic Z, Jovanovic T, Stankovic Z, Marinkovic N (1985) Abnormal megakaryocytopoiesis in the Belgrade laboratory rat. Blood 65:60–64

Recent Developments in Toxicology:
Trends, Methods and Problems
Arch. Toxicol., Suppl. 14, 157–159 (1991)
© Springer-Verlag 1991

Effects of Snake Venoms on Tissue-Type Plasminogen Activator Release

H.-P. Klöcking, and A. Hoffmann

Institute of Pharmacology and Toxicology, Medical Academy Erfurt, Nordhäuser Str. 74, 5010 Erfurt, GDR

Introduction

Proteins from snake venoms may act on all parts of the haemostatic system in mammal organisms, i.e. also on the fibrinolytic system (Stocker and Meier 1989).

Previous investigations demonstrated that batroxobin, the enzyme from the venom of *Bothrops atrox* or *Bothrops moojeni*, respectively, caused an acute release of tissue-type plasminogen activator (t-PA) in the isolated perfused vascular preparation (pig ear) and in whole animals (rats) (Klöcking et al 1987, 1989). Therefore, further proteins from snake venoms were studied for their t-PA releasing effect in the isolated perfused pig ear.

Materials and Methods

The following proteins from snake venoms were used: ecarin (vials contain 50 ecarin units), a serine protease from *Echis carinatus* venom; protac®, a serine protease from *Agkistrodon c. contortrix* venom; factor X activator from Russel's viper venom (RVV-X); thrombocytin, a serine protease from Bothrops atrox venom; venzyme (ACTE), a serine protease from *Agkistrodon c. contortrix* venom.

All proteins from snake venoms were gifts from Pentapharm AG, Basel, Switzerland.

Acute t-PA release experiments: the t-PA release was studied in the isolated perfused pig ear. The test design and the determination of t-PA were described in detail previously (Klöcking et al 1976). To find out the percentage increase in t-PA release, the mean value of the t-PA activity of the first four fractions (100%) was related to that of eight fractions after starting the perfusion with test solution.

Statistical significance was calculated by Student's t-test; $p > 0.05$ was considered to be of no significance.

Table 1. Effect of snake venom protein on the t-PA release compared to thrombin

Substance	Concentration µmol/l	Number of experiments n	Increase in t-PA release %	Significance P
Venzyme	10	4	196 ± 120	< 0.001
	1	7	32 ± 22	n.s.
	0.1	5	32 ± 20	n.s.
Protac	10	3	82 ± 52	< 0.01
	1	4	20 ± 19	n.s.
	0.1	4	17 ± 16	n.s.
Thrombocytin	1	8	89 ± 56	< 0.002
	0.1	4	54 ± 18	< 0.001
	0.01	7	44 ± 26	± 0.01
Ecarin	0.1	3	187 ± 65	< 0.001
	0.01	7	83 ± 47	< 0.001
	0.001	5	48 ± 30	< 0.02
Thrombin	0.01	21	191 ± 108	< 0.001
control		9	14.5 ± 12	

Results and Discussion

Already at a concentration of 1 nmol/l perfusion solution, ecarin caused a significant increase in t-PA release which could be further enhanced in a dose-dependent manner (Table 1).

At a concentration of 0.1 µmol/l, ecarin was able to increase the t-PA activity by 190%. Thrombocytin, protac®, and venzyme were less potent by several orders of magnitude. An increase in t-PA release by 80% was caused by ecarin at a concentration of 0.01 µmol/l perfusion solution. To obtain the same result, concentrations of 1 µmol/l perfusion solution were required in case of thrombocytin, and 10 µmol/l if protac® or venzyme were used.

The snake venom preparations used act on different parts of the haemostatic system. Ecarin activates prothrombin (Morita and Iwanaga 1981), while protac activates protein C (Stocker et al 1987). Venzyme shows a procoagulant activity by catalyzing the rapid release of fibrinopeptide B and a much slower release of fibrinopeptide A from fibrinogen (Skimizu et al 1987).

Thrombocytin causes platelet aggregation and release of platelet constituents (Kirby et al 1974).

Additionally, these snake venoms have a t-PA releasing activity which was demonstrated in the isolated perfused pig ear. This fact is indicative of their direct action on the vascular endothelium.

With regard to the t-PA releasing effect, ecarin and the clotting enzyme thrombin are within the same order of magnitude (Klöcking et al 1988).

The tissue-type plasminogen activator, which is produced by vascular endothelial cells, is an important regulator of fibrinolysis, the so-called antithrombotic defence mechanism of the human organism.

On the other hand, in case of systemic action, acute t-PA release might induce enhanced bleeding risk under certain circumstances.

References

Furie BC, Furie B (1976) Coagulant protein Russell's viper venom. Meth Enzymol 45:191–205

Kirby EP, Niewiarowski S, Stocker K, Kettner Ch, Shaw E, Brudzynski M (1979) Thrombocytin, a serine protease from Bothrops atrox venom. 1. Purification and characterization of the enzyme. Biochemistry 18:3564–3570

Klöcking HP, Bock G, Drawert J, Hinsenbrock KP, Kaiser B, Sedlarik U (1976) Über die Freisetzung von Plasminogenaktivatoren. Folia Haematol 103:404–424

Klöcking HP, Hoffmann A, Markwardt F (1987) Release of plasminogen activator by batroxobin. Haemostasis 17:235–237

Klöcking HP, Hoffmann A, Markwardt F (1988) Influence of hirudin on the thrombin-induced release of tissue plasminogen activator from isolated perfused pig ear. Folia Haematol 115:110–112

Klöcking HP, Markwardt F, Güttner J (1989) On the mechanism of batroxobin-induced fibrinolysis. Pharmazie 44:504–505

Kornalik F, Blombeack B (1975) Prothrombin activation induced by ecarin – a prothrombin converting enzyme from Echis carinatus venom. Thromb Res 6:53–63

Morita T, Iwanaga S (1981) Prothrombin activator from Echis carinatus venom. Meth Enzymol 80:303–311

Shimizu A, Jimenez EC, Takagi J, Inada Y, Saito Y (1987) Venom from southern copperhead snake (Agkistrodon controtrix contortrix). I. Characterization of a protease that preferentially releases fibrinopeptide B. Toxicon 25:751–757

Stocker K, Meier J (1989) Schlangengiftproteine in der Hämostaseologie: Neuere Erkenntnisse. Folia Haematol 116:935–953

Recent Developments in Toxicology:
Trends, Methods and Problems
Arch. Toxicol., Suppl. 14, 160–165 (1991)
© Springer-Verlag 1991

Metabolic and Toxicological Studies on the Anticoagulant Rodenticide, Flocoumafen

G. E. Veenstra, D. E. Owen, and K. R. Huckle

Health, Safety and Environment Division, Shell Internationale Petroleum Maatschappij, P.O. Box 162, 2501 AN The Hague, The Netherlands

Introduction

Flocoumafen (STORM*), 4-hydroxy-3-[1,2,3,4-tetrahydro-3-[4-(4-trifluormethylbenzyloxy) phenyl]-1-naphthyl] coumarin (Fig. 1) is a 4-hydroxycoumarin-derived anticoagulant rodenticide.

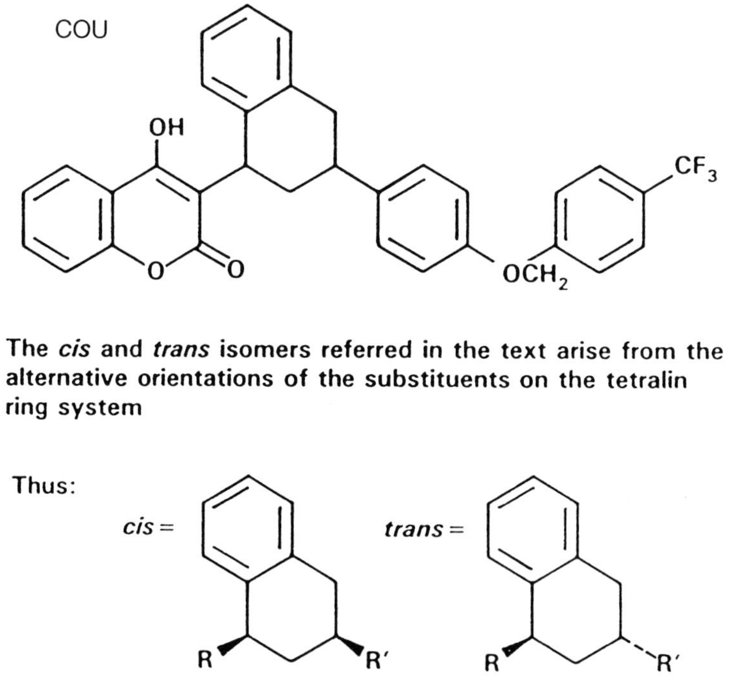

Fig. 1. Structure of flocoumafen

* STORM ist a Shell trademark

The novel structure of the side-chain on the 4-hydroxycoumarin moiety increases the activity of flocoumafen more than 100-fold against a wide range of rodent species when compared with warfarin. Flocoumafen therefore exhibits a high order of acute oral toxicity to most mammalian species. Toxicological effects, other than those associated with anticoagulant activity, have not been observed in a range of single and repeat dose studies.

Metabolic fate studies conducted in rodents (Huckle et al 1988, 1989a, 1989b) and non-rodent species (Huckle et al 1989c) show that flocoumafen is absorbed and is extensively distributed within the body, accumulating appreciably in the liver. The hepatic flocoumafen residue exhibits a long half life in rodents (in rats 220 days) and fowl and is not readily metabolised.

Flocoumafen and other second generation 4-hydroxycoumarin-based rodenticides are thought to act via the inhibition of vitamin K 2,3-oxide reductase (Thijssen & Baars 1987). This results in the depletion of vitamin K, an essential cofactor for the carboxylase that converts inactive precursors of the vitamin K-dependent clotting factors to their active forms (Suttie 1978, 1980).

In the present investigation, designed to develop effective antidotal therapy in non target mammalian species, the effects of repeated administration of vitamin K on anticoagulant recurrence and hepatic burden of the flocoumafen residue has been studied in the dog.

Materials and Methods

Four male and four female healthy beagle dogs were given a single oral dose of 0.5 mg/kg technical flocoumafen (94.9%, Sorex Ltd, Widnes UK) as a 100 mg/l solution in corn oil. Blood samples were taken from the femoral vein of each dog twice weekly and assayed for prothrombin time (PT) and activated partial thromboplastin time (APTT).

Regular examination for clinical signs of anticoagulation, and elevations in PT's or APTT's were used as a basis to initiate a 5 week antidotal treatment with vitamin K in eight of the dogs.

Two male and 2 female dogs were then allocated into one of two subgroups. Each subgroup received as initial treatment 7 days of therapy with vitamin K (Konakion 10, 1.1 ml ampoules, Roche, UK), consisting of daily subcutaneous injections of 2 or 5 mg/kg. A daily oral dose of either 2 or 5 mg/kg per day of a vitamin K (Konakion 10 tablets) was administered for the next 2 weeks, followed by 7 daily doses of either 1 or 2.5 mg/kg per day, reducing to 0.5 or 1.25 mg/kg per day for the final week of dosing.

Whole blood transfusion was applied (approx. 200 ml/dog) to dogs which showed advanced signs of anticoagulation. All dogs were held for $6\frac{1}{2}$ months after the end of the therapy to determine whether any subsequent anticoagulation occurred. At termination of the study the livers of all treated animals were taken and analysed for flocoumafen according to a previously described method (Huckle et al 1989c).

Results

Treatment phase

Clinical findings

Administration of a single dose of 0.5 mg/kg flocoumafen resulted in clear signs of anticoagulation in 3 male and 2 female dogs within 6–10 days of treatment (median 9 days). Those dogs (1 male and 2 females) that did not produce any clinical effects received a second dose of 0.5 mg/kg flocoumafen 7 weeks after the first dose. In this group 2 of 3 dogs showed clinical evidence of anticoagulation 7–8 days after the second dose.

The most common signs associated with treatment were pallor of gums, ears and conjunctival, and subdued behaviour. Coldness of the extremities and laboured breathing was observed in cases of severe anticoagulation.

Food consumption and body weight

Food consumption was usually reduced during an episode of anticoagulation. Body weights were not affected during any stage of the treatment, therapy and observation.

Haematology

Five days after administration of flocoumafen all dogs had elevated PT values (see Fig. 2). For those dogs which did not develop any clinical signs of anticoagulation maximum clotting times were recorded at day 8. By day 12 PT valued had returned to base line values and no further changes in clotting time were observed until 5 days after the second dose. APTT values followed the same general trend as the PT values (data not shown).

Therapeutic phase

Clinical signs

Treatment with vitamin K (2 or 5 mg/kg S.C.) was initiated immediately after observation of adverse clinical signs. In 3 of 7 animals no improvement on the condition of the animals was observed after one or two doses of vitamin K; a blood transfusion (approx. 200 ml) resulted in immediate improvement of the condition and full recovery within 24 hours. In the animals receiving vitamin K alone improvement from anticoagulation was rapid with complete recovery within 48 h.

During the remainder of the therapeutic phase (5 weeks) and during the observation period after completion of the vitamin K treatment (36 weeks) no further episodes of anticoagulation were observed.

Haematology

Vitamin K treatment resulted in a rapid decrease of the clotting times; within 48 hours after initiation of the therapy PT values had returned to base line values (Fig. 2).

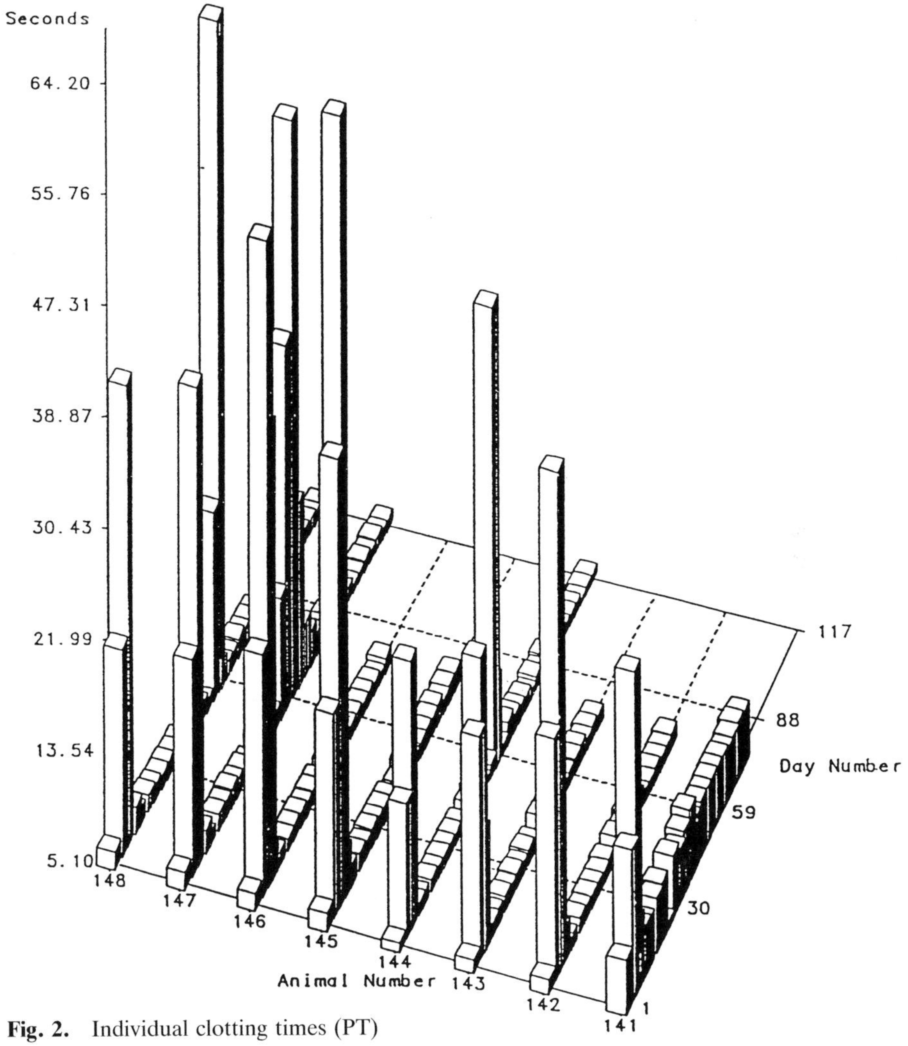

Fig. 2. Individual clotting times (PT)

No further elevations of clotting times were observed during or after therapeutic phase.

Liver Residues

Analysis of the livers for presence of flocoumafen at termination of the experiment revealed that 43 weeks after dosing approximately 8% of the administered dose was present in the liver in the animals which had received a single dose of 0.5 mg/kg flocoumafen. In the animals which were dosed twice with 0.5

Table 1. Concentration of Flocoumafen in liver tissue 43 weeks following acute oral administration of 0.5 mg/kg

	Total dose	Concentration (mg/kg)			total/liver
No	(mg)	cis	trans	total	(mg)
141M	5.8	0.72	0.47	1.19	0.46
142F	5.9	0.66	0.44	1.10	0.50
143M	6.7	0.70	0.62	1.32	0.52
144F*	13.2	0.94	0.60	1.54	0.62
145M	7.0	0.88	0.66	1.54	0.67
146F	6.6	0.86	0.50	1.36	0.55
147M*	12.2	1.02	0.59	1.61	0.65
148F*	12.5	1.00	0.60	1.60	0.69
Mean				1.41	0.58
SD				0.2	0.09

* Animal received 2 doses of 0.5 mg/kg
Limit of detection cis 0.01 mg/kg; trans 0.02 mg/kg

mg/kg flocoumafen approximately 5% of the total administered dose was recovered in the liver. The concentations of flocoumafen in liver tissue are presented in Table 1.

Conclusions

- There is no difference in antidotal response using either 2 or 5 mg/kg vitamin K per day.
- Assessment of clotting times, in particular PT values, is a useful diagnostic tool to establish a suspected exposure and to monitor response to therapy. However, extensions in clotting times are not always correlated with overt anticoagulant toxicity.
- There is significant individual animal variation in the length of time to develop clinical signs of intoxication and in the degree of elevation of clotting times. No changes were, however, detected within 6 days of exposure in any of the animals.
- The therapeutic regime used in this study is effective and suitable for dogs intoxicated with flocoumafen.
- Severe intoxication, as observed in gross haemorrhaging was successfully treated by whole blood transfusion.
- Flocoumafen is retained in the liver at concentrations similar to those in other species. The findings support the concept of the existence of a capacity-limited hepatic binding site for flocoumafen.

References

Huckle KR, Hutson DH, Warburton PA (1988) Elimination and accumulation of the rodenticide flocoumafen in rats following repeated oral administration. Xenobiotica 18:1465–1479

Huckle KR, Hutson DH, Logan CJ, Morrison BJ, Warburtn PA (1989a) The fate of the rodenticide flocoumafen in the rat: retention and elimination of a single oral dose. Pestic Sc 25:297–312

Huckle KR, Morrison BJ, Warburton PA (1989b) The percutaneous fate of the rodenticide flocoumafen in the rat: role of non-biliary intestinal excretion. Xenobiotica 19:63–74

Huckle KR, Warburton PA, Forbes S, Logan CJ (1989c) Studies on the fate of flocoumafen in the Japanese quail (Coturnix coturnix japonica). Xenobiotica 19:51–62

Suttie JW (1978) The fat soluble vitamins. In Handbook of Lipid Research, Vol. 2, ed. H. F. Deluca, Plenum Press, New York

Suttie JW (1980) Mechanism of action of Vitamin K: Synthesis of gamma-carboxyglutamic acid. Critical Reviews in Biochemistry 8:191–223

Thijssen HHW, Baars LGM (1987) Hepatic uptake and storage of Warfarin. The relation with the target enzyme vitamin K 2,3-epoxide reductase. J Pharmacol Exp Ther 243:1082–1088

Recent Developments in Toxicology:
Trends, Methods and Problems
Arch. Toxicol., Suppl. 14, 166–169 (1991)
© Springer-Verlag 1991

Influence of Natural Humic Acids and Synthetic Phenolic Polymers on Haemostasis

H.-P. Klöcking

Institute of Pharmacology and Toxicology, Medical Academy Erfurt,
Nordhäuser Straße 74, 5010 Erfurt, Germany

Introduction

Naturally occurring humic acids and synthetically prepared phenolic polymers belong to the group of polyanions. One of these compounds is heparin, which is well-known in human medicine for a long time already for being useful as directly acting anticoagulant substance (Verstraete and Machin 1980). Heparin prevents not only proliferation of already existing thrombi but also formation of new thrombi (Egbring 1983). Furthermore, heparin contributes to the release of tissue-type plasminogen activator (t-PA) from the endothelial cells of the vessel wall (Markwardt and Klöcking 1977), which plays an important role in dissolving thrombi. That prompted the investigation of naturally occurring humic acids as well as synthetic phenolic polymers with regard to their influence on plasminogen activator release and their anticoagulant effect.

Materials and Methods

The substances used were: sodium humate (Na-HS, m.w. 7900) and ammonium humate (NH_4-HS, m.w. 7500), both isolated from high-moor peat (collected near Rostock, Baltic Sea coastal region of the FRG) (Klöcking et al 1977); polymeric oxidation products from caffeic acid (KOP, m.w. 6500) and 3.4-hydroxyphenylacetic acid (3.4-DHPOP, m.w. 6700) were synthetized by oxidation of the basic compounds with sodium m-periodate (Helbig and Klöcking 1983).

Experimental animals: Conventional Uje: WIST male and female rats with a bodyweight of 285 ± 35 g (n = 26), male and female conventional Schön: ICR mice, weighing 28 ± 2 g (n = 30) and rabbits of either sex with a bodyweight of 2.8 ± 0.3 kg (n = 6).

The t-PA activity in rat plasma was determined according to Chmielewska and Wiman (1986). Bleeding time in mice was estimated according to Kaiser and Markwardt (1986). Thrombin time and partial thromboplastin time were determined with recalcified human and rabbit citrated plasma (Klöcking 1981).

Results and Discussion

The results of the *in vivo* study on acute t-PA release are compiled in Table 1. They represent the maximum increase in t-PA release expressed in International Units (IU) per ml plasma. Compared with the control, the administration of Na-HS, NH_4-HS or KOP at a dose of 5 or 10 mg/kg each caused a statistically significant increase in t-PA release.

These substances affect the coagulation system in direction of hypocoagulability. Doses, which twice prolonged the thrombin time (tt) and the partial thromboplastin time (ptt), are listed in Table 2. At a concentration of 6.5 μmol/l of KOP, or 3.4-DHPOP, respectively, the thrombin time is prolonged twice. The naturally occurring humic acid NH_4-HS showed two- to three-fold less potent anticoagulant effect. The partial thromboplastin time is prolonged 4–7 fold in the case of KOP, 3.4-DHPOP, NH_4-HS but only doubled at higher concentrations.

Table 1. Acute t-PA release in rats after intravenous bolus injection

Substance	Dose mg/kg	Maximal enhancement of t-PA release IU/ml plasma	Time of max. enhancement min	No. of expts n	Significance p
Saline	0.1 ml/100 g	1.3 ± 1.0	1	3	–
Na-HS	1	0.9 ± 0.8	1	3	n.s.
	5	6.2 ± 1.6	2	3	< 0.001
	10	21.6 ± 3.1	5	3	< 0.001
NH_4-HS	1	0.9 ± 1.0	1	3	n.s.
	5	10.6 ± 1.9	1	3	< 0.001
	10	12.9 ± 5.0	2; 5; 30	4	< 0.02
KOP	5	8.6 ± 3.6	2	3	< 0.02
	10	11.3 ± 1.7	2; 10; 30	4	< 0.001

Table 2. Anticoagulant effective concentrations of phenolic polymers (PP) in vitro (μmol/l)

PP	I_{eff}(TT)	I_{eff}(PTT)
3,4-DHPOP	6.5	29.5
KOP	6.5	48.0
Na-HS	21.0	92.5
NH_4-HS	14.5	82.5

Table 3. Influence of humic acids and phenolic polymers on bleeding time (min) in mice after intravenous injection

Substance	Concentration mg/kg	Bleeding time min	Number of experiments n
Na-HS	10.0	1.2 ± 0.9	5
NH_4-HS	10.0	0.9 ± 0.3	5
3.4-DHPOP	0.1	1.5 ± 0.7	3
	0.5	1.6 ± 0.1	3
	1.0	9.2 ± 3.2	3
	10.0	19.3 ± 9.4	3
KOP	0.1	3.1 ± 2.1	3
	0.5	2.1 ± 1.5	3
	1.0	3.7 ± 3.5	4
	10.0	15.7 ± 16.7	3
control (physiological saline)		1.4 ± 0.4	71

The relation between bleeding time and dose in mice after transsection of the tail was studied to evaluate the effect of the humic acids and phenolic polymers on primary haemostasis *in vivo*. Depending on the dose, the bleeding time was prolonged twice after a single intravenous injection of both phenolic polymers (Table 3). These haemorrhagic properties of KOP and 3.4-DHPOP may be related to their anticoagulant effect. A single intravenous injection of 10–20 mg/kg 3.4-DHPOP or KOP in rabbits prolonged the clotting time by 40–100%.

A potential anticoagulant effect as well as a tissue-type plasminogen activator releasing effect may explain the use of peat in treatment of thrombophlebitis described in the literature (Lachmann 1964).

References

Chmielewska J, Wiman B (1986) Determination of tissue plasminogen activator and its "fast" inhibitor in plasma. Clin Chem 32:482–485

Egbring R (1983) Neue Aspekte der Biochemie des Heparins und seine Bedeutung bei Prophylaxe und Therapie. Folia Angiol Suppl VIII:7–91

Helbig B, Klöcking R (1983) Darstellung und Charakterisierung von Huminsäure-Modellsubstanzen. Z Physiother 33:31–37

Kaiser B, Markwardt F (1986) Antithrombotic and haemorrhagic effects of synthetic and naturally occurring thrombin inhibitors. Thomb Res 43:613–620

Klöcking HP (1981) Erfassung von Hämostasestörungen durch Fremdstoffe. Ergebn exp Med 38:99–118

Klöcking R, Helbig B, Drabke P (1977) Eine verbesserte Methode für die Isolierung wasserlöslicher Huminsäuren und enzymatisch oxydierter o-Diphenolverbindungen. Pharmazie 32:297

Lachmann H (1964) Die Möglichkeiten der Heilmoor-Schwebstoff-Therapie. Bericht über den 8. Internationalen Kongreß für universelle Moor- und Torfforschung vom 5. bis 10. Oktober 1962 in Bremen (BRD). Linz a/D: Länderverlag GmbH, pp. 115–116

Markwardt F, Klöcking HP (1977) Heparin-induced release of plasminogen activator. Haemostasis 6:370–374
Verstraete M, Machin ST (1980) Clinical usage of heparin; present and future trends. Scand J Haematol, Suppl 36:1–188

Recent Developments in Toxicology:
Trends, Methods and Problems
Arch. Toxicol., Suppl. 14, 170–173 (1991)
© Springer-Verlag 1991

A Comparison of the Thrombogenic Activity of Prothrombin Complex Concentrates Using Stasis and Non-stasis Models

M. Richter, H.-P. Klöcking, G. Dornheim*, J. Güttner**, and R. Wulkow

Institute of Pharmacology and Toxicology, Medical Academy, Erfurt, Germany
* District Institute of Blood Transfusion, Suhl
** Institute of Microbiology and Experimental Therapy, Jena

Introduction

Patients treated with prothrombin complex concentrates (PCC) have frequently been found to develop thrombotic events (Kaspar 1975; Cederbaum 1976). Therefore, PCC must be regarded as potentially thrombogenic substances (Elödi 1977). In order to reduce the risk of thromboembolic complications the addition of heparin was recommended (Ménaché and Roberts 1975).

In the present study the influence of two adsorbent agents of different particle size on the thrombogenicity of PCC was evaluated. The two products obtained (without heparin) were both compared *in vitro* and *in vivo*. *In vivo* the venous stasis model according to Wessler et al (1959) and a non-stasis model of disseminated intravascular coagulation (Koth et al 1980) were used.

Materials and Methods

Prothrombin complex concentrates (PCC): Prepared either by adsorption to DEAE-Sephadex A-50 particle size 40–120 μm, Pharmacia Uppsala (PCC A) or by Molselect DEAE-50 particle size 100–320 μm, Reanal Budapest (PCC B) from human citrated plasma after separation of factor VIII by cryoprecipitation. Factor II, IX and X were determined as recommended by Behring-Werke, Marburg.

Experimental animals: Wistar rats (Schö/Wi) with a body mass of 290 ± 50 g, anaesthetized with 1.2 g ethyl urethane intraperitoneally.

In vitro procoagulant activity: Determination of thrombin generation time (TGt_{50}) was performed according to Sas et al (1975) and the determination of non-activated partial thromboplastin time (NAPTT) as described by Kingdon et al (1975), modified by Klöcking et al (1984).

In vivo thrombogenicity: Determination of thrombogenicity was done by means of the stasis model according to Wessler et al (1959). The thrombus size was graded by a score: 0 = fluid blood, 1 = a few macroscopic strands of fibrin, 2 = several small thrombi, 3 = two or more large thrombi, 4 = a single thrombus forming a cast of the isolated segment. For determination of thrombogenicity by means of the non-stasis model, rats received 100 mg of 4-aminomethylbenzoic acid per kg subcutaneously, and 30 min later 2.5 ± 0.4 MBq ^{125}I-fibrinogen per kg was given; after a further 30 min a 90 min-infusion of PCC via a jugular vein catheter started. The radioactivity over the lungs was continuously recorded according to Koth et al (1980). A significant increase in radioactivity was regarded as microthrombosis. After finishing the infusion the radioactivity in samples of lungs, kidney, spleen and liver was measured. Tissue sections were examined for fibrin thrombi according to Klöcking et al (1989).

Results and Discussion

In vitro coagulation assays (NAPTTR, TGt_{50}) revealed a significant difference in the procoagulant activity of PCC A compared with PCC B. The generally accepted limits for non-thrombogenic PCC (NAPTTR > 0.5; TGt_{50} > 10 min (Sas et al 1975; Prowse and Williams 1980) are not fully reached for PCC A. In contrast to that, PCC B meets these requirements (Table 1).

In the stasis model preparation B showed significant lower thrombogenicity than preparation A (Fig. 1). Mean effective doses were calculated from the percentage of thrombus formation in dependence of PCC doses administered (Lichtfield and Wilcoxon 1949). ED_{50} values for thrombus formation amounted to 200 factor IX U/kg (PCC B), and 15 factor IX U/kg (PCC A), respectively. In the non-stasis model using ^{125}I-fibrinogen pretreatment of animals and continous recording of radioactivity over the lungs the difference in thrombogenicity of PCC A and B could also be demonstrated, if higher doses of PCC than

Table 1. *In vitro* data of prothrombin complex concentrates

Assay	PCC A (n = 7)	PCC B (n = 8)
F IX (U/ml)	19.7 ± 3.0	18.0 ± 2.4
F II (U/ml)	21.9 ± 2.2	18.8 ± 2.1
F X (U/ml)	20.8 ± 1.9	18.0 ± 2.1
Protein (g/l)	15.1 ± 0.7	9.5 ± 1.4
NAPTTR*	0.63 ± 0.05	0.84 ± 0.08
TGt_{50} (min)	4.8 ± 0.5	28.9 ± 8.7

* ratio of non-activated partial thromboplastin time with and without PCC

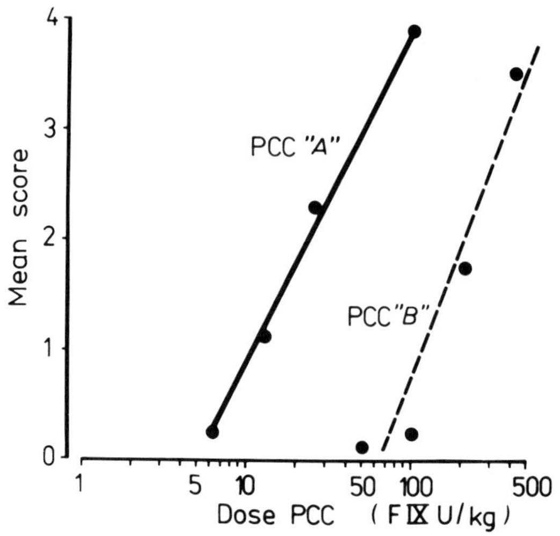

Fig. 1. Influence of two adsorbent agents on the thrombogenicity of prothrombin complex concentrates (PCC). PCC A: prepared by adsorption to DEAE-Sephadex A-50 particle size 40–120 μm. PCC B: prepared by adsorption to Mol-select DEAE-50 particle size 100–320 μm. n = 8 for each dose level of each PCC

Table 2. Microthrombosis in the non-stasis model following i.v. infusion of prothrombin complex concentrates

Dose F IX U/kg/h	Incidence PCC A	PCC B
250	2 / 6	0 / 10
500	3 / 4	0 / 5
1000	5 / 5	3 / 4
2000	–	5 / 5

in the stasis model were administered (Table 2). An increase in radioactivity indicating the occurrence of disseminated intravascular coagulation could be observed 52.7 ± 13.4 min after the PCC infusion was started. Microthrombosis was confirmed by histological examination and by measurement of radioactivity of tissue samples after finishing the infusion. The increase in radioactivity was 5–6 fold for lungs and kidney and about 2 fold for liver and spleen on average.

References

Cederbaum AI, Blatt PM, Roberts HR (1976) Intravascular coagulation with use of human prothrombin complex concentrates. Ann Intern Med 84:683–687

Elödi S (1977) Thrombogenic properties of prothrombin complex concentrates. Blood 50:961–965

Kaspar CK (1975) Thromboembolic complications (Task Force Report – International Society of Thrombosis and Haemostasis). Thromb Diath Haemorrh 33:640–644

Kingdon HS, Lundblad RL, Veltkamp JJ, Aronson DL (1975) Potentially thrombogenic materials in factor IX concentrates. Thromb Diath Haemorrh 33:617–631

Klöcking H-P, Kleßen C, Jablonowski C, Meerbach W, Dornheim G (1984) Untersuchungen zur Thrombogenität eines neuen Prothrombinkomplexkonzentrates. Fol Haematol 111:645–661

Klöcking H-P, Markwardt F, Güttner J (1989) On the mechanism of batroxobin-induced fibrinolysis. Pharmazie 44:504

Koth W, Nowak G, Markwardt F (1980) Monitoring of microthrombosis in experimental animals by continuous recording of ^{125}I-fibrin deposits and ^{52}Cr-labelled platelets in the lungs. Acta Biol Med Ger 39:157–162

Litchfield JT. Wilcoxon F (1949) A simplified method of evaluating dose-effect experiments. J Pharmacol Exp Ther 96:99–115

Menache D, Roberts HR (1975) Summary report and recommendations of the task force members and consultants. Thromb Diath Haemorrh 33:645–647

Prowse CV, Williams AE (1980) A comparison of in vitro and in vivo thrombogenic activity of factor IX concentrates using stasis (Wessler) and non-stasis rabbit models. Thromb Haemost 44:81–86

Sas G, Owens RE, Smith JK, Middleton S, Cash JD (1975) In vitro spontaneous thrombin generation in human factor IX concentrates. Br J Haematol 31:25–31

Wessler S, Reimer SM, Sheps MC (1959) Biologic assay of a thrombosis-inducing activity in human serum. J Appl Physiol 14:943–946

Recent Developments in Toxicology:
Trends, Methods and Problems
Arch. Toxicol., Suppl. 14, 174–178 (1991)
© Springer-Verlag 1991

Inhibition of Platelet Aggregation and Thromboxane A_2 Production by Chlorophenoxy Acid Herbicides

P. Ylitalo[1], M. Ylihakola[1], H. A. Elo[1], H.-M. Koponen[1], E. Seppälä[2], and H. Vapaatalo[2]

[1] Department of Pharmacology and Toxicology, University of Kuopio, P.O. Box 1627, SF-70211 Kuopio, Finland, and
[2] Department of Biomedical Sciences, University of Tampere, P.O. Box 607, SF-33101 Tampere, Finland

Introduction

In toxic doses chlorophenoxy acid herbicides cause hemorrhages in different organs of animals. The compounds are structurally related to some antiplatelet drugs, such as P-chlorophenoxyisobutyric acid (clofibric acid) and 4-[2-(benzene-sulfonamide)-ethyl]phenoxyacetic acid (sulotroban = BM 13.177). It has been found recently that chlorophenoxy acid herbicides inhibit human platelet aggregation (Elo et al. 1987). The mechanism of the antiplatelet action of these herbicides is not known, while sulotroban is an effective antagonist of thromboxane (Tx) A_2 (Patscheke et al. 1986). The present paper reports that some chlorophenoxy acid herbicides are potent inhibitors of human platelet aggregation and thromboxane production.

Material and Methods

Platelet-rich plasma (PRP, 300,000 platelet/mm^3) containing citrate as anticoagulant was prepared from healthy unmedicated nonsmokers (21–42 yrs). Platelet aggregation was induced at 37°C by 1.0–10 μM adenosine-5'-diphosphate (ADP, Sigma), 7.5.–30 μg/ml collagen (Sigma), or 0.6–2.0 μM U-46619, a TxA_2-receptor agonist (Upjohn). The phenoxy acid herbicides of minimum 99% purity were 2,4-dichlorophenoxyacetic acid (2,4-D) and 4-(2-methyl-4-chlorophenoxy)butyric acid (MCPB) (Riedel-de Haen AG, Hannover). All the concentrations refer to their water-soluble sodium salts.

The dose-response curves of platelet aggregation by each inducer, with and without 2,4-D or MCPB, were measured with a nine-channel photometer (FP-901 Chemistry Analyzer, Labsystems Ltd, Helsinki) equipped with a shaking thermoblock. After one-minute shaking, increasing concentrations of an inducer were added. The time course of the absorbance was counted every ten seconds for ten minutes. The anti-aggregatory effects of 2,4-D and MCPB were studied by increasing their concentrations from 0.004 to 1.0 mg/ml, and from 0.02 to 0.2 mg/ml, respectively. Immediately after the counting, the cuvettes were placed in ice-bath. The samples were centrifuged at 9,000 g for 1 min at 4°C. The platelet-free supernatant was separated, frozen with liquid nitrogen, and stored at −70°C. The concentrations of TxB_2, a stable metabolite of TxA_2, in the supernatant were measured by a direct radioimmunoassay (Seppälä et al. 1984). The percentages of the aggregation were calculated from the terminal plateau of the aggregation curves. The percentages of TxB_2 production by chlorophenoxy acids were calculated from the duplicates containing the vehicle alone.

The competitive antagonistic effect of 1.0 mg/ml 2,4-D versus the vehicle on the dose-response curve of platelet aggregation by the TxA_2-receptor agonist U-46619 was studied by increasing U-46619 concentrations from 0.3 to 4μM.

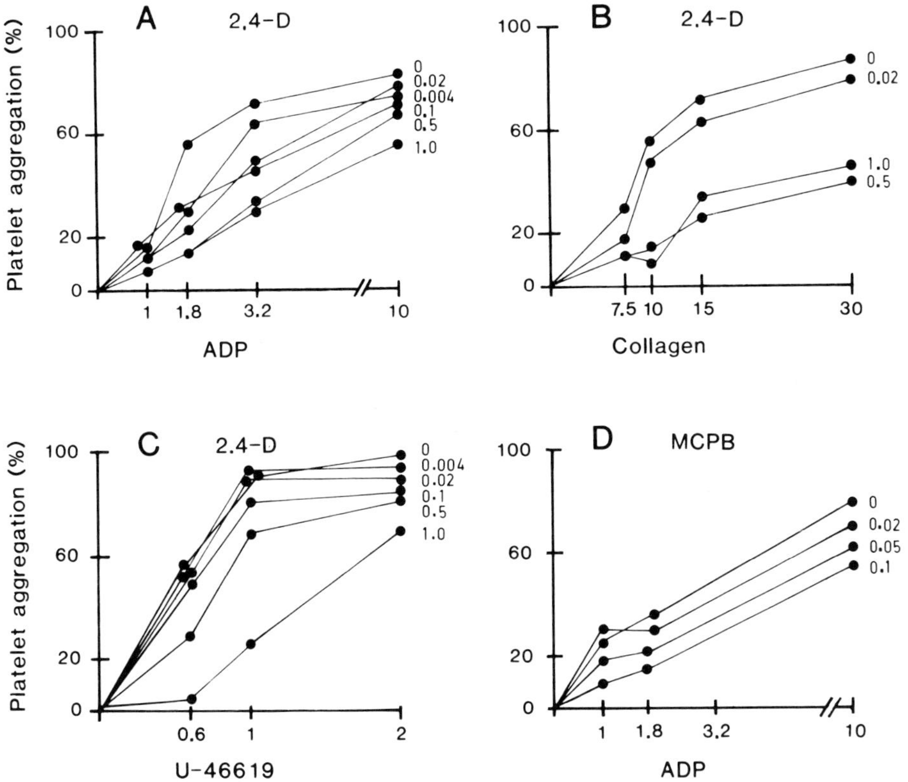

Fig. 1. The mean inhibitory effects of different concentrations of 2,4-dichlorophenoxyacetic acid (2,4-D) and 4-(2-methyl-4-chlorophenoxy)butyric acid (MCPB) on human platelet aggregation induced by adenosine-5'-diphosphate (ADP) (μM), collagen (μg/ml), or U-46619 (μM). 2,4-D (Graphs A–C) and MCPB (Graph D) concentrations (mg/ml) in platelet-rich plasma from 4 or more volunteers are given at the end of the curves

Results and Discussion

The herbicide concentrations used in the present study can occur in human blood during acute exposure, since in severe human intoxications the concentrations of chlorophenoxy acids have varied from 0.67 to 1.15 mg/ml (Nielsen et al. 1965; Prescott et al. 1979).

2,4-D inhibited, dose-dependently, the platelet aggregation induced by ADP, collagen or U-46619. MCPB also inhibited the aggregation by ADP (Fig. 1). The aggregation induced by the TxA_2-receptor agonist U-46619 was competitively inhibited by 1 mg/ml 2,4-D (Fig. 2). The shift of the aggregation curve to the right suggests that, at least in high concentrations, 2,4-D competes with U-46619 for the same receptors. The hypothesis of a TxA_2-receptor blockade by 2,4-D agrees with the findings of Patschke et al. (1986), who proposed that the anti-aggregatory effects of sulotroban, another phenoxyacetic acid derivative, are closely related to the competitive blockade of TxA_2-receptors in the platelets.

Both 2,4-D and MCPB reduced, dose-dependently, the production of TxB_2 in PRP during the aggregation process induced by ADP and collagen (Fig. 3). 2,4-D inhibited TxB_2 production in the same way also during the aggregation caused by U-46619. At the concentrations of 0.5 and 1.0 mg/ml, 2,4-D reduced TxB_2 to 0–20% of that in the parallel control vials with no herbicide. Although the molecular mechanism of platelet aggregation is complex and not fully

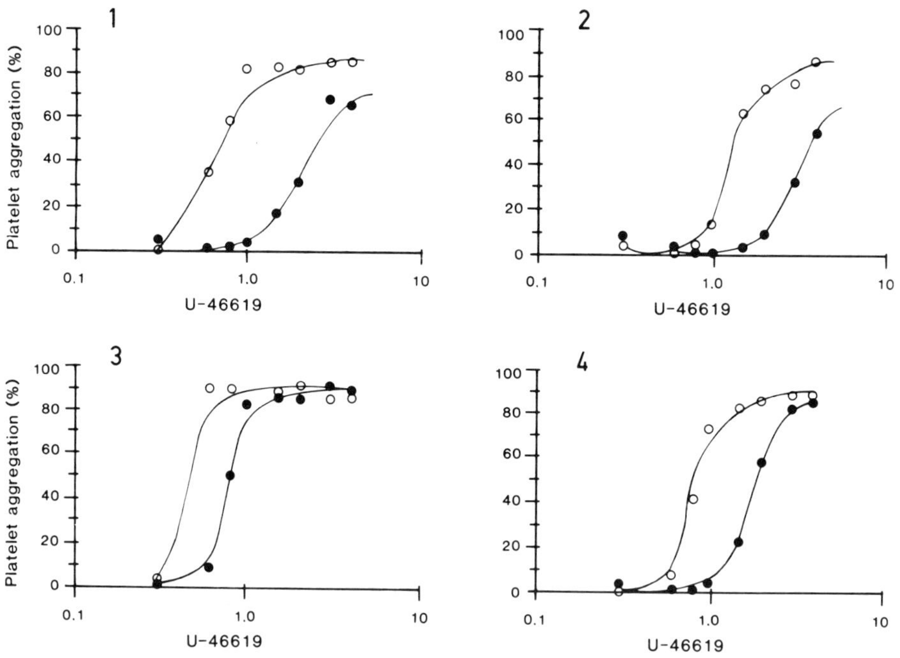

Fig. 2. The inhibitory effect of 1 mg/ml 2,4-dichlorophenoxyacetic acid (2,4-D) on platelet aggregation induced by increasing concentrations of the thromboxane A_2-receptor agonist U-46619 (μM) in platelet-rich plasma from 4 volunteers. Symbols: ○ = vehicle, ● = 2,4-D

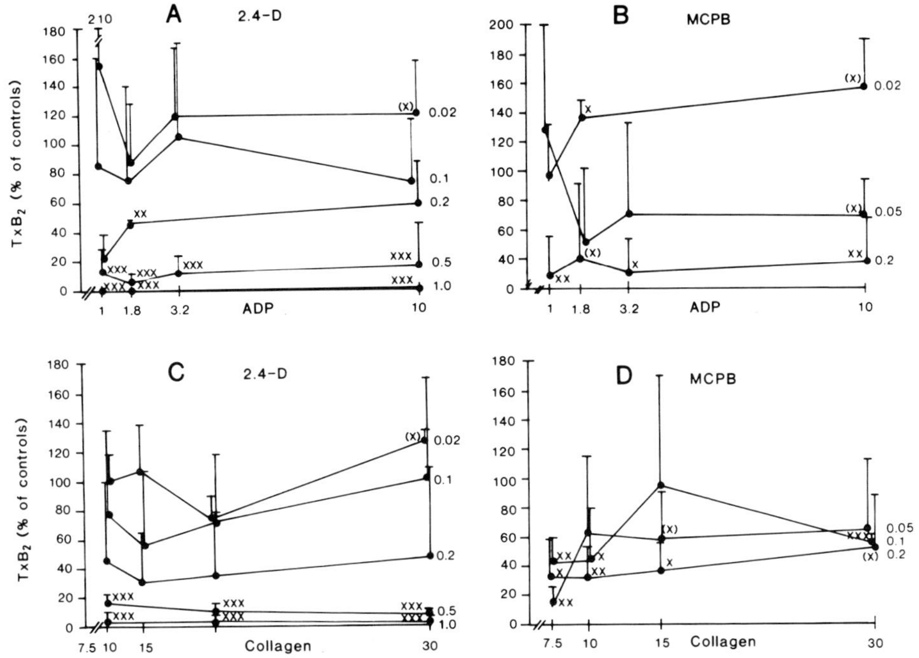

Fig. 3. The inhibitory effects (mean ± SD, n = 2–13) of different concentrations of 2,4-dichlorophenoxyacetic acid (2,4-D) and 4-(2-methyl-4-chlorophenoxy)butyric acid (MCPB) on thromboxane (Tx) B_2 production in human platelet-rich plasma during the platelet aggregation induced by adenosine-5'-diphosphate (ADP) (μM), and collagen (μg/ml). 2,4-D (A, C) and MCPB (B, D) concentrations (mg/ml) are given at the end of the curves. $^{(x)}p<0.1$, $^{x}p<0.05$, $^{xx}p<0.01$, $^{xxx}p<0.001$ (paired t-test) compared with the vehicle (= 100%)

understood, it is generally agreed that the TxA_2 system is involved in the ADP and collagen-induced platelet aggregation (Kinlough-Rathbone et al. 1977).

The present observations suggest that the antiplatelet effect of selected chlorophenoxy acid herbicides are closely related to the inhibition of the TxA_2 production of platelets, and that, at least in high concentrations, they also block TxA_2-receptors.

Grant. Huhtamäki Oy Leiras, Turku, Finland, and the Medical Research Council of The Academy of Finland.

References

Elo HA, Luoma T, Ylitalo P (1987) Inhibition of platelet aggregation by chlorophenoxyacetic acid herbicides. Xth International Congress of Pharmacology, Sydney, Australia. Abstracts, p. 456

Kinlough-Rathbone RL, Packham MA, Reimers HJ, Cazenave JR, Mustard JF (1977) Mechanism of shape change, aggregation and release induced by collagen, thrombin or A 23187. J Lab Clin Med 90:707–719

Nielsen K, Kampe B, Jensen-Holm J (1965) Fatal poisoning in man by 2,4-dichlorophenoxyacetic acid (2,4-D): determination of the agent in forensic materials. Acta Pharmacol Toxicol 22:224–234

Patschke H, Staiger C, Neugebauer G, Kaufmann B, Strein K, Endele R, Stegmeier K (1986) The pharmacokinetic and pharmacodynamic profiles of the thromboxane A-2 receptor blocker BM 13.177. Clin Pharmacol Ther 39:145–150

Prescott LF, Park J, Darrier I (1979) Treatment of severe 2,4-D and mecoprop intoxication with alkaline diuresis. Br J Clin Pharmacol 7:111–116

Seppälä E, Pora O, Metsä-Ketelä T (1984) A modified method for extraction and purification of prostaglandins with resin XAD-2. Prostagl Leukotr Med 14:235–241

Toxic Effects on Liver and Kidney

Recent Developments in Toxicology:
Trends, Methods and Problems
Arch. Toxicol., Suppl. 14, 181–184 (1991)
© Springer-Verlag 1991

The Effect of Paracetamol on the Hepatic Uptake of Microparticles *in Vivo* and *in Vitro*

S. Bühring, G. Hennighausen, and L. Jonas*

Institute of Pharmacology and Toxicology, University of Rostock, Leninallee 70, 2500 Rostock, and
* Institute of Pathology, University of Rostock, Strempelstraße 14, 2500 Rostock

Introduction

In the liver, Kupffer cells which are resident macrophages are mainly responsible for the elimination of most of the circulating foreign particles. The phagocytotic function is of importance for nonspecific and specific defence mechanisms and it influences the kinetic of drug carrier systems (Wisse 1979). Hepatotoxic compounds may disturb the phagocytotic activity of Kupffer cells. Hence the hepatic uptake of CC and MMS and the effect of paracetamol and DEM in vivo and in the isolated perfused rat liver was studied.

Materials and Methods

Female outbred rats (Rozt:Wistar, C, weighing 200–220 g) were treated intraperitoneally with paracetamol 2.3, 4.0 mmol/kg bw (Chemiekombinat Bitterfeld) and DEM 5.6 or 7.5 mmol/kg bw (synthetized in this laboratory by Dr. G. Kröning) 90 min before the determination of hepatic uptake.

Hepatic uptake in vivo: Blood samples were taken from the retroorbital plexus 1, 4, 7, 10 and 15 min after the injection of 20 mg colloidal carbon (Tusche No. 26, Robotron-Vertrieb Dresden, modified) into the tail vein. The concentration of carbon in the haemolyzed blood samples was determined spectrophotometrically at 650 nm. The MMS (Rotrop-HSA-MMS B1, Zentralinstitut für Kernforschung, Dresden) were also injected into the tail vein (0.17 mg MMS, 1.7 mBq). Blood samples were taken 30, 60, 90 and 120 sec after injection of MMS, the radioactivity was counted on a Gammacounter.

Abbreviations: CC, Colloidal carbon; DEM, Diethyl maleate; GSH, Reduced gluthathione; MMS, Tc-99m-labelled millimicrospheres of human serum albumin

Hepatic uptake in the isolated perfused rat liver: 10 min after a heparin injection (2500 I.E/kg) the rat livers were perfused via the portal vein by a recirculation technique with a modified Hanks'-Solution (1% gelatine, pH 7.4, 37°C, bubbling with 95% oxygen and 5% carbon dioxide) containing 0.1 ml carbon/100 ml or 1 ml MMS (10 mBq)/100 ml. After preperfusion of 5 min, samples were obtained from the recirculated perfusion medium at 0, 10, 15, 20, 40 and 60 min after starting the recirculated perfusion. 6 rats were treated intraperitoneally with 75 mg phenobarbital/kg b.w. for 3 d. Parecetamol (6 mmol/100 ml Hanks'-Solution) was given in the perfusion medium before starting the recirculated perfusion.

Determination of GSH in the liver: GSH was estimated as non-protein sulfhydryl content according Sedlak and Lindsay (1968).

Data analysis: The clearance was expressed as half life ($t_{1/2}$ in min) estimated by a kinetic program according the nonlineal approximation. The Student t-test was employed for statistical analysis.

Results

After treatment of rats with acute toxic doses of paracetamol the clearance of carbon was inhibited dose dependently, the clearance of MMS also decreased (Table 1).

Table 1. Effect of paracetamol (Par) on the hepatic uptake of CC and MMS in vivo

Treatment[a]	GSH (μmol/gliver)	Blood clearance[b] CC	MMS ($t_{1/2}$,min)
Control	5.6 ± 0.3 (12)	2.4 ± 0.2 (32)	0.7 ± 0.03 (8)
Par 2.3	4.0 ± 0.4 (3)	3.0 ± 0.3 (7)	–
Par 4.0	2.8 ± 0.3 (3)	9.3 ± 1.2 (6)*	1.0 ± 0.1 (8)*
DEM 5.6	1.2 ± 0.22 (7)*	4.3 ± 0.6 (7)*	1.1 ± 0.14 (8)*

[a] single administration (i.p.) of paracetamol and DEM in (mmol/kg b.w.) 90 min before determination of GSH and blood clearance
[b] values are the mean ± SE, number of animals in parentheses
* $p<0.05$, significance to the control group

The clearance function of the isolated perfused rat liver was decreased after pretreatment of rats with both paracetamol and DEM. Using livers of rats pretreated phenobarbital with 15 mg/kg for three days similar effects were observed, when paracetamol was used *in vitro* in the perfusion medium (Table 2). The inhibition of hepatic uptake of microparticles in vivo and in the isolated perfused rat liver was correlated with a decrease of GSH content in liver (Table 1 and 2). Hepatocytes were isolated with collagenase according Zimmermann et al (1987). The non-parenchymal cell fractions were pelleted (10 g, 5 min) and washed twice (300 g, 5 min). The non-parenchymal cell suspension with a

Table 2. Effects of paracetamol on the hepatic uptake of CC and MMS in the isolated perfused rat liver

Treatment[a]	GSH (μmol/gliver)	clearance[b] CC	MMS ($t_{1/2}$,min)
Control	4.2 ± 0.3 (17)	22.3 ± 1.5 (8)	16.0 ± 1.9 (9)
Par 4.0	2.0 ± 0.2 (6)*	42.2 ± 13.1 (4)*	33.4 ± 11.1 (2)*
Phenobarbital			
Par[c]	4.2 ± 0.7 (4)	43.9 ± 6.1 (6)*	–
DEM 7.5	0.5 ± 0.1 (9)*	45.8 ± 13.1 (4)*	28.8 ± 5.6 (5)*

a, b, * see table 1
c Paracetamol in vitro, 8 mmol/100 ml perfusion medium
Phenobarbital pretreatment 75 mg/kg i.p. 3 d

density of 1.06 gm cm^{-3} was freed from erythrocytes and cell debris by centrifugation through a visotrast (Fahlberg-List Magdeburg) density gradient (1.09 gm cm^{-3}).

The content of GSH was decreased in hepatocytes as well as in non-parenchymal cells isolated from livers of DEM-treated rats:

Hepatocytes: Control 182.9 nmolGSH/100 mg protein
DEM 37.3 nmolGSH/100 mg protein
Non-parenchymal cells:
Control 176.5 nmolGSH/100 mg protein
DEM <14.3 nmolGSH/100 mg protein

Discussion

The inhibition of the hepatic uptake of microparticles after treatment with paracetamol and DEM is correlated with a depletion of GSH. A decrease of the GSH content to a critical value may lead to loss of viability and failure of GSH dependent cellular functions such as phagocytosis (Orrenius 1985). It is probable that the phagocytotic activity of Kupffer cells is disturbed by depletion of GSH as indicated by the decrease of GSH in the non-parenchymal liver cell fraction. The isolated perfused rat liver is a suitable model for experimental studies of phagocytosis without humoral, nervous or hormonal influence from the organism. Electron microscopic studies with goldmarked MMS in the isolated perfused rat liver indicate the uptake of MMS and their digestion in the lysosomes by Kupffer cells. Acute toxic doses of paracetamol may impair the blood clearance of microorganisms and other phagocytotic processes involved in nonspecific and specific defence mechanisms.

Acknowlegements. The authors wish to thank Dr. G. Ester (Clinic of radiology, University Rostock) for estimation of Tc-99m MMS and Dr. Walzel (Institute of biochemistry, University Rostock) for marking the MMS with gold particles.

References

Hennighausen G (1989) Depletion of hepatic glutahtion in rats impairs phagocytosis in vivo. Arch Toxicol Suppl 13:326–329
Orrenius S (1985) Biochemical mechanisms of cytotoxicity, TIPS, FEST-suppl 15–20
Wisse E, Knook DL (1979) Progress in liver diseases. Vol. VI (Popper H, Schaffner F, eds). Grune & Stratton New York, 153–171
Sedlak J, Lindsay RH (1968) Estimation of total protein bound and non-protein suflhydryl groups in tissue with Ellman's reagent. Anal Biochem 25:192–205
Zimmermann T, Franks H (1987) Isolation and characterization of parenchymal cells from normal and cirrhotic rat liver. Cell Biochem Hunot 5:47–54

Recent Developments in Toxicology:
Trends, Methods and Problems
Arch. Toxicol., Suppl. 14, 185–187 (1991)
© Springer-Verlag 1991

Influence of Iron Chelators, 1,2-Dialkyl-3-Hydroxypyridin-4-Ones, on the Lipid Peroxidation and Glutathione Level in the Liver of Mice

V. Eybl[1], D. Caisová[1], J. Koutenský[1], and G. J. Kontoghiorghes[2]

[1] Dept. of Pharmacology, Medical Faculty Charles University, 30166 Pilsen, ČSFR, and [2] Dept. of Haematology, RHSM, London, UK

Introduction

Pretreatment with deferoxamine (DFO) protects male mice against the halo-alkane induced hepatoxicity. This effect is closely related to the ability of DFO to bind iron and to suppress iron-dependent lipid peroxidation which may be induced by various chemicals (Siegers et al 1988). In this paper the influence of some iron chelators of the group of 1,2-dialkyl-3-hydroxypyridin-4-ones on the lipid peroxidation (LP) and glutathione depletion was investigated. LP was induced by indium chloride, cadmium chloride or carbon tetrachloride.

Materials and Methods

The experiments were performed in male mice (ICR, SPF, 23–25 g b.w.). Indium chloride ($InCl_3$, purum, Fluka AG, Switzerland), cadmium chloride ($CdCl_2.2^1/2H_2O$, p.a., Lachema, ČSFR) and carbon tetrachloride (Lachema, ČSFR) were used. The 1,2-alkyl-3-hydroxypyrid-4-ones were synthesized (Kontoghiorghes et al 1987). The following derivatives were used: 1,2-dimethyl- (compound 1 = L_1), 1-ethyl-2-methyl- (compound 2), 1-propyl-2-methyl- (compound 3 = L_1NP_r), 1-methyl-2-ethyl- (compound 4), 1,2-diethyl-3-hydroxypyrid-4-one (compound 5) (Kontoghiorghes et al 1987).

The liver homogenates of mice pretreated with these chemicals were used. The doses and the ways of administration are indicated in the tables.

Lipid peroxidation expressed as amount of malondialdehyde (MDA nmol/g wet tissue) was measured in liver homogenates by the thiobarbituric acid test and the level of glutathione (GSH μmol/g wet tissue) with Ellman's reagent (Sedlak and Lindsay 1968; Uchiyama and Mikara 1978).

All results were statistically evaluated. The means ± SEM are given in the Tables.

Results and Discussion

The chelators administered s.c. were able to reduce the lipid peroxidation enhanced by indium pretreatment. The GSH-depletion in the liver was abolished (Table 1).

The compounds 1 (L_1) and 3 (L_1NP_r) were effective after the s.c. or p.o. administration as well. The direct binding of indium, a trivalent metal, is presumed. In the previous experiments the protective effect of these iron binding chelators in acute indium intoxication was proved (Eybl et al 1990). The chelators were also able to diminish the lipid peroxidation which was enhanced by cadmium pretreatment. This fact indicates that also iron might be involved in the cadmium induced lipid peroxidation (Table 2).

Table 1. Effect of chelators on indium induced lipid peroxidation and glutathione content in the liver of male mice 24 h after the i.p. administration of $InCl_3$. The chelators were administered immediately s.c. following the i.v. injection of $InCl_3$ (6 mg/kg) at a chelator: $InCl_3$ molar ratio of 30:1

Treatment	n	MDA nmol/kg[a] wet tissue	GSH μmol/g[a] wet tissue
1. Control	9	36.11 ± 1.75	6.81 ± 0.21
2. In only	12	50.08 ± 1.66	3.80 ± 0.27
3. In + Comp. 1	5	29.60 ± 0.81	6.22 ± 0.26
4. In + Comp. 2	5	40.60 ± 1.21	5.45 ± 0.02
5. In + Comp. 3	6	36.83 ± 2.23	6.58 ± 0.23
6. In + Comp. 4	7	37.14 ± 2.24	6.38 ± 0.24
7. In + Comp. 5	6	42.00 ± 1.13	5.63 ± 0.22

[a] Values represent the mean ± SEM

Table 2. Effect of chelators on indium or cadmium induced lipid peroxidation in the liver of male mice 24 h after the i.p. administration of $InCl_3$ or $CdCl_2$. The chelators were administered immediately s.c. following the i.v.injection of $InCl_3$ or $CdCl_2.2^1/2\ H_2O$ (always 6 mg/kg) at a chelator:$InCl_3$ molar ratio of 30:1

Treatment	n	MDA nmol/g[a]
A. Experiment with In		
1. Control	10	33.15 ± 1.09
2. In only	9	47.89 ± 1.04
3. In + Comp. 1 p.o.	6	29.50 ± 1.52
4. In + Comp. 1 s.c.	5	34.40 ± 2.42
5. In + Comp. 3 p.o.	7	35.50 ± 2.43
6. In + Comp. 3 s.c.	7	30.57 ± 0.95
B. Experiment with Cd		
7. Control	5	25.00 ± 0.55
8. Cd only	5	42.20 ± 2.01
9. Cd + Comp. 1 s.c.	5	29.40 ± 0.40
10. Cd + Comp. 3 s.c.	5	24.20 ± 1.47

[a] Values represent the mean ± SEM

Table 3. Effect of chelators on CCl_4 induced lipid peroxidation in the liver of male mice 12 h after the p.o. administration of CCl_4. Chelators were administered immediately following the i.p. administration of CCl_4. Doses: Comp. 1 = 110 mg/kg, comp. 3 = 113 mg/kg. CCl_4 (1% solution in sunflower oil) 10 mg/kg i.p.

Treatment	n	MDA nmol/g[a] wet tissue
A. Chelators administered s.c.		
1. Control	6	28.50 ± 2.14
2. CCl_4 only	6	66.83 ± 3.70
3. CCl_4 + Comp. 1 s.c.	6	45.50 ± 3.21
4. CCl_4 + Comp. 3 s.c.	6	70.83 ± 5.35
B. Comp. 1 administered s.c./p.o.		
5. Control	5	31.00 ± 2.26
6. CCl_4 only	5	67.80 ± 5.40
7. CCl_4 + Comp. 1 s.c.	6	43.67 ± 3.62
8. CCl_4 + Comp. 1 p.o.	6	109.00 ± 5.39
C. Comp. 3 administered s.c./p.o.		
9. Control	6	30.83 ± 1.47
10. CCl_4 only	6	64.83 ± 3.10
11. CCl_4 + Comp. 3 s.c.	6	68.00 ± 3.80
12. CCl_4 + Comp. 3 p.o.	6	99.17 ± 8.06

[a] Values represent the mean ± SEM

Compound 1 was effective in CCl_4 induced peroxidation only when administered s.c. Compound 3 was without protective effect (Table 3).

The results of the experiments show that the antioxidant activity and hepatoprotective effect of iron chelators used should be examined further. Particularly compound 1, which is used clinically in iron overload, might be valuable. The advantage is the possibility of p.o. administration.

References

Eybl V, Caisová D, Koutenská M, Koutenský J, Sýkora J, Smolíková V, Kontoghiorghes GJ (1990) Influence of 1,2-dialkyl-3-hydroxypyridin-4-one chelators on indium toxicity, retention and distribution in mice. Plzeň lék Sborn Suppl in print

Kontoghiorghes GJ, Sheppard L, Chambers S (1987) New synthetic approach and iron chelating studies of 1-alkyl-2-methyl-3-hydroxypyridin-4-ones. Arzneim Forsch 37:1099–1102

Sedlak J, Lindsay RH (1968) Determination of malonaldehyde precursor in tissues by thiobarbituric test. Analyt Biochem 25:192–205

Siegers CP, Steffen B, Younes M (1988) Antidotal effects of deferoxamine in experimental liver injury role of lipid peroxidation. Pharmacol Res Comm 20:337–343

Uchiyama M, Mikara M (1978) Estimation of total, protein-bound and non-proteins sulphydryl groups in tissue with Ellman's reagent. Analyt Biochem 86:271–278

Recent Developments in Toxicology:
Trends, Methods and Problems
Arch. Toxicol., Suppl. 14, 188–192 (1991)
© Springer-Verlag 1991

The Influence of Cytochrome P-450 Inducers on Carbonyl Reduction in Mouse Liver

E. MASER, B. HAHNEMANN, W. LEGRUM, U. OPPERMANN, and K. J. NETTER

Department of Pharmacology and Toxicology, Philipps-University Marburg, Lahnberge, D-3550 Marburg, FRG

Introduction

Carbonyl compounds (aldehydes and ketones) are widely distributed in nature and have a broad spectrum of biological functions, i.e. as metabolic intermediates, specific effectors or xenobiotic substrates. Many of them may interfere with the physiological metabolism and, therefore, must be eliminated. In mammals the reduction of xenobiotic aldehydes and ketones to their respective alcohol metabolites plays the most important role prior to their elimination from the body (El Masry et al 1956). The enzymes involved are called carbonyl reductases and are subdivided into ketone and aldehyde reductases, based on their substrate specificity and inhibitor sensitivity (Felsted and Bachur 1980; Sawada et al 1981). In general, aldehyde reductases do not reduce ketones and are inhibited by barbiturates. In contrast, ketone reductases, which reduce many aldehydes, are insensitive to barbiturates, but are inhibited by the flavonoid quercitrin.

In previous investigations it was shown that metyrapone (MPon), the potent cytochrome P-450 inhibitor, is reduced to metyrapol (MPol) in liver microsomes by a phenobarbital (PB) sensitive aldehyde reductase, and in liver cytosol by a quercitrin (Qu) sensitive ketone reductase. Moreover, the two enzymes differ in cofactor requirement and immunological homology (Maser 1989). The microsomal MPon-reductase was purified to homogeneity showing a molecular weight of 34 kDa (Maser and Netter 1989).

The objective of this study was to investigate whether xenobiotic inducers of the oxidative biotransformation also influence the reductive metabolism of carbonyl compounds. Therefore, female as well as male C57BL/6J Han mice were pretreated with a set of previously described inducers of the coumarin hydroxylase locus (Hahnemann et al 1989).

Abbreviations: Metyrapone (MPon), Metyrapol (MPol), Phenobarbital (PB), Quercitrin (Qu), Pyrazine (PN), Pyrazole (PL), Aminotriazole (AT), Cobalt (Co), Indium (In), 3-Methylcholanthrene (3-MC)

Methods

Female and male C57BL/6J Han mice, 7 weeks old, were used throughout this study. The pretreatments were as follows: pyrazine (PN), pyrazole (PL) and aminotriazole (AT) were injected intraperitoneally at a dose of 200 mg/kg b.wt. daily for two days. The heavy metals cobalt (Co) and indium (In) were injected subcutaneously at a dose of 60 mg $CoCl_2 \cdot 6\,H_2O$/kg b.wt. for two days and 120 mg $In_2(SO_4)_3 \cdot 5\,H_2O$/kg b.wt. for only one day, respectively. The classic cytochrome P-450 inducers phenobarbital (PB; 80 mg/kg b.wt. for two days) and 3-methylcholanthrene (3-MC; 30 mg/kg b.wt. for two days) served as references. Control mice received an aliquot of isotonic saline for two days. All substances were dissolved in isotonic saline exept 3-MC, which was dissolved in arachis oil. Preparation of microsomes and determination of cytochrome P-450 were performed according to Legrum et al (1979). Protein was assayed according to Lowry et al (1951).

Enzyme activity studies and determination of MPon and MPol by HPLC were performed according to Maser and Legrum (1985).

Results

In general, pretreatment of mice with the selected xenobiotics, generates a higher MPol formation in the hepatic cytosol, whereas it is lowered in microsomes (Fig. 1 a–d). This finding serves as an additional differentiating feature of the two MPon-reducing enzymes in the respective liver fractions.

The most prominent differences are observed after the N-containing heteroaromates pyrazine (PN) and pyrazole (PL). In detail, PL decreases aldehyde reductase activity in microsomes of males to 54% (Fig. 1a) and in that of females to 75% (Fig. 1b). PN enhances the specific activity of the cytosolic ketone reductase to about 160% in both sexes (Fig. 1 c–d). However, AT, Co, In and the classic cytochrome P-450 inducers PB and 3-MC had a weaker or no effect.

Quercitrin and phenobarbital used *in vitro* as diagnostic inhibitors of the carbonyl reductases in livers of PN or PL pretreated male mice reveal: the microsomal aldehyde reductase activity, which is decreased after PL-pretreatment, shows a scarcely altered PB-inhibition, whereas the Qu-inhibition of the ketone reductase, induced after PN-pretreatment, is clearly stronger compared with that of respective control experiments (Table 1). As expected, PB does not affect the cytosolic ketone reductase and Qu does not inhibit the microsomal aldehyde reductase.

Discussion

Carbonyl reductases comprise a family of enzymes which share typical features such as monomeric structure with a molecular weight of about 30–40 kDa, dependence on NADPH as cofactor and insensitivity to pyrazole, the potent inhibitor of alcohol dehydrogenase. In addition to their substrate specificity, up to now, they are subdivided by their inhibitor susceptibility into PB-sensitive aldehyde reductases and Qu-sensitive ketone reductases. The ketone MPon was shown to be reduced in mouse liver microsomes by an aldehyde reductase (Maser and Netter 1989) and in the respective cytosol by a ketone reductase (Oppermann and Maser 1990).

In this investigation it becomes obvious that inducers of the cytochrome P-450 mediated oxidative metabolism are able to influence also the reductive metabolism of carbonyl compounds. However, the substances do not turn out to be exclusively inductive, as shown for the cytosolic ketone reductase (Fig. 1 c–d), since in contrast a lowered enzyme activity in the case of the microsomal aldehyde reductase is observed (Fig. 1 a–b). Up to now the family of carbonyl reductases is subdivided reflecting their specificity towards substrates and their

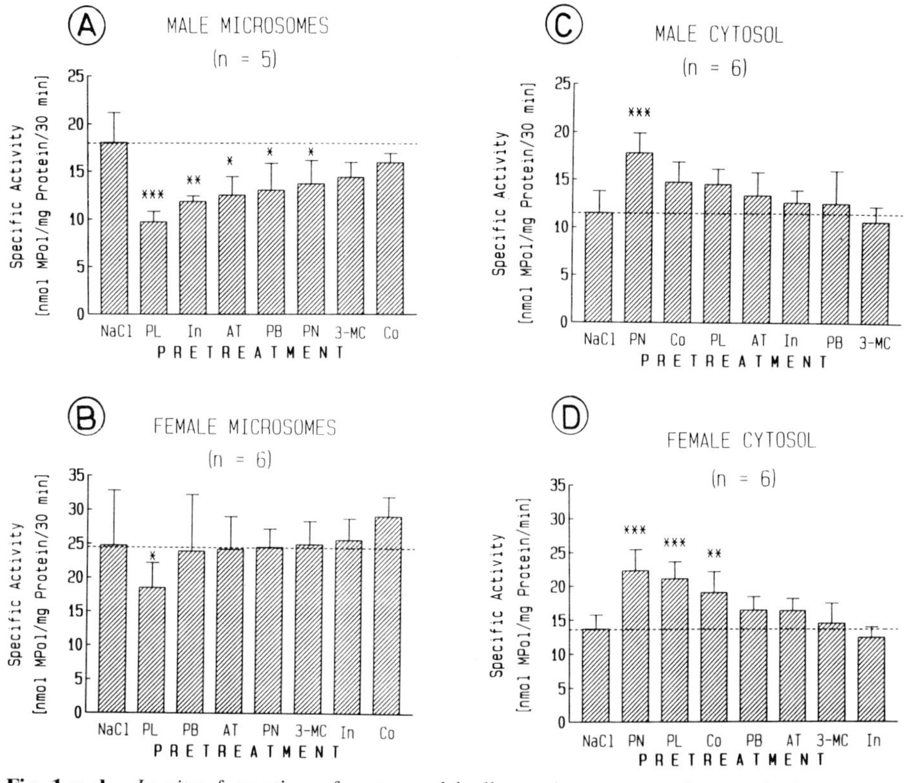

Fig. 1 a–d. *In vitro* formation of metyrapol in liver microsomes and cytosol after *in vivo* pretreatment of female and male C57BL/6J Han mice with cobalt (Co), indium (In), pyrazine (PN), pyrazole (PL), aminotriazole (AT), phenobarbital (PB) and 3-methylcholanthrene (3-MC). The reductase activity was assayed in the standard incubation mixture containing 50 mM sodium phosphate buffer, pH 7.4, a NADPH-regenerating system and 4.6 mM metyrapone

Table 1. Relative inhibition of microsomal and cytosolic carbonyl reductases by diagnostic inhibitors

	Inhibitor			
	PB		Qu	
Pretreatment	Ms	Cyt	Ms	Cyt
None	15.6	–	–	40.2
PL	12.9	–	–	52.1
PN	16.1	–	–	51.7

The enzyme activity was assayed in the standard incubation mixture containing 50 mM sodium phosphate buffer, pH 7.4, a NADPH-regenerating system, 4.6 mM metyrapone and inhibitor concentrations of 1 mM, respectively. The percent of inhibition was calculated from uninhibited control experiments. (n = 6; for control activities see Fig. 1 a–d)
Ms: microsomes
Cyt: cytosol
–: no inhibition observed

sensitivity towards inhibitors. On the basis of these results it can be deduced that further subdivision might be possible based on the susceptibility towards "diagnostic inducers" administered *in vivo,* e.g. N-containing heteroaromates. A combination of *in vivo* pretreatment of mice with PN or PL and subsequent quantification of the *in vitro* inhibition by PB and/or Qu shows that the expected enhancement of inhibition of the cytosolic enzyme indeed occurs. This may serve to characterize the presumably newly synthesized apoprotein.

References

El Masry AM, Smith JN, Williams RT (1956) Studies in detoxication. 69. The metabolism of alkylbenzenes: n-propylbenzene and n-butylbenzene with further observations on ethylbenzene. Biochem J 64:50–56

Felsted RL, Bachur NR (1980) Mammalian carbonyl reductases. Drug Metab Rev 11:1–60

Hahnemann B, Kühn B, Heubel F, Legrum W (1989) Selective induction of the coumarin hydroxylase by N-containing heteroaromatic compounds. Arch Toxicol Suppl 13:297–301

Legrum W, Stuehmeier G, Netter KJ (1979) Cobalt as a modifier of microsomal monooxygenase in mice. Toxicol Appl Pharmacol 48:195–204

Lowry OH, Rosebrough NJ, Farr AL, Randall RJ (1951) Protein measurement with the Folin phenol reagent. J Biol Chem 193:265–275

Maser E (1989) Characterization of microsomal and cytoplasmic metyrapone reducing enzymes. Arch Toxicol Suppl 13:271–274

Maser E, Legrum W (1985) Alteration of the inhibitory effect of metyrapone by reduction to metyrapol during the metabolism of methacetin in vivo in mice. Naunyn-Schmiedeberg's Arch Pharmacol 331:283–289

Maser E, Netter KJ (1989) Purification and properties of a metyrapone-reducing enzyme from mouse liver microsomes – This ketone is reduced by an aldehyde reductase. Biochem Pharmacol 38:3049–3054

Oppermann U, Maser E (1990) Cofactor requirement and inhibitor profiles of carbonyl reducing enzymes in kidney of rodents. Naunyn-Schmiedeberg's Arch Pharmacol Suppl 341:R8

Sawada H, Hara A, Hayashibara M, Nakayama T, Usui S, Saeki T (1981) Microsomal reductase for aromatic aldehydes and ketones in guinea pig liver. Purification, characterization, and functional relationship to hexose-6-phosphate dehydrogenase. J Biochem 90:1077–1085

Recent Developments in Toxicology:
Trends, Methods and Problems
Arch. Toxicol., Suppl. 14, 193–196 (1991)
© Springer-Verlag 1991

Influence of Inducers of Cytochrome P-450 on Dealkylation of 7-Alkoxycoumarins

D. MÜLLER, TH. GROH, and F. K. SPLINTER

Institut für Pharmakologie und Toxikologie der Friedrich-Schiller-Universität,
Löbderstraße 1, O-6900 Jena, Germany

Introduction

P-450 dependent monooxygenation, which is responsible for detoxication and toxication of many xenobiotics, can be induced by many substances. Depending on the inducer, different P-450 forms are induced, which differ in their substrate specificity and in their involvement in the formation of toxic and carcinogenic metabolites. Therefore in toxicological investigations a sensitive and specific detection of the extent and type of induction is important. A simple detection is possible by measuring the metabolism of model substrates, e.g. ethylresorufin O-deethylation depends specifically on P-450 forms induced by β-NF and other polycyclic aromatic hydrocarbons (Guengerich et al 1982), whereas pentylresorufin O-depentylation is considerably, but not specifically, induced by PB (Lubet et al 1985). Among 7-alkoxycoumarin derivatives 7-ethoxycoumarin was mostly used as a model substrate; its deethylation is induced by β-NF and PB (Guengerich et al 1982). The aim of this study was to find 7-alkoxycoumarins and 7-alkoxy-4-methylcoumarins whose 7-dealkylation depends more specifically on certain P-450 forms.

Material and Methods

Preweanling male rats were injected intraperitoneally 1 × 50 mg/kg b.w. β-NF (Serva Heidelberg) or on 3 consecutive days 3 × 60 mg/kg b.w. PB (AWD Dresden), controls received saline. The animals stayed with their mothers; who

Abbreviations: AC-DA/AMC-DA: 7-alkoxycoumarin/7-alkoxy-4-methylcoumarin O-dealkylation; BC/BMC: 7-butoxycoumarin/7-butoxy-4-methylcoumarin O-debutylation; β-NF: β-naphthoflavone; EC/EMC: 7-ethoxycoumarin/7-ethoxy-4-methylcoumarin O-deethylation; MC/MMC: 7-methoxycoumarin/7-methoxy-4-methylcoumarin O-demethylation; P-450: cytochrome P-450; PB: phenobarbital; PC/PMC: 7-propoxycoumarin/7-propoxy-4-methylcoumarin O-depropylation; PhC/PhMC: 7-phenylmethoxycoumarin/7-phenylmethoxy-4-methylcoumarin O-debenzylation received a standard diet and tapwater ad libitum. 24 hours after the last pretreatment (18 days after birth) liver 10,000 g supernatant was prepared. O-dealkylation of 7-alkoxy- and 7-alkoxy-4-methylcoumarins (synthesized by F. K. Splinter) was determined as described for 7-ethoxycoumarin-O-deethylation (Müller 1990). The concentration of all substrates in the incubation assay was 400 μmol/l.

Results

Fig. 1 shows the influence of β-NF on AC-DA and AMC-DA. EC, PC and BC rates are increased by a factor of about 5, whereas MC and PhC rates are induced to a lesser extent. If these 7-alkoxycoumarins contain an additional 4-methyl group the dealkylation rates (MMC, EMC, PMC, BMC, PhMC) are not influenced by β-NF at all.

The influence of PB on the above-mentioned reactions is demonstrated in Fig. 2. MC, EC, PC and BC are induced by a factor of about 4, whereas PhC shows a 15-fold increase following PB administration. The induction of the O-dealkylation rates of the 4-methyl derivatives by PB is generally smaller (factor 2–3 for MMC, EMC, PMC, BMC and 4 for PhMC).

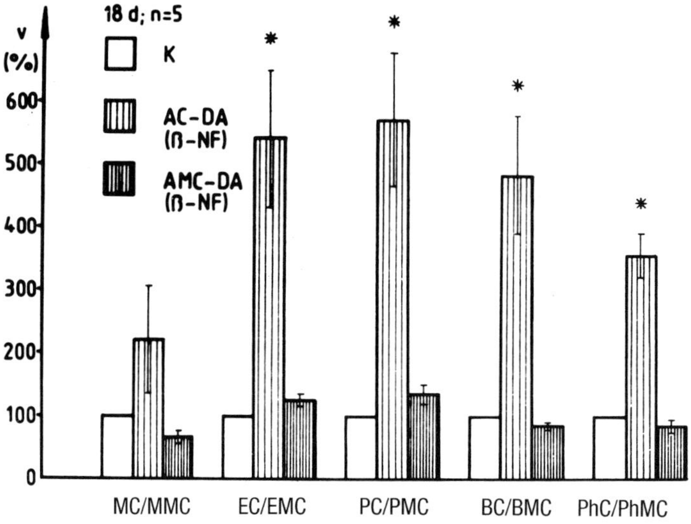

Fig. 1. Influence of a single administration of 50 mg β-naphthoflovane (β-NF)/kg b.w. on 7-alkoxycoumarin/7-alkoxy-4-methylcoumarin O-dealkylation (AC-DA/AMC-DA) in rat liver 10,000 g supernatant. K = controls (= 100%). O-dealkylation of 7-methoxycoumarin/7-methoxy-4-methylcoumarin = MC/MMC; of 7-ethoxycoumarin/7-ethoxy-4-methylcoumarin = EC/EMC; of 7-propoxycoumarin/7-propoxy-4-methylcoumarin = PC/PMC; of 7-butoxy-coumarin/7-butoxy-4-methylcoumarin = BC/BMC; of 7-phenylmethoxycoumarin/7-phenyl-methoxy-4-methylcoumarin = PhC/PhMC. Significant differences (Student's t-test, $P \leq 0.05$) to the control value are marked by *

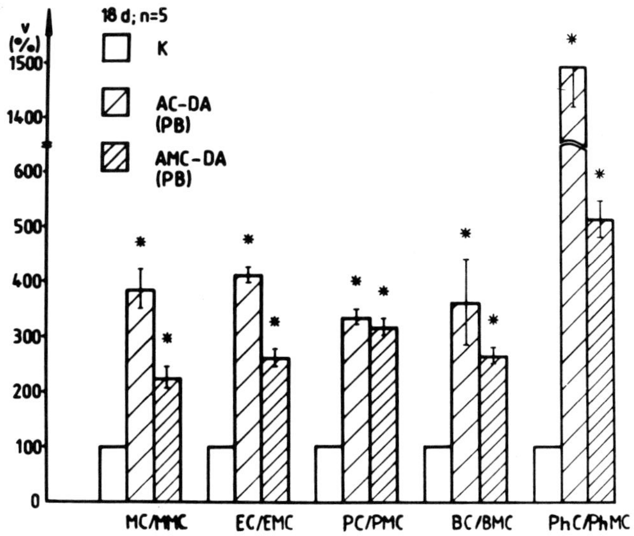

Fig. 2. Influence of 3 administrations of 60 mg phenobarbital (PB)/kg b.w. on 7-alkoxycoumarin/7-alkoxy-4-methylcoumarin O-dealkylation (AC-DA/AMC-DA) in rat liver 10,000 g supernatant. For further explanations cf. legend of Fig. 1

Discussion

The results show that the 7-O-dealkylation rate of all 7-alkoxycoumarin derivatives (without 4-methyl group) can be induced by β-NF as well as by PB. The extent of induction, however, differs. EC, PC and BC are considerably induced by β-NF and to a smaller extent by PB. In older rats these differences in the inducing effects of both inducers are much more pronounced (Müller 1990). MC and PhC rates, however, are preferentially increased by phenobarbital. PhC proved to be the most sensitive reaction for the detection of an induction by PB (induction factor 15). The inducibility by PB, however, is not specific, since there is also a certain induction by β-NF. In the presence of a 4-methyl group in the 7-alkoxycoumarin molecule 7-O-dealkylation rate of all derivatives cannot be induced by β-NF; there is only an induction by PB. The highest induction by PB is seen with PhMC (induction factor 5). Thus PhC can be recommended for a sensitive detection and PhMC for a specific and also relatively sensitive detection of induction by PB. So far pentylresorufin O-depentylation was mostly used for this purpose because of its strong induction by PB (Lubet et al 1985). This reaction has some disadvantages in comparison with PhMC. It is also induced by β-NF. Moreover, control activities of PhMC are more exactly measureable due to the higher fluorescence of the metabolite.

Among the 10 reactions investigated, there was no specific induction by β-NF. Thus ethylresorufin is the more specific substrate of the β-NF-induced P-450 form than any of these alkoxycoumarins.

References

Guengerich FP, Dannan GA, Wright ST, Martin MV, Kaminsky LS (1982) Purification and characterization of liver microsomal cytochromes P-450: electrophoretic, spectral, catalytic, and immuno-chemical properties and inducibility of eight iso-zymes isolated from rats treated with phenobarbital or β-naphthoflavone. Biochemistry 21:6019–6030

Lubet RA, Mayer RT, Cameron JW, Nims RW, Burke MD, Wolff TH, Guengerich FP (1985) A rapid and sensitive assay for measuring induction of cytochrome(s) P-450 by phenobarbital and other xenobiotics in the rat. Arch Biochem Biophys 238:43–48

Müller D (1990) Influence of xenobiotics on drug metabolism and its sensitive detection. Exp Pathol 39:187–194

Recent Developments in Toxicology:
Trends, Methods and Problems
Arch. Toxicol., Suppl. 14, 197–200 (1991)
© Springer-Verlag 1991

The Effect of Monoclonal Antibodies on Dipyrone-Induced Rat Liver Cytochrome P450

H. Kraul[1], M. Pasanen[2], and O. Pelkonen[2]

[1] Institute of Clinical Pharmacology, Friedrich Schiller University, Bachstraße 18, O-6900 Jena, Germany, and
[2] Department of Pharmacology and Toxicology, University of Oulu, SF-90220 Oulu, Finland

Introduction

Dipyrone is used to estimate cytochrome P450(P450)-dependent biotransformation in humans especially via phenobarbital(PB)-inducible isozymes. However, there exist discrepancies concerning specificity.

One possibility to examine the involvement of different P450-isozymes is to explore the ability of dipyrone to induce different P450-forms. These P450-forms can be further characterized by use of specific antibodies towards P450.

Material and Methods

Antibodies: The monoclonal antibodies (Mab) used in studies were a generous gift of S. S. Park and H. V. Gelboin (Laboratory of Molecular Carcinogenesis, National Institutes of Health, Bethesda, MD 20892, U.S.A.). Mab 1-7-1 towards 3-methylcholanthrene (MC)-induced form of rat liver P450 (Park et al 1982) and Mab 2-66-3 to PB-induced form of rat liver P450 (Park et al 1984) were available. As a control, an ascites fluid (HyHel-9) containing non-specific IgG was used.

All other chemicals were of the highest grade available.

Abbreviations: MC, 3-methylcholanthrene; PB, phenobarbital; AHH, aryl hydrocarbon (benzo(a)pyrene) hydroxylase; EROD, 7-ethoxyresorufin O-deethylase; PROD, 7-pentoxyresorufin O-depentylase; ECOD, 7-ethoxycoumarin O-deethylase; P450, cytochrome P-450; Mab, monoclonal antibody

Pretreatment of experimental animals: 80-day-old male Wistar rats received 200 mg dipyrone/kg b.w. (Novalgin®, Hoechst Veterinär GmbH, Germany) daily i.p. for 3 (n = 5) or 7 days (n = 5). Control animals (n = 4) were administered 0.9% sodium chloride, the solvent for dilution. Animals were killed 24 hours after the last administration.

Enzyme activities: Liver microsomal fractions were isolated by differential centrifugation. P450 content was determined by the method of Omura and Sato (1964). 7-Ethoxycoumarin O-deethylase (ECOD) activity was assayed according to Aitio (1978), aryl hydrocarbon hydroxylase (AHH) activity was measured according to Nebert and Gelboin (1968). The activities of 7-ethoxyresorufin O-deethylase (EROD) and 7-pentoxyresorufin O-depentylase (PROD) were determined by the endpoint method of Burke et al (1977).

In vitro inhibition experiments were performed with Mab 2-66-3 and Mab 1-7-1 (4 parts microsomal protein/1 part Mab).

Statistical analysis: The data represent the mean ± SD. Significant differences of a sample compared with control was determined using Wilcoxon rank sum test ($p < 0.05$).

Results and Discussion

Most of the enzyme activities measured, with the exception of AHH activity, were significantly enhanced after 3 as well as after 7 days pretreatment with dipyrone (Table 1). The induction was virtually complete after 3 days pretreatment. It seems that different P450-forms are involved in induction efficacy caused by affecting PB-inducible (PROD) and MC-inducible (EROD) P450-forms as well as unspecific parameters (ECOD, P450). However, what is remarkable is the dominant effect on PROD activity (31.6- and 44.4-fold induction), which is suitable to characterize PB-inducible P450-forms (Lubet et al 1985).

To distinguish between different induced P450-forms, monoclonal antibodies were incubated with reaction volume (Fig. 1). PROD activity in induced animals was clearly inhibited by PB-specific Mab 2-66-3 (about 10 to 15% remaining activity), Mab 1-7-1 did not cause significant diminution. Surprisingly, EROD activity was also inhibited by Mab 2-66-3 but not by Mab 1-7-1. Therefore, the increased EROD activity caused by dipyrone treatment seems to be the result of affecting PB-inducible P450-forms, although ethoxyresorufin has been declared to be a specific substrate for MC-inducible forms (Burke et al 1985). AHH activity was neither induced nor influenced by Mabs.

Results show the predominant role of PB-inducible P450-forms (P450IIB) in induction efficacy of dipyrone. The increase in other enzyme activities as

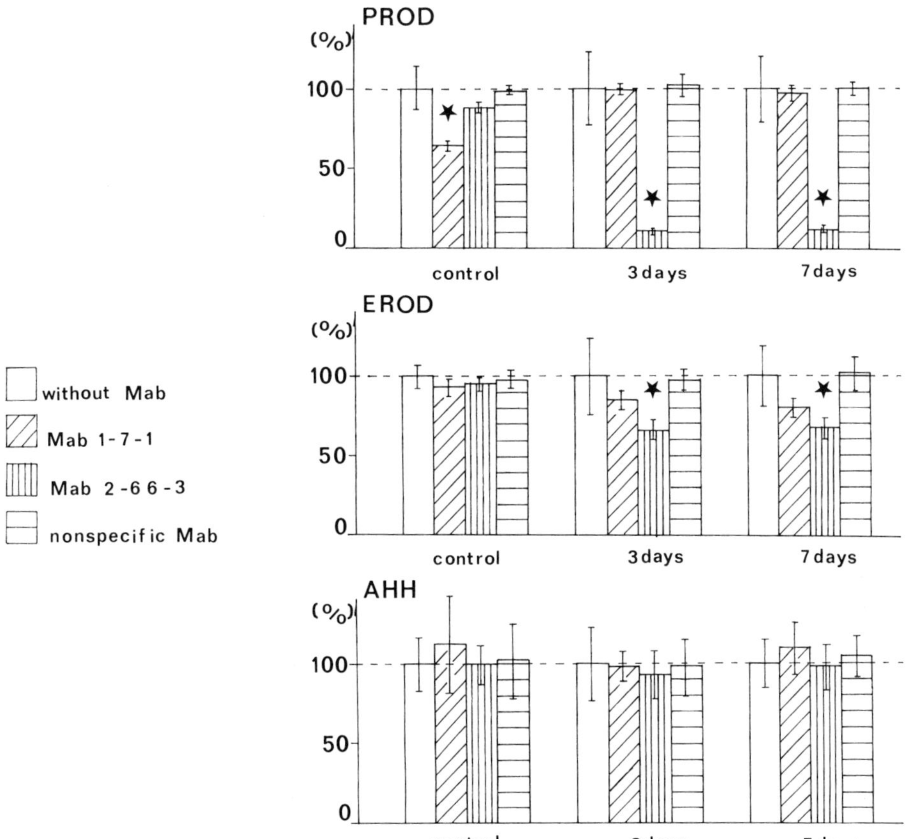

Fig. 1. Effects of two Mabs on activities of PROD, EROD and AHH by hepatic microsomes from untreated (control) and dipyrone-induced (3 and 7 days) rats. Data are expressed as percent of control incubation (without Mab). Significant inhibition is indicated by an asterisk

Table 1. Effect of dipyrone administration on rat liver monooxygenase activities

treatment	microsomal protein (mg/ml)	P450 (nmol/mg protein)	PROD	EROD	ECOD	AHH
control	14.09 ± 1.36	0.77 ± 0.13	0.024 ± 0.003	0.904 ± 0.062	0.600 ± 0.077	0.341 ± 0.056
3 days	15.67 ± 1.31 (1.11)	1.11 ± 0.11* (1.44)	0.781 ± 0.205* (31.6)	1.609 ± 0.388* (1.78)	1.287 ± 0.335* (2.14)	0.346 ± 0.063 (1.06)
7 days	16.27 ± 0.77* (1.15)	1.21 ± 0.19* (1.57)	1.100 ± 0.250* (44.4)	1.677 ± 0.316* (1.86)	1.571 ± 0.214* (2.62)	0.284 ± 0.044 (0.83)

Note. The rats were treated as described under Material and Methods. Each value represents the mean ± SD of four (control) or five (3 and 7 days treatment) rats. Enzyme activities are given as nmol/mg microsomal protein/min. The numbers in parenthesis denote the degree of induction compared with controls. Significant differences from control values: * $p < 0.05$.

well as the inhibition by Mabs (for example in EROD activity) seems to be rather the result of overlapping substrate specificities than of other P450-enzymes.

References

Aitio A (1978) A simple and sensitive assay of 7-ethoxycoumarin deethylation. Anal Biochem 85:488–491

Burke MD, Prough RA, Mayer RT (1977) Characteristics of a microsomal cytochrome P-448-mediated reaction ethoxyresorufin O-deethylation. Drug Metab Disp 5:1–8

Burke MD, Thompson S, Elcombe CR, Halpert J, Haaparanta T, Mayer RT (1985) Ethoxy-, pentoxy- and benzyloxyphenoxazones and homologues: A series of substrates to distinguish between induced cytochromes P-450. Biochem Pharmacol 34:3337–3345

Lubet RA, Mayer RT, Cameron JW, Nims RW, Burke MD, Wolff T, Guengerich FP (1985) Dealkylation of pentoxyresorufin: A rapid and sensitive assay for measuring induction of cytochrome(s) P-450 by phenobarbital and other xenobiotics in the rat. Arch Biochem Biophys 238:43–48

Nebert DW, Gelboin HV (1968) Substrate inducible microsomal aryl hydrocarbon hydroxylase in mammalian cell culture. I. Assay and properties of the induced enzyme. J Biol Chem 243:6242–6249

Omura T, Sato R (1964) The carbon monoxide-binding pigment of liver microsomes. I. Evidence for its hemoprotein nature. J Biol Chem 239:2370–2378

Park SS, Fujino T, Miller H, Guengerich FP, Gelboin HV (1984) Monoclonal antibodies to phenobarbital-induced rat liver cytochrome P-450. Biochem Pharmacol 33:2071–2081

Park SS, Fujino T, West D, Guengerich FP, Gelboin HV (1982) Monoclonal antibodies that inhibit enzyme activity of 3-methylcholanthrene-induced cytochrome P-450. Cancer Res 42:1798-1808

Recent Developments in Toxicology:
Trends, Methods and Problems
Arch. Toxicol., Suppl. 14, 201–204 (1991)
© Springer-Verlag 1991

Antipyrine Metabolism in Coke Oven Workers

P. Arvela, O. Pelkonen, L. Pyy*, E. Sotaniemi**, K. Vähäkangas, and E. Yrjänheikki*

Department of Pharmacology and Toxicology, University of Oulu,
** Department of Internal Medicine, Oulu University Central Hospital, and
* Institute of Occupational Health, Oulu, Finland

Introduction

Antipyrine test has been widely used to study the effects of environmental factors on drug metabolism and liver capacity both in animals and man (Vesell 1978). Antipyrine clearance has also been used in testing workers exposure to mixtures of chemicals (Dössing 1982). Chemical industry workers, alcoholics and healthy controls have been compared in terms of their antipyrine metabolizing capacity earlier (Arvela et al 1988). It was concluded that although the large interindividual variability in antipyrine metabolism reduce the usefulness of the test in small groups, it still may be an useful test on an individual basis. In the present study the antipyrine test was evaluated in 6 individual coke oven workers before and after one year's work in an atmosphere rich of various polycyclic aromatic hydrocarbons (PAH) which are known to induce drug metabolizing capacity of the liver.

Methods

Antipyrine was purchased from Aldrich Co. Three antipyrine metabolites, OHA, NORA and HMA, were a generous gift of Dr. M. Eichelbaum, Dr. Margarete-Fischer-Bosch Institute of Clinical Pharmacologg., Stuttgart. Experiment protocol was approved by the Ethical Committee of the Medical Faculty University of Oulu. Informed consent was obtained from the volunteers involved before starting the study.

6 healthy workers (numbers 1, 4 and 6 were smokers) were given antipyrine orally 20 mg/kg once before they started their work at the coke oven plant and the second time after one year of permanent work in the same conditions. The

Abbreviations: OHA, 4-hydroxyantipyrine; NORA, norantipyrine; HMA, 3-hydroxymethyl-antipyrine; PAH, polycyclic aromatic hydrocarbons

PAH concentrations from the personal dust samples were also measured twice, in 1988 and 1989 with the HPLC-method described by Jongeneelen et al (1987). Blood samples were collected at 1, 3 and 8 hours and urine for 24 hours after the antipyrine intake. The antipyrine concentrations in blood were measured by GLC according to Prescott et al (1973) and the three main metabolites (OHA, NORA and HMA) in urine according to Eichelbaum et al (1981). The pharmacokinetic parameters were calculated as described earlier (Pelkonen et al 1985). Urine cotinine levels were measured with a commercial kit (Diagnostic Products Corp. Los Angeles, CA 90015).

Results and Discussion

The average PAH concentrations in personal dust samples are shown in Fig. 1. In spite of a relative high exposure to various PAHs during one year, no clear changes in antipyrine metabolism were seen (Fig. 2). Only a small decrease in the antipyrine half-life in 5 individual workers was observed. Even the urinary excretion of OHA, the formation of which is catalyzed by the cytochrome P450 involved in PAH metabolism varied from a 50% decrease to nearly 25% increase in individual workers (Fig. 3). It is known that smoking habits can effect the metabolism of antipyrine. Arvela et al (1988) earlier reported that cessation of smoking for one month can decrease the excretion of the three main metabolites of antipyrine to half of the smokers normal value (Arvela et al 1988). The smoking habits of the coke oven workers were confirmed by measuring their urine cotinine levels and the average cotinine concentration of the three smokers remained at the same level and the non-smokers had clean urine in their both measured samples.

The negative results indicate that the workers may have adapted to the conditions of the high PAH-atmosphere. The same kind of adaptation has been

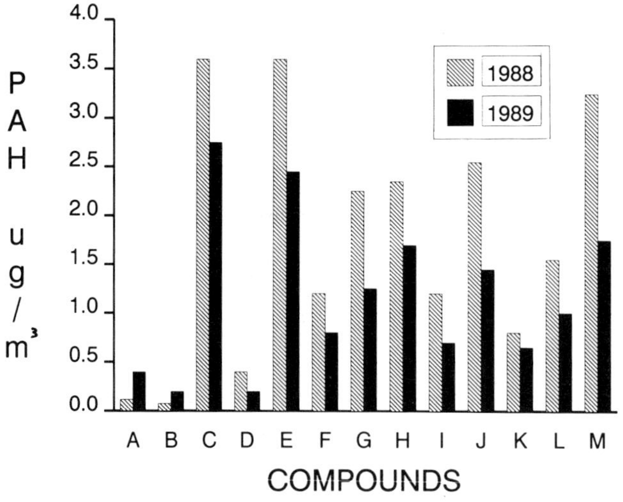

Fig. 1. The average PAH concentrations in personal dust samples collected in year 1988 and 1989 during a shiftwork. A = Nitrofluoranthene, B = Fluoranthene, C= Phenantrene, D = Anthracene, E = Pyrene, F = Benzofluoranthene, G = Chrysene, H = Benzo(e)pyrene, I = Benzofluoranthene, J = Benzo(a)pyrene, K = Dibenzanthracene, L = Benzo(ghi)pyrene, M = Dust

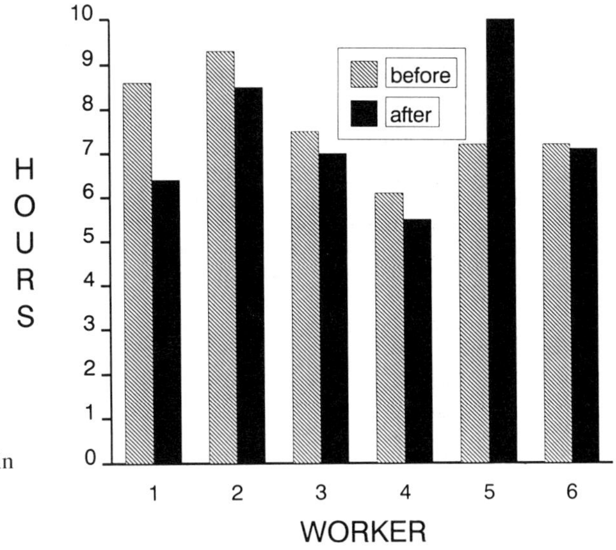

Fig. 2. Antipyrine half-life in 6 workers before start and after one year work in coke-oven conditions

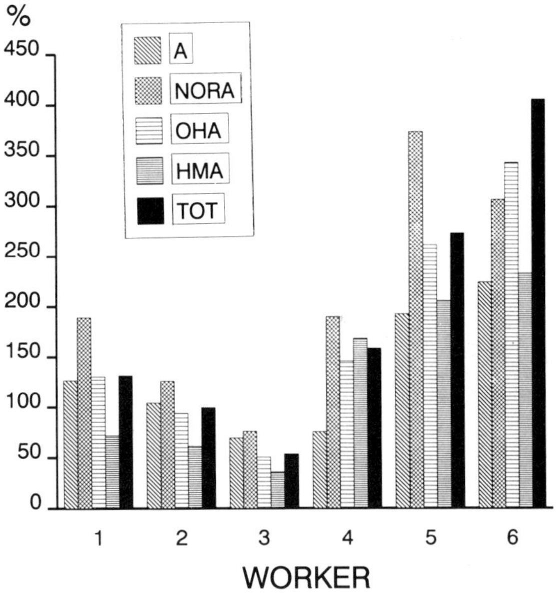

Fig. 3. Urinary excretion of antipyrine (A) and its three main metabolites norantipyrine (NORA), hydroxyantipyrine (OHA), hydroxymethylantipyrine (HMA) and total amount of A + OHA + NORA + HMA (TOT) excretion after working 1 year in the coke oven plant. The values are given as per cent of control values which were determined before the work was started

described by Sotaniemi et al (1984) in diabetic patients during the treatment with various drugs. It is also possible that exposure to PAHs through inhalation does not change the metabolic capacity of the liver and thus reduce the usefulness of antipyrine-test to measure this kind of occupational exposures.

Grant. Sigfrid Juselius Foundation, Finland.

References

Arvela P, Pelkonen O, Puurunen J, Sotaniemi E (1988) Antipyrine Metabolism in Chemical Industry Workers as Compared with Alcoholics and Healthy Controls. In Environmental and Occupational Hazards No 8:345–349

Dössing M (1982) Changes in Hepatic Microsomal Enzyme Function in Workers Exposed to Mixtures of Chemicals. Clin Pharmacol Ther 32:340–346

Eichelbaum M, Sonntag B, Dengler HJ (1981) HPLC Determination of Antipyrine Metabolites Pharmacol 23:192–202

Jongeneelen FJ, Anzion RBM, Henderson PT (1987) Determination of Hydroxylated Metabolites of Polycyclic Aromatic Hydrocarbons in Urine. J Chromatography 413:227–232

Pelkonen O, Sotaniemi EA, Ahokas JT (1985) Coumarin 7-hydroxylase Activity in Human Liver Microsomes. Properties of the Enzyme and Interspecies Comparisons. Br J clin Pharmac 19:59–66

Prescott LF, Adjepon-Yamoah KK, Roberts E (1973) Rapid gas-liquid chromatographic estimation of antipyrine in plasma. J Pharm Pharmac 25:205–207

Sotaniemi EA, Arranto AJ, Salmela PI, Stengård JH, Pelkonen RO (1984) Hepatic microsomal enzyme activity in patients with non-insulin dependent diabetes mellitus (NIDDM) and its clinical significance. Acta endocrinol suppl 262:125–129

Vesell E (1978) Genetic and Environmental Factors Affecting Drug Disposition in Man. Clin Pharmacol Ther 22:659–679

Recent Developments in Toxicology:
Trends, Methods and Problems
Arch. Toxicol., Suppl. 14, 205–208 (1991)
© Springer-Verlag 1991

Influence of Bile Acids on the Development of Hepatic Transport of Organic Anions

A. Barth

Institute of Pharmacology and Toxicology, Friedrich Schiller University Jena, Löbderstraße 1, O-6900 Jena, Germany

Introduction

Transfer of xenobiotics and endogeneous compounds from blood into bile is immature at birth. Bile acids play a dominant role in bile formation and are able to influence the biliary excretion of various drugs (Gregus and Fischer 1979). Maturation of bile acid excretion was found to be on the 20th day of life, whereas the biliary excretion of exogeneous organic anions matures between the 20th and 30th days of life in rats (Barth et al 1977, 1986). Despite choleretic effect taurocholic acid did not enhance the low biliary transport capacity for indocyanine green in 20 day old rats (Barth et al 1981). It was the aim of the present experiments to elucidate the effects of other bile acids on the development of the biliary excretion process. In 20 and 60 day old rats the biliary excretion of eosin was determined after simultaneous administration of eosin and bile acids.

Materials and Methods

Male Wistar rats were used. The animals were kept under conventional conditions. Under urethane anaesthesia (1.2 g/kg ip) the common bile duct was cannulated. Eosin (100 mg/kg iv, VEB Laborchemie Apolda) and eosin plus bile acids as sodium salts [100 mg/kg iv, cholic acid (Reanal Budapest), tauroglycocholic acid (Merck Darmstadt), deoxycholic acid (VEB Berlin-Chemie), chenodeoxycholic acid (Feinbiochemica Heidelberg)] were given into the femoral vein in a volume of 1 ml/kg aqueous solution. Bile was collected for 1 hour after dye administration. Bile was measured gravimetrically, the optical density was measured spectrophotometrically at 518 nm. The standard eosin curve was not influenced by bile acids.

Student's t-test was used for statistical analysis.

Table 1. Bile flow in 20 and 60 day old rats 1 hour after administration of eosin (E, 100 mg/kg b.m. iv) alone or eosin together with bile acids (100 mg/kg b.m. iv, sodium salt of cholic acid (CA) tauroglycocholic acid (TG-CA), deoxycholic acid (DOC) and chenodeoxycholic acid (CDOC). Arithmetic means and S.E.M., * significant differences from eosin treated controls, $p \leq 0.05$

Treatment	Bile flow (μl/kg × min) 20 d	n	60 d	n
E	44.1 ± 3.7	12	63.4 ± 1.7	5
E + CA	70.0 ± 4.6*	8	89.5 ± 6.4*	6
E	43.4 ± 4.2	5	64.2 ± 6.2	4
E + TG-CA	52.6 ± 8.6	5	79.0 ± 7.9	4
E	44.1 ± 3.7	12	60.9 ± 3.8	4
E + DOC	38.9 ± 7.1	7	42.7 ± 3.5*	4
E	54.5 ± 5.8	3	64.2 ± 6.2	4
E + CDOC	14.5 ± 3.2*	3	49.2 ± 3.8	3

Results and Discussion

Among the trihydroxylated bile acids cholic acid especially enhanced the bile flow in young and adult rats. The tauroglycocholic acid showed little choleretic potency. Both dihydroxy bile acids diminished the bile flow, deoxycholic acid in adults, chenodeoxycholic acid more pronounced in 20 day old rats (Table 1).

Cholic acid was able to increase the biliary eosin excretion in immature 20 day old rats without reaching adult values (Fig. 1). But the conjugated tauroglycocholic acid and both dihydroxylated bile acids, deoxycholic and chenodeoxycholic acid, strongly inhibited the biliary dye excretion (Fig. 1, 2). In adult rats it is known that bile acids influence the biliary excretion of organic anions differently and independent of their choleretic potency (Gregus and Fischer 1979). According to the literature, bile acids are able to activate hepatocytes normally not involved in the transport of xenobiotics (Vonk et al. 1978). In immature 20 day old rats, cholic acid may act in this way, but in adult ones the higher bile acid dependent bile flow after cholic acid administration had no effect on the dye excretion (Fig. 1). Tauroglycocholic, deoxycholic and chenodeoxycholic acid were ineffective in increasing the low eosin excretion capacity in 20 day old rats, on the contrary, these bile acids inhibited the excretion process (Fig. 1, 2). Experiments with liver slices demonstrated toxic effects of dihydroxy bile acids on eosin release *in vitro* (Radegast and Kopplin 1988). Schölmerich et al. (1981) found membrane damage and enzyme release in isolated hepatocytes after incubation with chenodeoxycholic acid. Gregus et al. (1982) demonstrated inhibition of biliary excretion of organic anions by dihydroxy bile acids related to toxic effects on respiration in liver mitochondria. The inhibiting effect of deoxycholic and chenodeoxycholic acid on the eosin excretion also in young rats presented here may be due to toxic action on energy

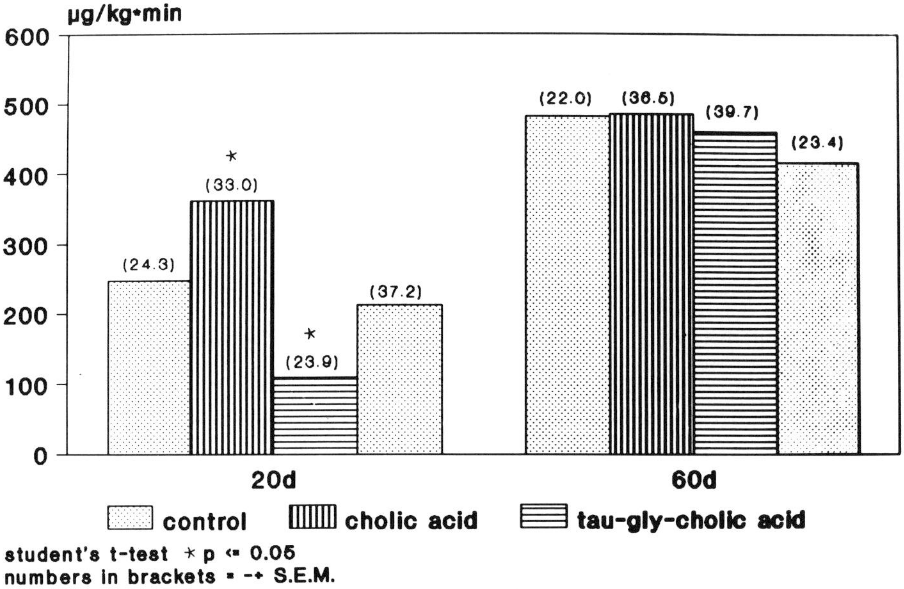

Fig. 1. Biliary excretion of eosin in 20 and 60 day old male Wistar rats 1 hour after iv administration of 100 mg eosin/kg b.m. (individual control to each bile acid trial) and eosin plus 100 mg/kg b.m. sodium salt of cholic acid or tauroglycocholic acid (tau-gly-cholic acid). Arithmetic means, n = 6–12, * significant differences from controls

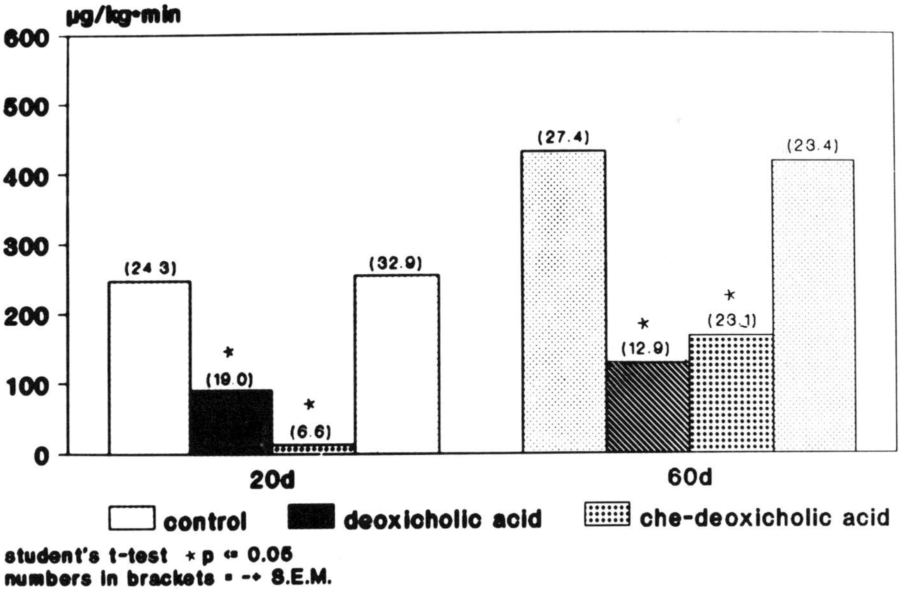

Fig. 2. Biliary excretion of eosin in 20 and 60 day old male Wistar rats 1 our after iv administration of 100 mg eosin/kg b.m. (individual control to each bile acid trial) and eosin plus 100 mg/kg b.m. sodium salt of deoxycholic acid or chenodeoxycholic acid (che-deoxycholic acid). Arithmetic means, n = 6–12, * significant differences from controls

supply for active excretion process. The inhibiting effect of tauroglycocholic acid only in 20 day old rats may be due to the glycocholic component and demonstrated the sensitiveness of young animals. In conclusion, changes in bile flow after administration of bile acids do not correlate with changes of biliary dye excretion in young and adult rats. The hepatobiliary transport of exogeneous compounds is dependent on age and bile acid excretion. The low excretion capacity in young rats could not be abolished by bile acid administration.

References

Barth A, Zaumseil J, Klinger W (1977) Gallenfluß und Gallensäureausscheidung bei männlichen Wistarratten (Jena) verschiedenen Alters. Z Versuchstierk 19:26–35

Barth A, Gregus Z, Klinger W, Varga F (1981) The influence of bile acids on biliary excretion of indocyanine green in rats of different ages. Acta biol med germ 40:1723–1728

Barth A, Klinger W, Hoppe H (1986) Hepatic elimination kinetics of organic anions in rats: developmental aspects and influence of phenobarbital. Archs int Pharmacodyn Ther 283:16–29

Gregus Z, Fischer E (1979) Effect of sodium taurocholate on hepatic uptake and biliary excretion of organic anions in rats. Archs int Pharmacodyn Ther 240:180–192

Gregus Z, Varga F, Fischer E (1982) Effect of bile acids on the biliary excretion of amaranth and the respiration of liver mitochondria. Acad Scient Hungarica 59:89–97

Radegast S, Kopplin U (1988) Eosintransport in Rattenleberschnitten – Einfluß von Alter, Phenobarbital und Gallensäuren. Med Diss F.S.U. Jena

Schölmerich J, Schmidt K, Kremer B, Becher MS, Gerok W (1981) Different effect of taurolithocholate and chenodeoxycholate on structure and function of isolated hepatocytes. Klin Wschr 59:655–668

Vonk RJ, Jekel PA, Meijer DKF, Hardonk MJ (1978) Transport of drugs in isolated hepatocytes. The influence of bile acids. Biochem Pharmacol 27:397–405

Recent Developments in Toxicology:
Trends, Methods and Problems
Arch. Toxicol., Suppl. 14, 209–212 (1991)
© Springer-Verlag 1991

Comparative Nephrotoxicity Studies of Cisplatin and its Iodinized Derivative in Rats

D. Appenroth, S. Kessler, K. Winnefeld, and H. Bräunlich

Institute of Pharmacology and Toxicology, FSU, Löbderstraße 1, O-6900 Jena, Germany

Introduction

Cisplatin, cis-diammine-dichloroplatin (CP) is a widely used anticancer drug. But its curative effect is connected with nephrotoxicity. Reaction products of CP and low molecular weight proteins are involved in the toxic effect. Passage through membranes and binding to proteins are determined among other things by physicochemical properties of the substance. Therefore it was decided to investigate the nephrotoxicity of a CP derivative which is very similar in structure, but differs in lipophilicity. This iodinized derivative, cis-diammine-diiodoplatin (CP-J_2), was determined to have a higher n-octanol/water partition coefficient than CP (0.25 and 0 respectively).

Methods

Experiments were done with adult female Wistar rats (colony bread, conventional conditions, 6 animals per group). CP was registered by atomic absorption spectrometry. All methods used are described in detail by Appenroth et al (1990a). CP and CP-J_2 were administered intraperitoneally. Data were expressed as mean values $\pm$ S.E.M. Statistical evaluations were done with Student's t-test.

Results and Discussion

Figure 1 shows urinary volume and protein excretion after 6 mg CP/kg, the equimolar and double the equimolar dose of CP-J_2. It becomes obvious that CP nephrotoxicity develops during the first 3 days after the administration. Following the equimolar dose of CP-J_2 no significant deviation from the control range appeared. If double of the equimolar dose of CP-J_2 was administered nephrotoxicity occurs a little more rapidly, but the maximum nephrotoxicity was the same as that with CP and was observed at the same time after the injection. It can be concluded that the double the equimolar dose of CP-J_2 is followed by the same nephrotoxicity as CP. This was verifed by blood urea nitrogen concentrations (not shown).

Figure 2 shows the kinetic behaviour of CP and CP-J_2. It is obvious that the Pt-serum level declines equally after both CP complexes. However, the Pt concentration after CP-J_2 is significantly higher. It is hypothesised that this is due to better resorption from the intraperitoneal depot because of its higher lipophilicity. The higher serum level of CP-J_2 causes enhanced glomerular filtration which is visible as a higher urinary excretion (Fig. 2). CP can enter the cell by reabsorption from the tubular lumen and by active secretion from the peritubular site. In particular the reabsorption depends on lipophilicity. Consequently it was expected that there would be higher Pt concentrations in the kidney after the administration of CP-J_2. However, no difference exists between these complexes (Fig. 2). From morphological studies it was known that CP destroys the luminal membrane which normally acts as a barrier for reab-

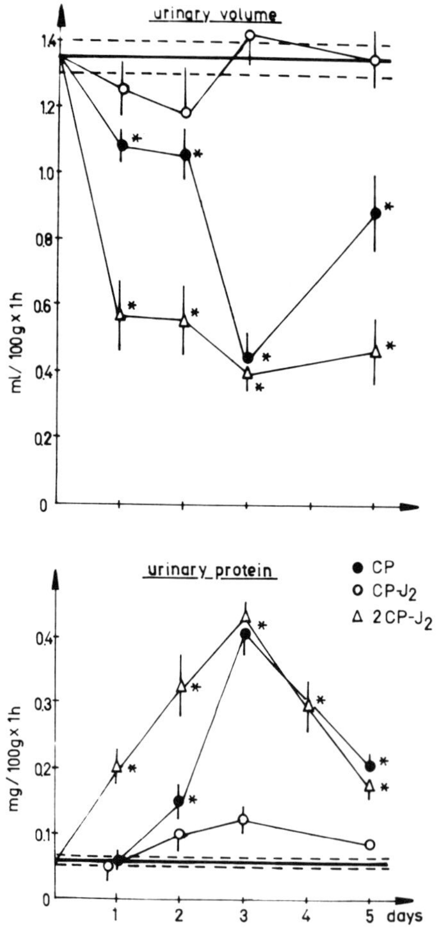

Fig. 1. Urinary volume and protein excretion during 5 days after 6 mg CP/kg and the equimolar dose of CP-J_2 and the double of it (2 CP-J_2) compared with control (====). * statistically significant differences between treated and control rats ($p \leq 0.05$)

sorption (Appenroth et al 1990b). Because of its low nephrotoxicity this is highly unlikly following the equimolar dose of CP-J_2. Therefore it was supposed that the destruction following CP administration facilitates the reabsorption more than the greater lipophilicity of CP-J_2. However, the same Pt-level in the kidney was found. Nevertheless nephrotoxicity of CP is higher. The hydrolysis process of CP running inside the cell is required to allow fast reactions with, for instance, proteins, RNA, and DNA with the consequence of activity and nephrotoxicity. It can be speculated that the binding of iodine to the Pt is not as labile as that of chlorine. More likely, because of the very big atomic radius, iodine prefers the trans-position. It is known also that trans-platin does not act against tumours and possesses much lower toxicity than CP. Finally the lower nephrotoxicity of the iodinized derivative has not been proved despite the same Pt concentration inside the target organ.

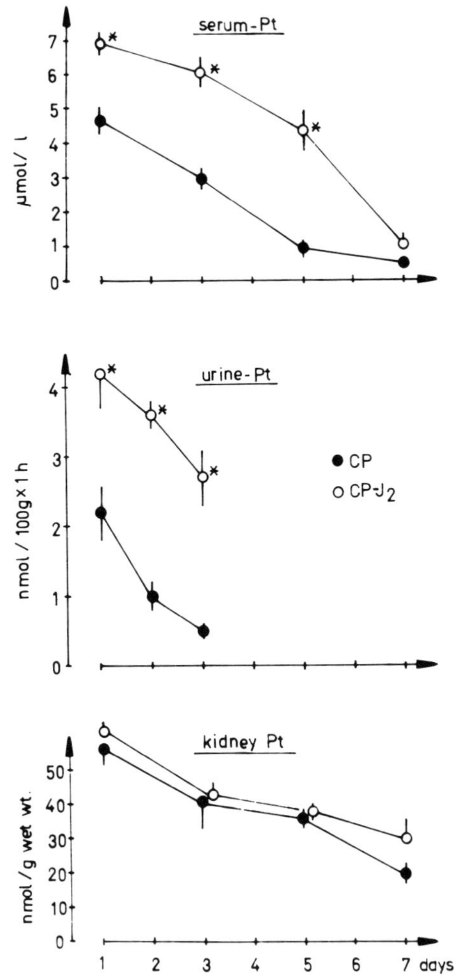

Fig. 2. Pt concentrations in serum, urine, and kidney during 7 days after CP and CP-J_2.
* statistically significant differences between CP and CP-J_2 treated rats ($p \leq 0.05$)

Acknowledgement. For the synthesis of CP and CP-J_2 we thank Dr H.-P. Schröer from the Institute of Microbiology and Experimental Therapy of the Academy of Sciences of the GDR, Jena.

References

Appenroth D, Gambaryan S, Bakhteeva V, Winnefeld K, Schröter H, Bräunlich H (1990a) Triidothyronine (T_3) increases cisplatin nephrotoxicity in young and adult rats. J Appl Toxicology 10:395–399

Appenroth D, Gambaryan S, Gerhjardt S, Kersten L, Bräunlich H (1990b) Age dependent differences in the functional and morphological impairment of the kidney following cisplatin administration. Exp Pathol 38:231–239

Recent Developments in Toxicology:
Trends, Methods and Problems
Arch. Toxicol., Suppl. 14, 213–217 (1991)
© Springer-Verlag 1991

Inhibitory Effects of a Platelet Activating Factor-Antagonist 48740RP in Cyclosporin-Induced Nephrotoxicity in the Rat as Assessed by Renal Glutathione Metabolism Parameters

F. Massicot, C. Pham Huy, C. Martin, M. Thevenin, J.-M. Warnet, and J. R. Claude

Laboratoire de Toxicologie, Faculté de Pharmacie, 75270 Paris Cédex 06, France

Introduction

Cyclosporin A (CsA) is a potent immunosuppressive agent that has improved renal allograft survival. However, together with its powerful immunosuppressive effect, CsA therapy is often accompanied by structural, biochemical and functional abnormalities – especially nephrotoxicity (Thomson et al 1984). It was shown that CsA has peroxidative properties and thus may lead to nephrotoxicity (Inselman et al 1988). So, it was suspected that CsA affects kidney GSH metabolism parameters (Mayer and Cockett 1988). Recently, it was shown that nephrotoxicity was associated with an increased renal oxidized glutathione content (Massicot et al 1991).

Platelet activating factor (PAF) is an endogenous phospholipid which exerts a wide range of biological effects including renal injury (Braquet et al 1987). Recent studies indicate that PAF is released during acute allograft rejection. The impairment of renal function evoked by CsA is also observed after PAF infusion, suggesting that PAF is involved in CsA-induced nephrotoxicity.

Different PAF-acether antagonists were developed. One of these, 48740RP, 3-(3-pyridyl)-1H,3H-pyrido(1,2-c)thiazole-7-carboxamide (Rhône-Poulenc), selectively antagonizes the effects of PAF. Its effects on kidney glutathione content and glutathione related enzyme activities were investigated in the rat.

Abbreviations used: CsA, cyclosporin A; GSH, reduced glutathione; GSSG, oxidized glutathione; PAF, Platelet Activating Factor; 48740 RP, 3-(3-pyridyl)-1H,3H-pyrido(1,2-c)thiazole-7-carboxamide (Rhône-Poulenc)

Materials and Methods

Animals: Rats (180–200 g) from Janvier (Le Genest-St Isles – France) were housed in individual metabolic cages. Each rat was maintained on a standard diet with free access to water.

Drugs and chemicals: Cyclosporin A (CsA) (Sandoz) was supplied in castor oil (Cremophor EL). CsA solution was diluted with 0.15 M NaCl for injection (8.63 mg dissolved in 1 ml). 48740RP, hydrochloride, was dissolved in 0.9% NaCl at a final concentration of 5.6 mg/ml.

Treatment: Groups of five animals were treated for 3 days either with CsA (50 mg/kg/day; sc) alone or with 48740RP (15 mg/kg/d; po). Five animals were also treated with CsA plus 48740RP. Control group was given a corresponding volume of Cremophor EL.

A timed 24 h-urine was collected on day 1 (before treatment) and 4 (24 h after the last injection). Animals were anesthetized with pentobarbital on the fourth day and right auricle of heart was excized for blood collection. The liver and kidney were immediately perfused in situ with heparinized isotonic saline and quickly removed and weighed.

Analytical methods: Serum: urea and creatinine were measured by means of a SMAII Technicon analyzer. *Urine:* alanine aminopeptidase (AAP) activity was determined according to Diener 1981. *Kidney cortex:* Reduced glutathione (GSH) and oxidized glutathione (GSSG) were measured in the acid soluble portion of kidney homogenates according to Griffith (1980). Kidney cytosolic fractions were prepared by differential sedimentation to measure the glutathione-related enzyme activities: reductase (GSSG-Rd) (Carlberg and Mannervic 1985), peroxidase (GSH-Px) (t-butyl hydroperoxide as substrate) (Gunzler and Flohe 1985). All enzyme activities were expressed relative to protein content according to Bradford (1976).

Statistical analysis: All the results are expressed as mean ± S.E.M. and the significance of the differences was assessed by Student's t-test, p values smaller than 0.05 being considered as significant.

Results

As reported in Table 1, treatment of rats for 3 days with 50 mg/kg CsA resulted in higher serum urea nitrogen (59%) and creatinine (11%) levels than in

Table 1. Effect of cyclosporin A plus 48740RP on serum BUN and creatinine and urine AAP activity in kidney cortex (mean ± S.E.M.)

	Serum		Urine: AAP	
	BUN	Creatinine	0h	72 h
Cremophor	9.7 ± 1.3 (5)	39.5 ± 0.9 (4)	0.12 ± 0.03 (5)	0.20 ± 0.10 (4)
CsA	15.4 ± 2.3 (5)*	43.8 ± 1.5 (5)*	0.09 ± 0.02 (4)	0.20 ± 0.04 (4)
48740RP	7.8 ± 0.09 (5)	35.4 ± 2.1 (5)	0.09 ± 0.01 (5)	0.09 ± 0.02 (5)
CsA + 48740RP	14.7 ± 1.0 (5)	41.2 ± 2.8 (5)	0.10 ± 0.03 (5)	0.11 ± 0.02 (5)

CsA: cyclosporin A; BUN: Blood urea nitrogen (mmol/l); Creatinine (μmol/l); AAP: alanine aminopeptidase (IU/24 h). Number of animals in brackets.
Significance of differences from Cremophor (serum) or pretreatment values at 0h (urine AAP activity): * $p < 0.05$

controls. Concerning urinary alanine aminopeptidase (AAP), the results are equivocal because of an unexpected increase with a large S.E.M. of the cremophore treated animals. Nevertheless, the enzyme level in CsA-treated animals is significantly increased, but in this case with a low S.E.M. As shown in table 2, CsA treatment results in a marked increase in GSSG concentration (120%; $p < 0.001$) and GSH (77%; $p < 0.01$) associated with an increase in GSH peroxidase activity (36%; $p < 0.01$) as compared with the controls. In rats treated by 48740RP alone, all biological determinations were similar to those of the controls (Table 1 and 2).

After co-treatment with CsA plus 48740RP, animals show serum creatinine and BUN levels not significantly different from CsA-treated group, but there is no increase in urine AAP activity at 72 h. Furthermore, co-treatment resulted in decreased GSSG and GSH concentrations (40%; $p < 0.05$ and 29%; $p < 0.01$, respectively) associated with a decrease in glutathione-peroxidase and in glutathione-reductase (18%; $p < 0.05$ and 16%; NS; respectively) activities relative to CsA group (Table 2).

Discussion

The present findings confirm that CsA treatment (50 mg/kg/d; sc) once daily for 3 days to rats results in increased BUN and serum creatinine level (Ryffel and Mihatsch 1986). CsA administration produces in the renal cortex a marked increase in GSSG and in GSH levels. The increase in GSSG level should be correlated with the rise in glutathione-peroxidase activity. In contrast, glutathione-reductase activity is very similar to that of the controls and cannot account for the elevated level of renal GSH concentration. The present increase in GSSG concentration is consistent with a previous study in the rabbit showing that increased GSSG may be related to a CsA-induced oxidative stress (Massicot et al) and is also consistent with the postulated role of lipid

Table 2. Effect of cyclosporin A plus 48740RP on glutathione and on GSH-Peroxidase and GSSG-reductase activities in rat kidney cortex (mean ± S.E.M.)

	Cremophor EL	CsA	48740RP	CsA + 48740RP
GSSG	0.05 ± 0.01	0.11 ± 0.02 b	0.05 ± 0.01	0.07 ± 0.01*
GSH	1.42 ± 0.07	2.51 ± 0.15 a	1.21 ± 0.04	1.77 ± 0.16**
GSH-Px	1.01 ± 0.05	1.38 ± 0.10 a	0.96 ± 0.08	1.11 ± 0.08*
GSSG-Rd	0.40 ± 0.02	0.43 ± 0.04	0.40 ± 0.03	0.36 ± 0.01

GSSG: oxidized glutathione; GSH: reduced glutathione (μmole/g tissue); GSH-Px: glutathione-peroxidase μmole/min/mg protein; GSSG-Rd: glutathione-reductase μmole/min/mg protein. Five animals per group. Significance of differences from Cremophor EL: a: $p < 0.01$; b: $p < 0.001$ and from CsA: * $p < 0.05$; ** $p < 0.01$

peroxidation in CsA nephrotoxicity (Inselman et al 1988). The marked increase in GSH level in response to CsA treatment is an unexpected observation. It was reasonable to suspect that cyclosporin could decrease GSH level. However, in the present study, both GSH and GSSG are increased. When a rapid production of GSSG is occurring, the net result of GSSG formation is a loss of intracellular GSH (Deneke and Fanburg 1989) which is followed by a GSH resynthesis as a rebound effect (D'Souza 1988). The rise in kidney GSH in response to CsA may be due to an increased rate of production or to an inhibition of the degradation of the tripeptide. Mayer and Cockett (1988) attribute it to an enhanced uptake of GSH as the result of alterations caused by CsA in the renal transport system. The mechanism of GSH elevation and whether this elevation is deleterious or not remain to elucidate. Whatever it may be, elevation of cellular glutathione (GSH) in response to CsA-treatment may process from a CsA-induced oxidant injury.

In this study, the PAF antagonist 48740RP alone does not alter kidney function since 48740RP induces no changes in BUN, serum creatinine level and urinary AAP activity as compared with the controls. Furthermore, 48740RP exerts no effect on kidney GSSG and GSH levels nor glutathione-related enzyme system.

Treatment of the CsA-treated animals with 48740RP does not prevent the increase in serum BUN and creatinine levels, except for urine AAP activity, but attenuates glutathione metabolism variations due to CsA toxicity. So it is suggested that 48740RP although not being able to prevent completely CsA nephrotoxicity, produces an attenuation of the renal glutathione metabolism disturbances in relation to the postulated CsA-induced lipid peroxidation, and could thus be an useful new drug for minimization of CsA-induced nephrotoxicity.

Acknowledgements. We wish to thank l'Association pour le Développement de la Recherche en Toxicologie Expérimentale, 4 av. de l'Observatoire, 75006 Paris-France for financial support.

References

Bradform MM (1976) A rapid and sensitive method for the quantitation of microgram quantities of protein utilizing the principle of protein-dye binding. Anal Biochem 72:248–254

Braquet P, Touqui L, Shent Y, Vargaftig BB (1987) Perspective in Platelet activing factor research. Pharmacol Reviews 39 (2):97–145

Carlberg I, Mannervik B (1985) Glutathione reductase. Meth Enzymol 113:484–490

Deneke SM, Fanburg BL (1989) Regulation of cellular glutathione. Am J Physiol 257 (1):L163–L173

Diener U, Knoll BL, Rautenstrauch OR, Wisser H (1981) Urinary excretion of N-acetyl-β-D-glucosaminidase and alanine aminopeptidase in patients receiving amikacine or cis-platinum. Clin Chim Acta 112:149–157

Gunzler WA, Flohe L (1985) Glutathione peroxidase. In: Greenwald RA (ed). CRC Press Boca Raton Florida (CRC handbook methods for oxygen radical research):285–290

Inselman G, Blank M, Baumann K (1988) Cyclosporine A induced lipid peroxidation in microsomes and effects on active and passive glucose transport by brush border membrane vesicles of rat kidney. Res Commun Chem Pathol Pharmacol 43:417–423

Lefort J, Sedivy P, Desquand S, Randon J, Coeffier E, Maridonneau-Parini I, Floch A, Benveniste J, Vargaftig BB (1988) Pharmacological profile of 48740RP, a PAF-acether antagonist. Eur J Pharmacol 150:257–268

Massicot F, Thevenin M, Martin C, Warnet JM, Dutertre H, Claude JR (submitted) Effects of cyclosporin A on kidney glutathione metabolism and cytochrome P-450 content in the rabbit. Pharmacol Res

Mayer RD, Cockett ATK (1988) Cyclosporin-mediated increase in kidney glutathione and effects on glutamyl-cycle enzymes. J Biochem Toxicol 3:213–221

Ryffel B, Mihatsch J (1986) Cyclosporine nephrotoxicity. Toxicol Pathol 14:73–82

Thomson AW, Whiting PH, Simpson JG (1984) Cyclosporine: Immunology, toxicity and pharmacology in experimental animals. Agent and Actions 15 (3/4):306–327

Recent Developments in Toxicology:
Trends, Methods and Problems
Arch. Toxicol., Suppl. 14, 218–221 (1991)
© Springer-Verlag 1991

The Urinary Excretion of Specific Proteins in Workers Exposed to Lead

A. Vyskočil[1], Z. Fiala[1], J. Šalandová[2], A. Popler[2], E. Ettlerová[3], and S. Emminger[4]

[1] Department of Hygiene, Medical Faculty, Charles University, Šimkova 870, 500 38 Hradec Králové, Czechoslovakia,
[2] District Institute of National Health, Kyjevská 44, 532 03 Pardubice, Czechoslovakia,
[3] Computer Centre, Medical Faculty, Charles University, Šimkova 870, 500 38 Hradec Králové, Czechoslovakia, and
[4] Department of Occupational Hygiene, Regional Hygiene Station, Balbínova 821, 500 31 Hradec Králové, Czechoslovakia

Introduction

There is some evidence that lead nephropathy is associated with prolonged occupational exposure to lead. Effects of lead on kidney function occur at blood lead levels (PbB) of about 60 μg/100 ml (Ibels and Poolock 1986). The effects could involve both the renal tubules and glomerular-vascular apparatus (Hammond 1977; Ibels and Poolock 1986). In this paper, renal function, both glomerular and tubular was studied in adults who had severe occupational exposure to lead.

Methods

The study was conducted on two groups of male subjects working in a glassworks. The exposed group consisted of 45 workers occupationally exposed to lead. It was compared with a group of 23 age-matched males with only administrative functions in the same plant.

Exposure to lead of all workers was monitored over a whole workday by collecting particles > 0.5 μg from the breathing zone with membrane filters Synpor II (Czechoslovakia) connected to personal sampling pump with constant flow rate 2 l/min. The content of lead of the filters was measured by atomic absorption spectrometry after wet ashing with nitric acid and hydrogen peroxide. Exposure to lead was also assessed by measuring the whole blood lead concentration using atomic absorption spectrometry (Fernandez 1975).

One "spot" urine sample per subject was collected in the middle of a typical work week at the end of the work day. All the samples were collected over phosphate buffer pH 7.4. The concentration of β_2-microglobulin (β_2-m) and albumin were determined by the RIA tests (ÚVVVR Prague, Czechoslovakia). Creatinine was measured by the method of Jaffé (Henry et al 1974), lysozyme by the method of Litwack (1955), glucose by the Oxochromglucose BIOLA-Test (Lachema, Czechoslovakia) and lactate dehydrogenase (LDH) by the UV-test (Sevac, Czechoslovakia).

Results and Discussion

Table 1 shows that control and exposed subjects were well matched for the listed criteria. There was a considerable variation of exposure to lead at the workplaces examined. The level of lead in air was on average 11 times higher than Czechoslovak TLV (0.05 mg/m^3).

In Table 2 the exposed workers were segregated into four groups according to their PbB concentrations. The only parameters of renal function in the

Table 1. Characteristics of the control and exposed group

	Control	Exposed	P
Number volunteered	27	57	–
Number retained for study	23	45	–
Age/year/Mean ±SD	42 ± 10	39 ± 9	NS
Number of smokers[4]	11 (48)	23 (51)	NS
Regular alcohol consumption[1, 4]	7 (46)	12 (27)	NS
Regular analgesics consumption[2]	2	2	NS
Non-occupational exposure[3]	0	0	NS
Duration of employement/year/Mean ± SD	17 ± 8	14 ± 7	NS
Duration of lead exposure/year/			
Mean ± SD	–	12 ± 7	–
Range	–	2–31	
Concentration of Pb in air/mg/m^3/			
Mean ± SD	–	0.55 ± 0.47	–
Range	–	0.05 ± 1.50	
Concentration of Pb in blood/μg/100 ml/			
Mean ± SD	18 ± 6	69 ± 18	–
Range	8–35	29–118	

[1] More than five glasses of wine, beer or liquor per week
[2] More than 1 tablet per week
[3] More than 1 h/week domestic utilisation of organic solvents
[4] In parentheses are given percentages of number retained for study
NS: not statistically significant: Student's t-test or chi-square test with Yates correction

Table 2. Urinary excretion of albumin, β_2-m. lysozyme, LDH and total proteins in control workers (PbB ≤ 40 μg/100 ml) and workers exposed to lead segregated into four groups according to their PbB concentrations

	Concentrations of lead in blood (μg/100 ml)				
	Control		Exposed		
	≤ 30 (n=22)	31–50 (n=14)	51–70 (n=14)	71–90 (n=8)	>90 (n=10)
Albumin (mg/g creat)[a]	11.0 (1.7–36.0)	17.3 7.6–46.0)	17.3 (5.3–79.5)	19.7[c] (6.9–79.7)	17.8[c] (8.1–50.7)
β_2-m (μg/g creat)[a]	42.4 (5.7–157.69)	46.8 (9.6–137.1)	57.1 (10.2–342.9)	43.9 15.3–111.3)	91.4[c] (21.0–296.0)
Lack of Lysozyme (mg/g creat)					
LDH (U/g creat)[a]	10.9 (4.6–27.6)	12.1 (4.5–22.7)	10.6 (3.6–87.9)	8.4 (3.9–14.4)	12.3 (5.4–21.4)
Totalproteins (mg/g creat)[b]	148 ± 15	148 ± 21	157 ± 24	126 ± 12	124 ± 13

[a] Geometric mean (range)
[b] Arithmetic mean ± SEM
[c] Value differs significantly from control ($p < 0.05$) by Student's t-test
n: number of persons; creat: creatinine

Table 3. Prevalence of abnormal values[a] in the urinary excretion of proteins and enzymes in lead-exposed workers and in their controls

Urinary parameter	Control	Exposed	P
Albumin	1/22	10/35	<0.05
β_2-m	1/22	2/43	NS
Lysozyme	1/22	1/44	NS
LDH	1/22	1/44	NS
Total proteins	2/21	3/42	NS

[a] Higher than the 95th percentile of the values found in the control group
NS: not statistically significant, chi-square test with Yates correction

exposed workers which differed significantly from the control were albumin in the groups with PbB > 71 μg/100 ml and β_2-m in the group with PbB > 90 μg/100 ml.

Glomerular permeability appeared to be affected as evidenced by increased albumin excretion and the prevalence of its increased values (Table 3). This study is in agreement with previous studies (Hong et al 1980; Wedeen et al 1979). It did not show any changes in markers of tubular dysfunction (lysozyme, LDH, glucose) with exception of β_2-m which was increased only in the group with PbB > 90 μg/100 ml. These results are in agreement with other studies (Hong et al 1980; Wedeen et al 1979; Vacca et al 1986).

Results of this study suggest that tubular proteinuria is not an early manifestation of lead nephrotoxicity. In agreement with conclusions of Hammond (1977) adults seem to experience tubular damage only at very high levels of exposure (PbB > 90 μg/100 ml). Glomerular permeability is affected in workers with PbB > 70 μg/100 ml, the generally accepted "safe" upper limit.

References

Fernandez FJ (1975) Micromethod for lead detection in whole blood by atomic absorption with use of graphite furnace. Clin Chem 212:558–561

Hammond PB (1977) Exposure of humans to lead. Ann Rev Pharmacol Toxicol 17:197–214

Henry RJ, Cannon DC, Winhelman JW (1974) Clinical chemistry: Principles and Technics, 2nd Ed, New York, Harper and Row, 552–554

Hong CD, Hanenson IB, Lerner SI, Hammond PB, Pesce AJ, Pollak VE (1980) Occupational exposure to lead: Effects on renal function. Kidney Int 18:489–494

Ibels LS, Poolock Ca (1986) Lead intoxication. Med Toxicol 1:387–410

Litwack G (1955) Photometric determination of lysozyme activity. Proc Soc Exp Biol Med 39:401–403

Vacca CV, Hines JD, Hall III PV (1986) The proteinuria of industrial lead intoxication. Environm Res 41:440–446

Wedeen RP, Mallik DK, Batuman V (1979) Detection and treatment of occupational lead nephropathy. Arch Intern Med 139:53–57

Miscellaneous Toxic Effects

Recent Developments in Toxicology:
Trends, Methods and Problems
Arch. Toxicol., Suppl. 14, 225–228 (1991)
© Springer-Verlag 1991

Reduction of Arsenic Trioxide Toxicity in Mice by Repeated Treatment with Glucose

F.X. Reichl, H. Kreppel, L. Szinicz, B. Fichtl, and W. Forth

Walther Straub-Institut für Pharmakologie und Toxikologie der LMU München, Nußbaumstraße 26, W-8000 München 2, FRG

Introduction

Glucose treatment improved the survival and symptomatology in mice poisoned with As_2O_3. In these experiments, irrespective of the treatment, the dying animals exhibited an almost complete depletion of glycogen in the liver (Reichl et al 1990). In order to obtain more insight into the mechanisms involved, postmortem changes of glucose and glycogen content in brain and muscle were investigated in animals acutely poisoned with As_2O_3 and treated with glucose or saline.

Materials and Methods

Animals: Male white NMRI mice (ITF, Tuttlingen, FRG), were fed standard diet (Alma H 1003, Botzenhardt KG Kempten, FRG) and water ad libitum. The animals were allowed to acclimatize for 7 days before the experiment.

Chemicals: Arsenic trioxide (As_2O_3, Merck, Darmstadt, FRG) dissolved in saline; 5% glucose solution (Glucosteril 5%, Fresenius AG, Bad Homburg, FRG). All chemicals were analytical grade products.

Experimental protocol: The animals were allotted randomly to groups (n = 10). Group 1 received As_2O_3, 12.9 mg/kg s.c., without further treatment. Group 2 received As_2O_3 as above and 15 min later and at two hourly intervals for the whole experimental period (52 h) saline, 10 ml/kg. Group 3 was treated as group 2 but received 5% glucose (10 ml/kg) instead of saline. Three control groups without As_2O_3 (n = 5) received only glucose (group 4) or saline (group 5) as described above or remained without any treatment (group 6). The animals were constantly observed during the experiment. Mice which died before the end of the experiment underwent autopsy and tissue sample removal immediately after death. Survivors were decapitated after the 52 hour experimental period and tissue samples were removed immediately. For the measurement of glucose and glycogen, muscle tissue and brain were frozen in liquid nitrogen and stored for later analysis at –76°C. The content was measured enzymatically according to Bergmeyer (1984).

Statistical evaluation: Statistical significance of the differences was tested by the Wilcoxon-Mann-Whitney test ($\alpha = 0.05$, two tailed) (Sachs 1978).

Results

All control animals survived the 52 h experimental period. The proportion survival was in group 1, 2, and 3: 0/10, 1/10, and 6/10, respectively. In untreated mice, the glycogen and glucose content in the brain was 1.37 and 0.57, and in the muscle 6.5 and 1.4 (μg/g organ), respectively. In the brain of

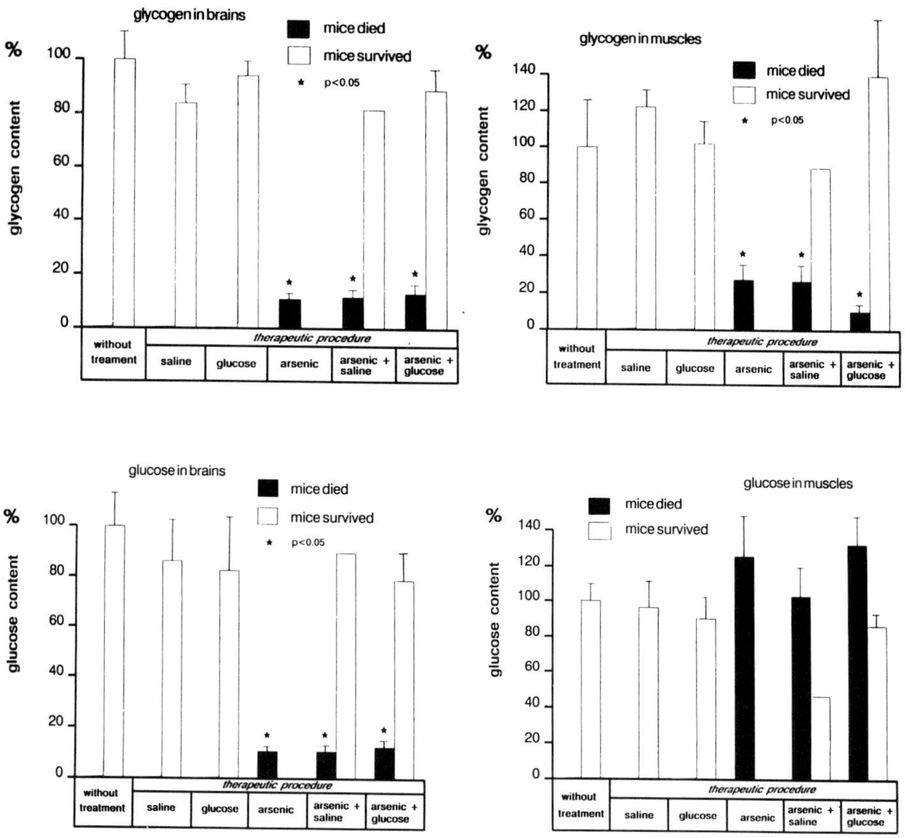

Fig. 1. Glycogen and glucose content in brain and muscle of mice receiving As_2O_3, treated by various therapeutic procedures. Values for animals which died during the test (■) and the surviving ones (□) are shown (100% = mice without any treatment). (*) significantly different from animals without any treatment

non-survivors, As_2O_3 treatment resulted in a decrease of the glycogen content by about 90%, irrespective of the treatment (Fig. 1). A similar finding was observed in the muscles, where a decrease by 80%–90% was observed (Fig. 1). The glucose content of the brain was reduced by about 90% in non-survivors receiving As_2O_3, however the same groups of animals exhibited similar or higher glucose content in muscle than the untreated controls (Fig. 1).

In the mice, poisoned with As_2O_3 and post-treated with glucose or saline, which died, almost no change (increase) was seen in the glycogen content of the brain or muscle.

Discussion

In the brain (CNS) the glucose content remained low in non-survivors in spite of glucose treatment. The reason might be that other tissues may utilise other energy sources whereas the brain is mainly dependent on glucose and utilises all the glucose available. The reason why peripheral tissues (muscles) exhibit an increase in glucose content after As_2O_3 poisoning may be that pyruvate dehydrogenase inhibition, the "biochemical lesion" caused by trivalent arsenicals (Peters 1955), not only causes gluconeogenesis inhibition (Szinicz and Forth 1988) but also results in inhibited carbohydrate utilisation. In agreement with this assumption, an increase in pyruvate and lactate was observed in trivalent arsenical poisoning (Peters 1955; Reichl et al 1988).

In summary the data presented seem to be in good agreement with the proposal of the earlier work (Szinicz and Forth 1988) that hypoglycemia, following glycogen depletion and gluconeogenesis inhibition, represents a vital problem of acute trivalent arsenical poisoning and that glucose substitution should be considered for treatment.

References

Bergmeyer HU (1984) Methods of enzymatic analysis; metabolites 1: Carbohydrates, 6. Verlag Chemie, Weinheim. Physiol 197:37–40

Peters RA (1955) Biochemistry of some toxic agents. 1. Present state of knowledge of biochemical lesions induced by trivalent arsenical poisoning. Bull John Hopkins Hosp 97:1–20

Reichl FX, Szinicz L, Kreppel H, Forth W (1988) Effect of arsenic on carbohydrate metabolism after single or repeated injection in guinea pigs. Arch Tox 62:473–475

Reichl FX, Szinicz L, Kreppel H, Fichtl B, Forth W (1990) Effect of glucose in mice after acute experimental poisoning with arsenic trioxyde (As_2O_3). Arch Tox, in press

Sachs L (1978) Angewandte Statistik; statistische Methoden und ihre Anwendung. Springer-Verlag, Berlin, Heidelberg, New York

Szinicz L, Forth W (1988) Effect of As_2O_3 on gluconeogenesis. Arch Tox 61:444–449

Recent Developments in Toxicology:
Trends, Methods and Problems
Arch. Toxicol., Suppl. 14, 228–230 (1991)
© Springer-Verlag 1991

Effect of Oral Treatment with Bal, DMPS or DMSA Arsenic in Organs of Mice Injected with Arsenic Trioxide

B. Schäfer, H. Kreppel, F. X. Reichl, B. Fichtl, and W. Forth

Walther Straub-Institut für Pharmakologie und Toxikologie, Nußbaumstraße 26,
8000 München 2, FRG

Introduction

The dithiols DL-2,3-dimercaptopropanol (British Anti-Lewisite, BAL; Peters et al 1945), DL-2,3-dimercaptopropanesulfonate (DMPS; Petrunkin 1956) and meso-2,3-dimercaptosuccinic acid (DMSA; Liang et al 1957) are effective arsenic antidotes. The possibility of oral application to reduce the As_2O_3 toxicity is proved for DMPS and DMSA, but not for BAL. Why BAL should not be given orally has not been discussed (Aposhian 1983; lit. in press). On the other hand, in one paper so far (Kreppel et al 1989) oral treatment with BAL has been shown to be effective in reducing lethality in acute As_2O_3 poisoning. Hence, the effectiveness of BAL in hydrophilic and in lipophilic solvent, DMPS, and DMSA in reducing the arsenic content of tissues after oral administration was investigated.

Material and Methods

Male white NMRI-mice (ITF, Tuttlingen, FRG) were fed standard diet (Alma H 1003, Botzenhardt KG, Kempten, FRG) and water *ad libitum*. The animals were allowed to acclimatize for 3 days before the experiment. Mice (n = 6/group) were randomly allotted to groups. They were injected s.c. with arsenic trioxide (8.4 mg/kg) containing a tracer dose of 76-As (III). As_2O_3 (99.99% purity, Merck, Darmstadt, FRG) was dissolved in saline. The injection volume was 0.1 ml/10 g bodyweight. 30 minutes later the animals were given dithiols (0.7 mmol/kg) via a gastric tube, controls received equal volumes of saline (0.9%). The following dithiols were used: DMPS (95% purity, Heyl & Co, Berlin, FRG) dissolved in saline; DMSA (98% purity, EGA-Chemie, Steinheim/Albuch, FRG) dissolved in saline by adjusting to pH 5.6; BAL (pure, Serva, Heidelberg, FRG) dissolved in saline or in peanut oil (Sigma, Deisenhofen, FRG). All solutions were freshly prepared before use. 0.5 h, 2 h, 4 h, 6 h, and

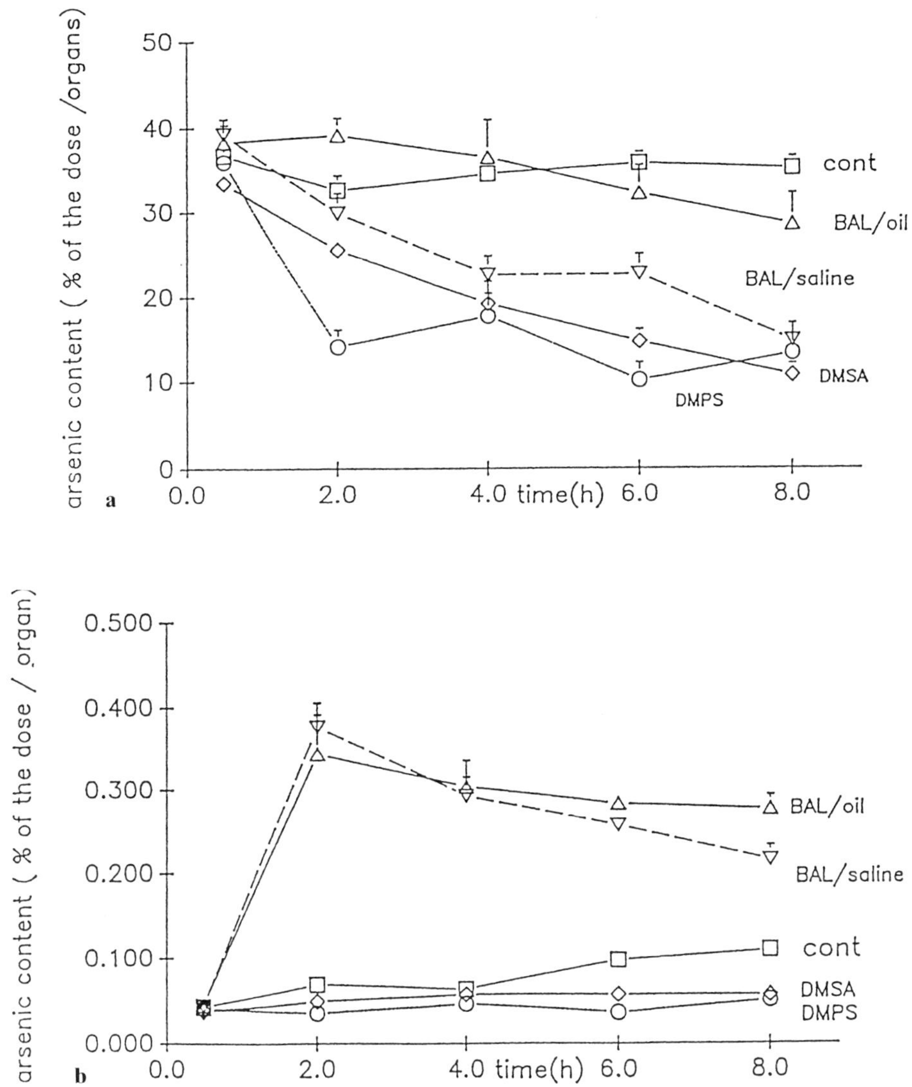

Fig. 1a, b. Influence of 0.7 mmol/kg DMPS, DMSA, BAL/oil, BAL/saline, and saline (controls), given orally, on the total arsenic content of organs (**a**) and brain (**b**) of mice (n = 6/group), mean ± SEM) injected with 8.4 mg/kg $^{76}As\text{-}As_2O_3$ s.c. Tissues investigated: heart, lungs, liver, spleen, kidneys, testes, skin (20% b.w.), muscle (40% b.w.), brain and blood (6.7% b.w.) 0.5 h, 2 h, 4 h, 6 h, and 8 h after the arsenic injection

8 h after the arsenic injection the animals were sacrified under ether. Blood was obtained by heart puncture. Heart, lungs, liver, spleen, both kidneys, small intestine (with faeces), large intestine (with faeces), testes, an aliquot of the back skin (contralateral to the injection site), an aliquot of the skeletal muscle of the right hind leg and the brain were removed. The radioactivity in blood

and organs was measured using a Gamma-counter (Gammamatic I, Kontron, Eching, FRG). The arsenic content in blood, skin and muscle was calculated assuming that these organs represent 6.7, 20 and 40%, respectively, of the bodyweight.

Statistical comparisons between treatment groups were performed using non-parametric methods: Wilcoxon-Mann-Whitney test, Kruskal-Wallis test.

Results and Discussion

DMPS and DMSA were significantly more effective in most organs (heart, lungs, liver, spleen, kidneys, testes, muscle, skin, brain and blood) in reducing the arsenic content than was BAL in saline or BAL in oil. BAL in oil was the least effective. In the first two hours DMPS was the most effective antidote (Fig. 1). BAL in saline reduced the arsenic content in most organs investigated more effectively than BAL in oil. The differences in reducing the arsenic content in tissues between BAL in oil and BAL in saline might be caused by the vehicle of administration influencing the absorption and/or the metabolization of BAL in the gastrointestinal tract. In brain and testes, however, the oral application of BAL in oil as well as in saline caused a marked accumulation of arsenic from 2 hours on which remained elevated during the whole experimental period (8 h). DMPS and DMSA significantly reduced the arsenic content in these organs (Fig. 2). A comparison of parenteral (Kreppel et al in press) and oral administration of the dithiols shows a similar sequence of efficacy in reducing the arsenic content in organs. The data presented in this paper confirm the conclusions of other authors that the recommendation of BAL as drug of choice in poisoning by inorganic arsenic needs to be carefully re-evaluated.

References

Aposhian HV (1983) DMSA and DMPS-water soluble antidotes for heavy metal poisoning. Ann Rev Pharmacol Toxicol 23:193–215

Aposhian HV (in press) Biochemical toxicology of arsenic. Rev Biochem Pharmacol

Kreppel H, Reichl FX, Mückter H, Fichtl B, Forth W (1989) Influence of oral treatment dithiol compounds on the survival rate of mice poisoned with arsenic. Proc 6th Int Trace El Symp, Leipzig 4:1039–1045

Kreppel H, Reichl FX, Szinicz L, Hunder G, Forth W, Fichtl B (in press) Arsenic content in organs of mice acutely poisoned with arsenic trioxide and treated with BAL, DMPS, or DMSA. Plzen Med Rep

Liang YI, Chu CC, Tsen YL, Ting KS (1957) Studies on antibilharzial drugs. VI. The antidotal effects of sodium dimercaptosuccinate and BAL-glucoside against tartar emetic. Acta Physiol Sin 21:24–32

Peters RA, Stocken LA, Thompson RHS (1945) British Anti-Lewisite (BAL). Nature 156:616–619

Petrunkin VE (1956) Synthesis and properties of dimercapto derivates of alkylsulfonic acids. Ukr Khem Zh 22:603–607

Recent Developments in Toxicology:
Trends, Methods and Problems
Arch. Toxicol., Suppl. 14, 231–235 (1991)
© Springer-Verlag 1991

Cyanide Antagonism in a Rodent Model

R. BHATTACHARYA, K. JEEVARATNAM, S. K. RAZA*, and S. DAS GUPTA

Division of Pharmacology and Toxicology,
* Division of Synthetic Chemistry, Defence Research and Development Establishment,
Tansen Road, Gwalior-474002, India

Introduction

The prophylactic experimental approach to counter cyanide poisoning involves pretreatment with methemoglobin producers viz. sodium nitrite ($NaNO_2$) and aminophenols (Kiese and Weger 1965). 4-Dimethylaminophenol (DMAP), a rapid methemoglobin former was suggested to give greater protection than nitrite (Kiese and Weger 1969). However, its clinical application is limited because of its toxicity (Szinicz et al 1979) and inability to offer a sustained protection for longer duration, when personnel are at risk to cyanide exposure while negotiating an industrial disaster. A protracted methemoglobinemia is presumed to offer sustained protection against cyanide poisoning. Hence, in the present study, the efficacy of sustained methemoglobinemia was studied in rats as a result of co-administration of $NaNO_2$ (slow methemoglobin producer) and DMAP in appropriate doses. A reduction in DMAP dosages is also envisaged based on methemoglobin kinetics.

Methods

Male Wistar rats (100 ± 10 g), obtained from the animal house of this establishment and fed on standard Lipton feed, were used in this study. All animals were fasted over night prior to experimentation while water was allowed ad libitum.

Toxicology: The acute (24 h) LD_{50} of potassium cyanide (KCN), $NaNO_2$ and DMAP administered subcutaneously (s.c.) was determined by Dixon's up and down method (1965). The prophylactic efficacy of methemoglobin generating agents alone or in combination with sodium thiosulfate ($Na_2S_2O_3$) was evaluated by determining the protection index (PI).

Biochemical parameters: In vivo methemoglobin estimation in blood was carried out at various time intervals by the modified method of Evelyn and Malloy

(1938) as described by Bright and Marrs (1982). At specific time interval after administration of KCN the rats pretreated with the antidotes were killed by cervical dislocation and the brain cytochrome oxidase activity determined following the method of Cooperstein and Lazarow (1951). Protein was estimated by the Biuret method (Gornall et al 1948).

Statistics: Data obtained were statistically analysed employing paired Student's t-test and the level of significance was set at $p<0.05$.

Materials: DMAP.HCl (~99% purity) synthesised by the chemistry division of this establishment was characterized by IR, MR and Mass spectrometer. The other chemicals used were AR grade obtained commercially: $NaNO_2$ (Merck), KCN (Ferrack) and $Na_2S_2O_3$. $5H_2O$ (Qualigens). All the injectable solutions were prepared in normal saline.

Results

Acute s.c. LD_{50} values of KCN, $NaNO_2$ and DMAP were 0.12, 1.4 and 0.90 mmol/kg respectively. $NaNO_2$ when injected at three different doses viz. 1.45, 0.73 and 0.36 mmol/kg produced a methemoglobin profile that corresponded with a dose dependant decline in protection against KCN (data not given). Transformation of 30–40% of hemoglobin into methemoglobin was chosen as the optimal level because it may be tolerated in experiments on humans and it has a therapeutic effect in experimental cyanide poisoning in mice and dogs (Kiese and Weger 1965). $NaNO_2$ at 0.36 mmol/kg dose evoked this effect with a peak methemoglobin (29%) around 45 min post treatment. DMAP at equimolar dose (0.36 mmol/kg) produced maximal methemoglobin (44%) at 10 min which progressively declined to normal by 120 min. The methemoglobin profile of $NaNO_2$ was significantly different from that of DMAP (Fig. 1). Pretreatment time of $NaNO_2$ and DMAP were chosen to coincide with their peak methemoglobin time i.e. 45 min and 10 min respectively. The Protection Index (PI) defined here as the ratio of LD_{50} of cyanide in protected rats to the LD_{50} in rats given only saline, was 1.7 in $NaNO_2$ as against 4.2 in DMAP pretreated rats. Incorporation of $Na_2S_2O_3$ at a dose recommended for prophylactic studies in mice (Isom and Way 1976) augmented the PI to 6.5 and 11.9 for $NaNO_2$ and DMAP respectively. Co-administration of $NaNO_2$ and DMAP in split dose combination *in vivo* ($NaNO_2$ and DMAP either at 0.18 mmol/kg each, 1:1, or 0.27 and 0.09 mmol/kg respectively, 3:1) resulted an almost similar methemoglobin kinetic profile with a sustained optimal level of methemoglobinemia (Fig. 1). Pretreatment of $NaNO_2$ and DMAP co-administration resulted in a substantial PI that persisted up to 60 min (Fig. 2). Incorporation of $Na_2S_2O_3$ in the pretreatment regimen further accentuated the protection. KCN ($^1/_2$ LD_{50}) inhibited 75% of brain cytochrome oxidase activity (Fig. 3). However, prior treatment prevented this inhibition to 50% within a span of 20 to 60 min pretreatment. $Na_2S_2O_3$ augmented the protection to 75%.

Discussion

In therapy of cyanide poisoning intravenous (i.v.) administration of the antidote is necessary for the rapid detoxification of cyanide. However, when personnel are at risk to exposure to any source contaminated with cyanide, prophylaxis by the oral or subcutaneous route is desirable (Marrs et al 1984). An effective prophylaxis is accomplished only when a desired optimal level of methemoglobinemia is persistent for a considerable period. The data presented primarily deals with a sustained protection achieved against cyanide intoxication through co-administration (s.c.) of $NaNO_2$ and DMAP in rats. DMAP (–10 min) alone offered a greater protection (PI 4.2) than in combination. However, the kinetic profile suggested the inability of the above regimen to offer a sustained protection over a period, as evident from the rapid decline in the methemoglobin level by 30 min (Fig. 1). Moreover, the adverse side effects of DMAP (Szinicz et al 1979) warranted a reduction in the dosage of the same. Co-administration of $NaNO_2$ and DMAP (3:1) resulted in a persistent methemoglobinemia with a correspondingly sustained protection against cyanide over a prolonged period. Efficacy of this regimen is also substantiated by the fact that pretreatment of $NaNO_2$ and DMAP co-administration with $Na_2S_2O_3$ significantly prevented the inhibition of brain cytochrome oxidase enzyme. Moreover, *in vitro* results (data not given) showed a prolonged persistence of methemoglobinemia at higher levels in human erythrocytes as compared with those of rats ($NaNO_2$, 2.1 μmol and DMAP, 0.69, μmol per gm of hemoglobin *in vitro*) because of the low activity of methemoglobin reductase in humans (Agar and Harley 1972). It may be speculated that a judicious lower dose of DMAP in combination with $NaNO_2$ be envisaged to offer a sustained prophylaxis against cyanide poisoning in human.

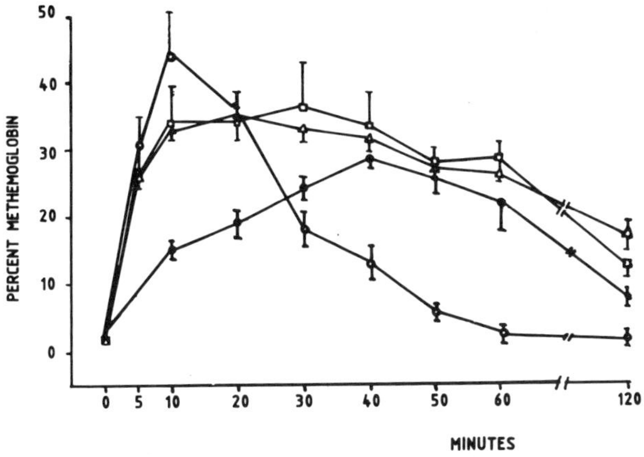

Fig. 1. Methemoglobin profile in rats following subcutaneous administration of $NaNO_2$ or DMAP alone or in split dose combination. The vertical bars indicate SE from the mean values for six rats. $NaNO_2$ (0.36 mmol/kg), (●), DMAP (0.36 mmol/kg), (○), $NaNO_2$ (0.18 mmol/kg) + DMAP (0.18 mmol/kg), (□), $NaNO_2$ (0.27 mmol/kg) + DMAP (0.09 mmol/kg), (△)

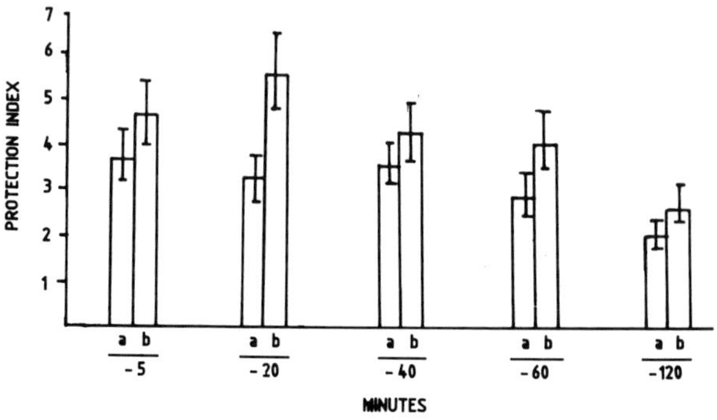

Fig. 2. Protection index in various pretreatment regimens at different time intervals. Vertical bars denote fiducial limits at 95% confidence level.
a) $NaNO_2$ 0.27 mmol/kg + DMAP 0.09 mmol/kg
b) $NaNO_2$ 0.27 mmol/kg + DMAP 0.09 mmol/kg + $Na_2S_2O_3$ (6.3 mmol/kg, ip, –15 min)

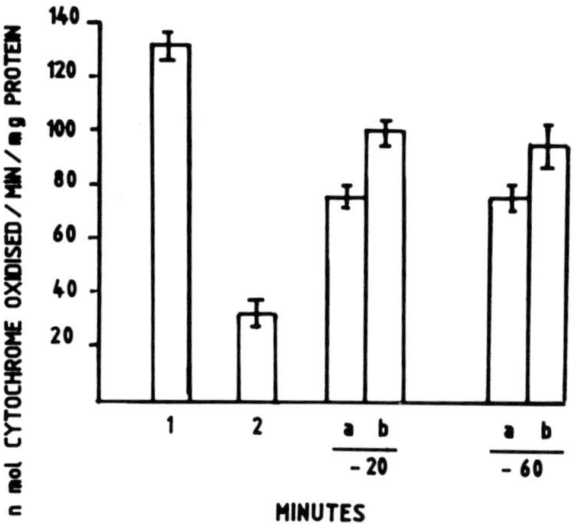

Fig. 3. Brain cytochrome oxidase activity following cyanide administration in unprotected and protected rats. Vertical bars indicate SE from the mean values for four to six rats.
1 Control
2 KCN $\frac{1}{2}$ LD_{50}
a) $NaNO_2$ 0.27 mmol/kg + DMAP 0.09 mmol/kg
b) $NaNO_2$ 0.27 mmol/kg + DMAP 0.09 mmol/kg + $Na_2S_2O_3$ (6.3 mmol/kg, ip, –15 min)

Acknowledgements. Authors are thankful to Brig. (Dr.) K. M. Rao, Director, Defence R & D Establishment, Gwalior for his keen interest in this investigation and Mr. Mitthan Lal for secretarial assistance.

References

Agar NS, Harley JD (1972) Erythrocytic methemoglobin reductases of various mammalian species. Experientia 28:1248–1249

Bright JE, Marrs TC (1982) A model for the induction of moderate levels of methemoglobinemia in man using 4-dimethylaminophenol. Arch Toxicol 50:57–64

Cooperstein SJ, Lazarow A (1951) A microspectrophotometric method for the determination of cytochrome oxidase. J Biol Chem 189:665–670

Dixon WJ (1965) The up and down method for small samples. J Am Statist Assoc 60:967

Evelyn RA, Malloy HT (1938) Micro determination of oxyhemoglobin, methemoglobin and sulfhemoglobin in a single sample of blood. J Biol Chem 126:655–662

Gornall AG, Bardawill CJ, David MM (1948) Determination of serum proteins by means of the biuret reaction. J Biol Chem 177:751–766

Isom GE, Way JL (1976) Lethality of cyanide in the absence of liver cytochrome oxidase. Biochem Pharmacol 25:605–608

Kiese M, Weger N (1965) The treatment of experimental cyanide poisoning by hemiglobin formation. Arch Toxicol 21:89

Kiese M, Weger N (1969) Formation of ferrihaemoglobin with aminophenols in the human for the treatment of cyanide poisoning. Eur J Pharmacol 7:97–105

Marrs TC, Scawin J, Swanston WW (1984) The acute intravenous and oral toxicity in mice, rats and guinea pigs of 4-dimethylaminophenol (DMAP) and its effects on haematological variables. Toxicology 31:165–173

Szinicz L, Weger N, Schneiderhan W, Kiese M (1979) Nephrotoxicity of aminophenols: effect of 4-dimethylaminophenol on isolated rat kidney tubules. Arch Toxicol 42:63–73

Recent Developments in Toxicology:
Trends, Methods and Problems
Arch. Toxicol., Suppl. 14, 236–239 (1991)
© Springer-Verlag 1991

The Quantitative Determination of Cadmium in Turkish Tobacco

N. Oto[1], and S. Duru[2]

[1] National Poison Center, Ankara, Turkey, and
[2] Pharmaceutical Toxicology College of Pharmacy Hacettepe University, Ankara, Turkey

Introduction

Cadmium is one of the metals which is not normally found in the pure form in nature. It is also widely used in various fields of industry. The most important toxicological feature of cadmium, on human body, is known to be its very long half-life (Fleisher 1974; Yost 1974; Bernard 1986; Page 1973). Cadmium originates from various sources, and reaches human body by food, beverages, drinks, air, and tobacco (Sherlock 1984).

Since tobacco has a widespread abuse in the world and Turkey is one of the countries where cigarette smoking is a very common habit, it was aimed to analyse the levels of cadmium in tobacco, harvested from various regions of Turkey: Aegean, Black Sea, Marmara, South-East and Eastern Anatolia. Measurements were carried out on three different grades of tobacco (Grade A, Grade B, Kappa Grade) produced during the last 5 years (over the period of 1983 to 1987), in order to investigate the cadmium exposure of Turkish people through smoking.

Sample preparation

2 grams of dried tobacco powders were weighted in a porcelain crozet. Samples were taken into an ash-furnace and the temperature was elevated to 450°C with 75°C increments. Samples were kept in the furnace at 450°C for 24 hours. At the end of this period all the samples were burned to whiteness (the colour observable by naked-eye). The ash was then taken out of the furnace and cooled to room temperature. The ash samples were dissolved in 0.7 ml concentrated nitric acid (Merck). This solution was evaporated to dryness at 270°C. The remaining residuals were dissolved in 0.06 M HCI and the solution was transferred into 10 ml measuring tubes (Piscator 1976; Ure 1976; Van der Lee 1987).

Table 1. Distribution of cadmium levels in tobacco with respect to regions in Turkey

	cadmium concentration measured on wet weight ($\mu g/g \pm$ standard error)
Easthern Anatolia	1.239 ± 0.205, n=14
South-East Anatolia	1.253 ± 0.168, n=13
Marmara Region	1.561 ± 0.330, n=13
Aegean Region	0.800 ± 0.125, n=14
Black Sea Region	1.463 ± 0.254, n=12

The cadmium levels of tobacco samples were measured by means of a flame atomic absorption spectrometer. The concentration of cadmium in 1 gram of sample was calculated.

Materials and Methods

The determination of cadmium was performed on an Instrumentation Laboratory 457 Spectrometer. The parameters of measurement set-up are as follows:
Light Source: Cadmium cathode lamp
Wave Length: 228.8 nm
Supply Current: 3 mA
Slot Width: 1 mm
Fuel: Acetylene
Ambient: Air

In this study; nitric acid (65%), hydrochloric acid (37%), cadmium powder were obtained from Merck Co.

The statistical analysis used were: Kruskall-Wallis Variance Solution and Linear Regression Analysis to analyse the data.

Results and Discussion

In this study, cadmium levels (calculated from the wet weight) ranged from 0.800 $\mu g/g \pm 0.125$ in Aegean Region to 1.561 $\mu g/g \pm 0.3$ in Marmara Region. In other regions the levels were found to lie within these extremes (Table 1).

The cadmium levels in tobacco samples from Aegean Region was significantly lower than that of the other regions ($p < 0.005$).

In addition, it was also found that differences between the cadmium levels of Grade A and Grade B ($p < 0.05$) and Grade A and Kappa Grade ($p < 0.001$) tobacco was statistically significant (Fig. 1).

A clear difference was not observed in the cadmium levels of tobacco samples taken from the products of consecutive years from the same region.

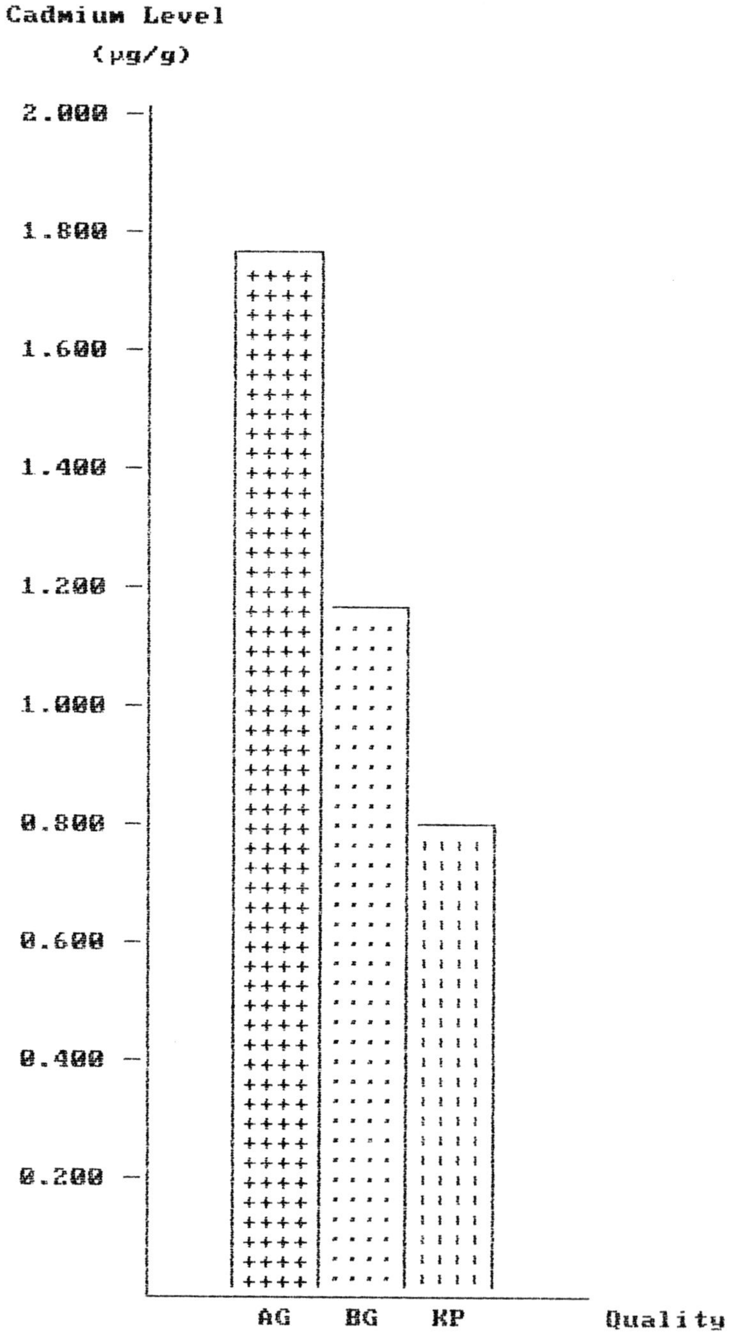

Fig. 1. Cadmium levels in tobacco with respect to regions in Turkey

In this work, cadmium levels in tobacco, harvested from various regions of Turkey, was found to be higher than that of other regions. It was therefore intended to examine the soil in those areas where cadmium level in tobacco is higher, so that soil improvement work can be initiated.

References

Bernard A, Lauwerys R (1984) Cadmium in human population. Experientia 40:143–152

Fleisher M, Sarofim AF, Fasset DW, Hammond P (1974) Environmental impact of cadmium. Environ Hlth Persp 7:253–323

Page AL, Bingham FT (1973) Cadmium residues in the environment. Residue Rev 48:1–44

Piscator M, Kjellström P, LInd B (1976) Contamination of cigarette and pipe tobacco by cadmium–oxide dust. Lancet II:587

Sherlock JC (1984) Cadmium in foods and the diet. Experientia 40:152–156

Ure AM, Mitchell C (1976) The determination of cadmium in plant material and soil extracts by solvent extraction and atomic absorption with a carbon-rod atomizer. Analytica Chimica Acta 87:283–290

Van der Lee JJ, Temminghoff E, Houba VJG, Novozansky I (1987) Background corrections in the determination of cadmium and lead by flame. AAS in plant and soil samples with high Fe levels. Applied Spectroscopy 4:388–390

Yost KJ (1984) Cadmium, the environment and human health: an overview. Experientia 40:157–164

Recent Developments in Toxicology:
Trends, Methods and Problems
Arch. Toxicol., Suppl. 14, 240–245 (1991)
© Springer-Verlag 1991

Modifying Role of Plasma Glucose in Acute Carbon Monoxide Poisoning

D. G. Penney

Department of Physiology, Wayne State University, School of Medicine, Detroit, Michigan, USA

Introduction

Carbon monoxide as a poison is responsible for a large number of deaths and a great amount of morbidity. Hyperglycemia is reported to modify the outcome in global cerebral and spinal cord ischemia (LeMay et al 1987; Siemkowicz and Hansen 1978), cardiac arrest (Lundy et al 1987), and stroke (Pulsinelli et al 1983). This is of particular relevance to acute carboxyhemoglobinemia because hyperglycemia usually accompanies CO poisoning. The aim of this study was to evaluate the role of plasma glucose concentration in the outcome from acute severe CO poisoning in an animal model.

Methods

Levine-prepared, unanesthetized adult female rats (Charles River Breeding Labs) were used. The left common carotid artery and jugular vein were catheterized with PE-50 polyethylene tubing under ketamine-rompun anesthesia. The catheters emerged from the nape of the neck, providing external ports for withdrawal of blood, or infusion of agents. This procedure effectively occludes the vessels and increases the sensitivity of that side of the brain to CO (Laas et al 1983). The rats were exposed to CO for 90 min inside a large transparent plastic bag, then recovered unrestrained in room air. The gas mixture was made up by mixing air from a compressor (40 l/min) with CO from a commercial cylinder, as needed. The CO concentration was continuously monitored with a Beckman infrared analyzer. The carotid arterial blood pressure (BP) and heart rate were recorded on a Gould chart recorder. The rectal temperature was monitored with a Yellow-Springs Instrument Co. thermistor probe. Blood was withdrawn from the carotid cannula at intervals for determination of hematocrit (microhematocrit method), COHb (spectrophotometrically), glucose

Table 1. Criteria for scoring neurologic status

General Appearance (Ap):	
1	Normal
2	Fur slightly ruffled
3	Fur ruffled and slightly hunched over
4	Fur moderately ruffled; moderately hunched over; face appears broader
5	Fur very ruffled; extremely hunched over; face appears very broad
Posture (Po):	
1	Normal, all extremities under body
2	Leaning to right or left but both legs under body
3	Leaning to one side with left or right limbs extended
4	Lying on one side but with leg on one side touching floor
5	Completely on one side, no limbs touching floor
Shuffle (Sh):	Observation of front paw movements with hind legs off floor
1	Normal, moves straight right or left
2	Can move right or left, but tends to move one direction only
3	Just moves to one side, but can move both extremities
4	Just moves one front paw, no movement of other paw
5	Spins to one side or the other
Circling (C):	
1	Normal
2	Can move right or left, but moves mainly to one side or the other
3	Circles to right only, or to left only
4	Circles and walks over tail
5	Circles and falls to the right or left
Activity (Ac):	
1	Very active
2	Appears alert but less active than normal; sluggish
3	May be prodded into sustained activity, but drifts off; prefers to stand still
4	Active only while being stimulated
5	No spontaneous movement even when stimulated
Screen (Sc):	Ability to hang onto a vertical screen
1	Normal, does not fall off
2	Cannot change position, hangs on but progresses down screen, able to stop fall
3	Hangs on briefly, then falls, some attempts to stop fall
4	Continuous movement to floor
5	Falls immediately off, no attempts to hold on

(glucose oxidase method), and insulin (radioimmunoassay). Neurologic status (neurologic index = NI) was assessed before and after CO exposure by a system modified from Lundy et al (1985) (Table 1)

Results

Experiment I. In this group of rats body temperature (BT) (Table 2) and mean arterial blood pressure fell progressively during 90 min exposure to 2400 and 2700 ppm CO, the former to a greater extent at the higher CO concentration. The blood glucose concentration rose during initial CO exposure, then declined; glucose increased again after 2 hrs of room air recovery, also to a

Table 2. Responses of Levine-prepared, adult female rats exposed to carbon monoxide for 90 minutes, then allowed to recover from 90 to 330 minutes (4 hrs) in room air

CO Conc.	Body Temp. Change T_0-T_{90} (C°)	Body Temp. Recovery (%) T_{90}-T_{210}	T_{90}–T_{330}	Plasma Glucose Change T_{90}–T_{210} (mg/dl)	N.I. T_{330}	Survival (number/%)
2400 ppm	−3.4	65.4	88.8	+33.0	12.4	31/37, 83.8
	± 0.13 (33)	± 5.9 (29)	± 6.2 (30)	± 12.5 (29)	± 1.1 (28)	
2700 ppm	−4.3	8.6	48.8	+50.1	16.7	12/16, 75.0
	± 0.25 (12)	± 11.0 (10)	± 13.8 (9)	± 21.8 (10)	± 1.9 (10)	
P =	++	++	+	NS	NS	

Values are ± standard error of the mean.
Numbers in parenthesis are numbers of animals used.
Comparing the two treatments using Student's t-test: +, $P<0.05$; ++, $P<0.01$.

greater extent at the higher CO concentration. NI obtained after 4 hrs of recovery when BT and plasma glucose had renormalized, was positively correlated with the increase in glucose during the first 2 hrs of recovery at 2400 ppm CO ($r = 0.71$, $P<0.001$), but NI was greater at 2700 ppm than at 2400 ppm CO (Table 2). Mortality was higher and recovery of BT slower at the higher CO concentration. Recovery hyperglycemia was, in turn, correlated with the decline in glucose during the second half of CO exposure, suggesting a hyperglycemic rebound following hypoglycemia. Failure of ability to regain BT early in recovery was correlated with an increased NI later in recovery. Hematocrit increased 4.1% during CO exposure, then decreased 7.8% during recovery. Plasma insulin activity was depressed immediately following CO exposure, and increased during recovery.

The results suggest that neurologic damage and general morbidity in acute CO poisoning are directly related to the magnitude of the post-exposure hyperglycemic response, and also demonstrate dose-response relationships to CO concentration in terms of CO-induced hypothermia and mortality, and for post-exposure BT recovery, hyperglycemia, and neurologic damage.

Experiment II. In rats in this series, severe hyperglycemia (50% glucose solution administered by i.p. injection or jugular infusion, 6 ml/kg) or hypoglycemia (0.9% NaCl administered by i.p. injection or jugular infusion; 6 ml/kg) was induced prior to 90 min exposure to 2400 ppm CO. Plasma glucose was raised from 118 to 402 mg/dl prior to CO exposure in the first group. This resulted in an increased mortality rate (i.e. 52%), and during 4 hrs of room air recovery an impaired ability to regain BT, an increased release of lactate dehydrogenase activity into the plasma, and an increased NI as compared with the control group (no glucose or saline infusion) (Fig. 1). Plasma glucose concentration in hyperglycemic rats that died was elevated (non-significant) above survivors. Hypoglycemia which developed during CO exposure in the saline-infused

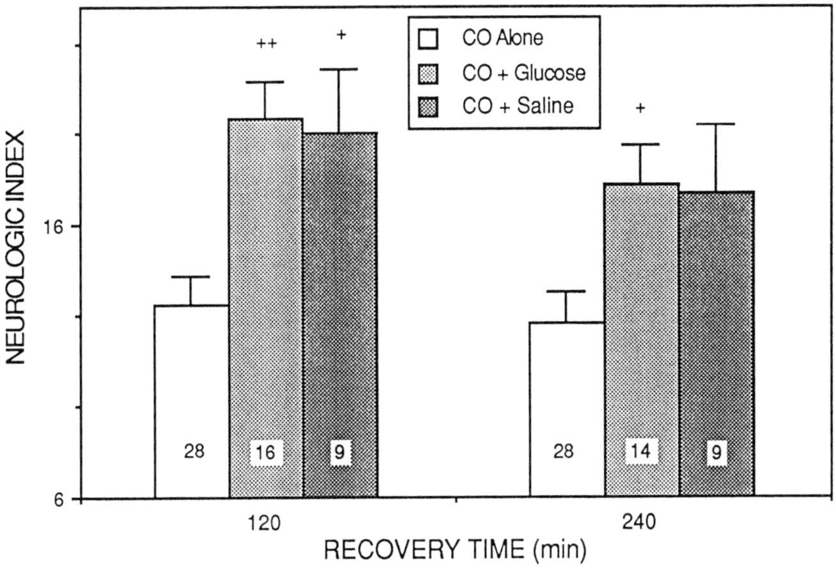

Fig. 1. Neurologic index obtained after 2 hrs and 4 hrs of room air recovery in Levine-prepared, adult female rats exposed to 2400 ppm carbon monoxide for 90 min. Only surviving rats are included; numbers indicated in each column. Compared with the controls (CO alone): +, $P<0.05$, ++, $P<0.01$

group (mean minimum glucose after 90 min, 44 mg/dl) resulted in an increased mortality rate (i.e. 44%), and during 4 hrs of room air recovery an impaired ability to regain BT and an increased NI as compared with the control group (Fig. 1). Plasma glucose concentration in the hypoglycemic and control rats that died was significantly depressed relative to survivors. BP fell far more rapidly during CO exposure in the hyperglycemic and hypoglycemic rats than in the controls. Plasma insulin activity was elevated during CO exposure in the hyperglycemic group as compared with the controls, but fell during recovery: Insulin activity remained low in the hypoglycemic group.

The results confirm the deleterious effect of elevated glucose during CO poisoning revealed in Experiment I. Whereas most rats in Experiment I which became hyperglycemic also experienced hypoglycemia late during CO exposure, no hypoglycemia occurred in the glucose-infused rats used here; thus, hypoglycemia could not have contributed to the observed neurologic damage. Thus, the association between high plasma glucose and neurologic damage suggested in Experiment I is confirmed as a cause and effect relationship by Experiment II. In contrast, induction of hypoglycemia alone during CO poisoning by the infusion of saline, has also been shown to produce increased morbidity and mortality of comparable severity to that produced by hyperglycemia.

Figure 2 summarizes the changes in BT, BP, heart rate, hematocrit, motor function, plasma glucose, etc. that are observed in the rat during acute CO poisoning, and the possible mechanisms involved. It is based upon this and other studies (Penney et al 1989).

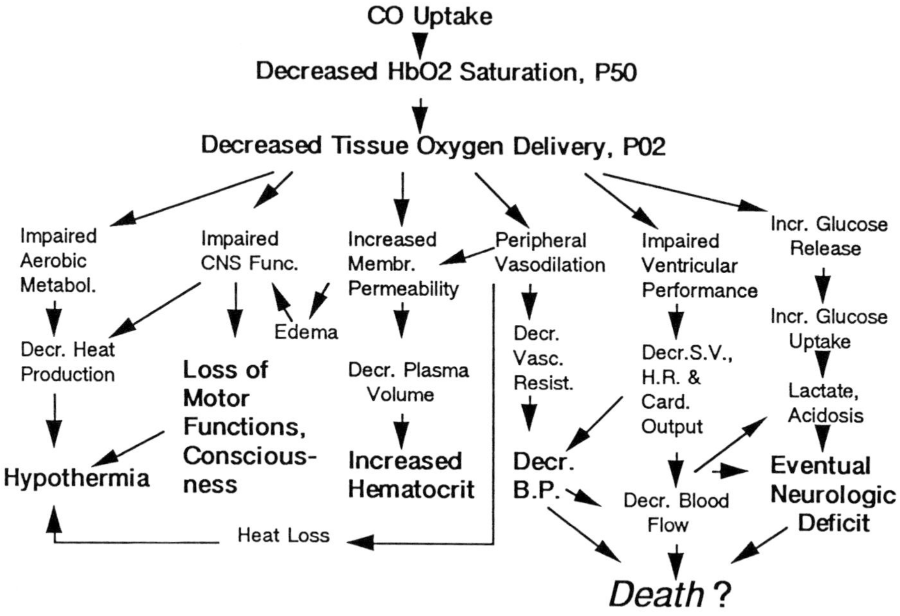

Fig. 2. Effects of acute severe carbon monoxide exposure in the rat and some of the presumed toxic mechanisms. Observed or measured parameters indicated in boldface at the bottom; mechanisms and intermediaries are indicated in smaller type. Abbreviations: B.P., blood pressure; CNS, central nervous system; CO, carbon monoxide; HbO2, oxyhemoglobin; H.R., heart rate; P50, partial pressure of oxygen at half-saturation of hemoglobin; PO2, partial pressure of oxygen; S.V., stroke volume

Discussion

As is well known, CO exerts its toxic effect by avidly binding to hemoglobin. Because brain blood flow has been shown to increase dramatically in animals breathing CO at all but the highest COHb levels (MacMillan 1975), the tissue effects of CO poisoning are probably somewhat different from those of cardiac arrest, global cerebral ischemia, etc., where blood flow ceases. Continued blood flow, even though a trickle, may be a worse condition than no blood flow if continued glycolysis and the resulting tissue acidosis are indeed culprits in producing brain damage (Rehncrona et al 1980), because glucose continues to be supplied and lactate continues to be generated. Incomplete ischemia has been shown to have a worse outcome than complete ischemia in non-CO animal models (Nordstrom et al 1976).

The results of these experiments are in agreement with a recent retrospective human study showing an association between neurologic deficit and elevated plasma glucose in people accidently or purposefully (ie. suicide) intoxicated with CO (Penney 1988). They suggest that emergency medical professionals should be increasingly concerned as to the glycemic status of CO-poisoning victims, and that swift measures should be taken to renormalize plasma glucose concentration.

References

Laas R, Igloffstein J, Meyerhoff S (1983) Cerebral infarction due to carotid occlusion and carbon monoxide exposure. I. Pathophysiological and neuropathological investigations. J Neurol Neurosurg Psychiatr 46:756–767

LeMay DR, Neal S, Neal S, Zelenock GB, D'Alecy LG (1987) Paraplegia in the rat induced by aortic cross-clamping: Model characterization and glucose exacerbation of neurologic deficit. Vascular Surg 6:383–390

Lundy EF, Dykstra J, Luyckx B, Zelenock GB, D'Alecy LG (1985) Reduction of neurologic deficit by 1,3-butanediol induced ketosis in Levine rats. Stroke 16:855–860

Lundy EF, Kuhn JE, Kwon JW, Zelenock GB, D'Alecy LG (1987) Infusion of five percent dextrose increases mortality and morbidity following six minutes of cardiac arrest in resuscitated dogs. J Crit Care 2:4–14

MacMillan V (1975) Regional cerebral blood flow of the rat in acute carbon monoxide intoxication. Can J Physiol Pharmacol 53:644–650

Nordstrom CH, Rehnchrona S, Siesjo BK (1976) Restitution of cerebral energy state after complete and incomplete ischemia of 30 min duration. Acta Physiol Scand 97:270–272

Penney DG (1988) Hyperglycemia exacerbates brain damage in acute severe carbon monoxide poisoning. Med Hypotheses 27:241–244

Penney DG, Verma K, Hull JA (1989) Cardiovascular, metabolic and neurologic effects of acute carbon monoxide poisoning in the rat. Toxicol Letts 45: 207–213

Pulsinelli WA, Levy DE, Sigsbee B, Scherer P, Plum F (1983) Increased damage after ischemic stroke in patients with hyperglycemia with or without established diabetes mellitus. Am J Med 74:540–544

Rehncrona S, Rosen I, Siesjo BK (1980) Excessive cellular acidosis: An important mechanism of neuronal damage in the brain. Acta Physiol Scand 110:435–437

Siemkowicz E, Hansen AJ (1978) Clinical restitution following cerebral ischemia in hypo-, normo- and hyperglycemic rats. Acta Neurol Scand 58:1–8

Recent Developments in Toxicology:
Trends, Methods and Problems
Arch. Toxicol., Suppl. 14, 246–248 (1991)
© Springer-Verlag 1991

Effect of Pyrazole on Dichloromethane Metabolism to Carbon Monoxide

D. Pankow, S. Kretschmer, and M. Weise

Institute of Industrial Toxicology, Martin Luther University, Franzosenweg 1a, O-4020 Halle/S

The reports of elevated carboxyhemoglobin (COHb) levels following exposure to dichloromethane (DCM) stimulated the research on the toxicology of this solvent. The metabolic pathway of DCM to carbon monoxide (CO) involves an oxygen insertion catalyzed by cytochrome P-450 (Gargas et al 1986), mainly by the isozyme cytochrome P-450 IIE1 (CYP2E1). Pyrazole (PYR) is known to depress the oxidative metabolism of DCM, but based upon the catalytic, spectral, and immunological properties it appears to induce CYP2E1 (Palakodety et al 1988). The effect of PYR on carboxyhemoglobinemia following DCM administration in rats was investigated.

Material and Methods

Male Wistar rats (body mass 259 ± 15 g) received a single dose of DCM, 6.2 mmol/kg (0.4 ml/kg) as a 10% solution in Oleum pedum tauri orally (p.o.). PYR was administered once or once daily for 3 days in physiologic saline by intraperitoneal (i.p.) injection. DCM was purchased from Riedel-de Haën, PYR from Merck-Schuchardt. Heparinized blood samples were taken from the retroorbital venous plexus. The COHb concentration in the blood was determined by means of the palladium-II-chloride method (Schmidt 1988); the DCM concentration in the blood by gas chromatography after extraction with 1,3-dimethylbenzene using trichloromethane as the internal standard. The values are given as the mean ± SEM (N = 3–7 rats per group). The Student t-test was used for statistical comparison.

Results

24 h after the last of 3 doses of PYR, 0.73 mmol/kg, the single DCM gavage produced COHb levels significantly higher than the controls receiving DCM

Abbreviations: CO = carbon monoxide, COHb = carboxyhemoglobin, CYP2E1 = cytochrome P-450 IIE1, DCM = dichloromethane, PYR = pyrazole

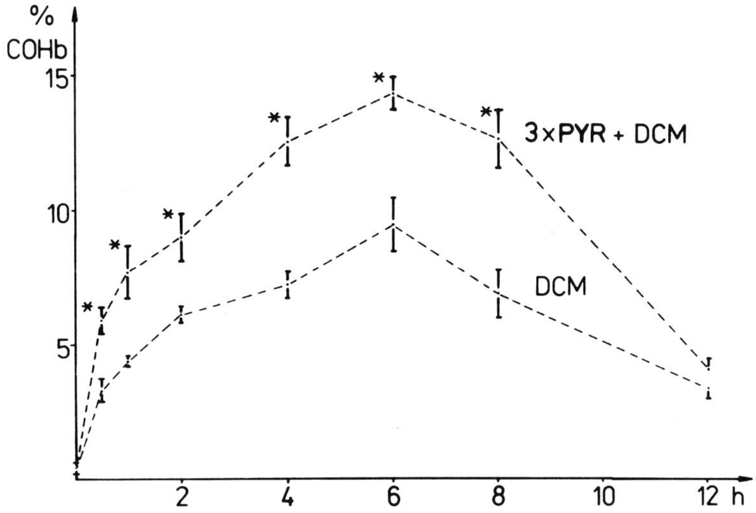

Fig. 1. Carboxyhemoglobin level in the blood of rats as a function of the time following administration of dichloromethane, 6.2 mmol/kg p.o., 24 h after completion of pretreatment with either saline solution, 1 ml/kg i.p., once daily for 3 days (DCM) or with pyrazole, 0.73 mmol/kg i.p., once daily for 3 days (3 PYR + DCM); asterisk: significantly different from COHb following DCM per se

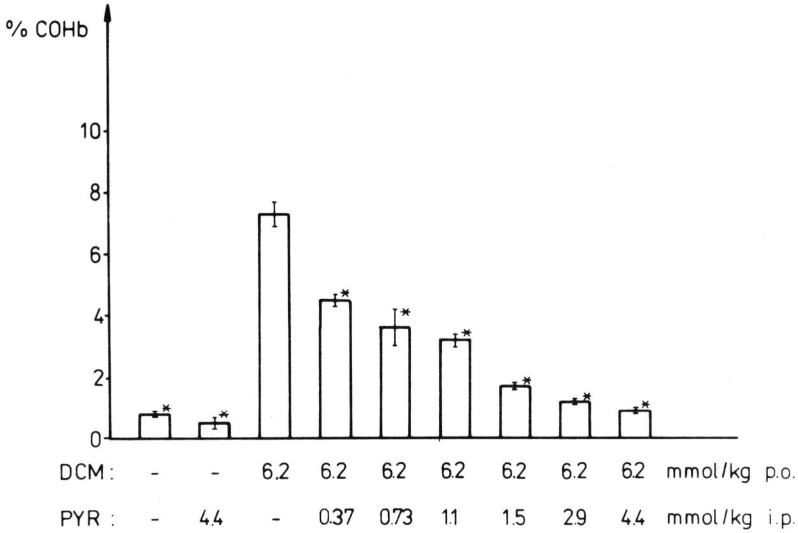

Fig. 2. Carboxyhemoglobin level in the blood of rats 6 hours after single simultaneous administration of dichloromethane (DCM) and pyrazole (PYR); asterisk: significantly lower than % COHb following DCM per se

after a similar pretreatment with saline (Fig. 1). The DCM concentration in the blood was lowered from 0.57 ± 0.06 to 0.42 ± 0.08 mmol/l at the half-life of DCM (about 170 min).

The rats received saline solution, 1 ml/kg, or PYR, 1.1, 1.5, or 2.9 mmol/kg, once daily for 3 days. 24 h after the last injection DCM was administered. The maximum COHb values were 9.0 ± 0.4, 12.5 ± 1.9 ($p < 0.05$), 10.8 ± 0.4 ($p < 0.05$), and 5.8 ± 0.3% ($p < 0.05$), respectively. Figure 2 summarises the effect of simultaneous administration of PYR and DCM on the COHb level. In comparison with the values after gavage of DCM *per se* the additional PYR administration produced a dose-dependent depression of the COHb formation.

Discussion

These results demonstrate that the metabolism of DCM to CO can be stimulated or inhibited by PYR. The effect depends on the interval between PYR and DCM administration and on the dose. This stimulation has not been described previously and is evident after suitable PYR pretreatment 24 h prior to DCM administration. The enhancement of COHb formation is consistent with a simultaneous reduction of the DCM blood level. It agrees with the observation that stimulation of DCM metabolism to CO is due to inducers of CYP2E1 such as isoniacid, ethanol, acetone and other solvents. The inhibition was observed following pretreatment with high PYR doses or following simultaneous administration of PYR and DCM. An inhibition has also been described after PYR pretreatment, 4.7 mmol/kg i.p., 30 min prior to DCM exposure (Gargas et al 1986; Kim and Carlson 1986). The inhibition may reflect the competition between PYR and DCM for oxidation by CYP2E1 as long as PYR is present in the blood. Studies in rats using PYR, 1.5 mmol/kg i.p., have suggested a first order elimination pattern with a half-life of 13.3 ± 0.9 h (Rydberg et al 1972). Another reason for the inhibition may be the hepatotoxic effect of PYR (Lieber et al 1970).

References

Gargas ML, Clewell III HJ, Andersen ME (1986) Metabolism of inhaled dihalomethanes in vivo: Differentiation of kinetic constants for two independent pathways. Toxicol Appl Pharmacol 82:211–223

Kim YC, Carlson GP (1986) The effect of an unusual workshift on chemical toxicity I. Studies on the exposure of rats and mice to dichloromethane. Fund Appl Toxicol 6:162–171

Lieber CS, Rubin E, DeCarli LM, Misra PS, Gang H (1970) Effects of pyrazole on hepatic function and structure. Lab Invest 22:615–621

Palakodety RB, Clejan LA, Krikun G, Feierman DE, Cederbaum AI (1988) Characterization and identification of a pyrazole-inducible form of cytochrome P-450. J Biol Chem 263:878–884

Rydberg U, Buijten J, Neri A (1972) Kinetics of some pyrazole derivates in the rat. J Pharm Pharmacol 24:651–652

Schmidt P (1988) Biologische Kontrollmethoden in der Arbeitsmedizin. Verlag Volk und Gesundheit, Berlin, 345–346

Recent Developments in Toxicology:
Trends, Methods and Problems
Arch. Toxicol., Suppl. 14, 249–253 (1991)
© Springer-Verlag 1991

Ethylene Oxide Distribution in Human Blood

B. Gansewendt, U. Foest, B. Marczynski, K. Golka, E. Hallier, and H. Peter

Institut für Arbeitsphysiologie an der Universität Dortmund, Ardeystraße 67, W-4600 Dortmund 1, FRG

Introduction

Ethylene oxide is a proven carcinogen in laboratory animals and a suspected carcinogen in humans (IARC 1985). Determination of the level of ethylene oxide in whole blood and alveolar air has been recommended for monitoring of occupational exposure (Brugnone et al 1985 and 1986), based on the report of a linear correlation between the concentration of the substance in ambient air, alveolar air and whole blood (DFG 1989). Others favour the determination of hydroxyethylated amino acids in hemoglobin as the most suitable biomonitor (ECETOC 1989; Farmer et al 1986). This is however only possible if no enzymatic metabolism of the substance takes place in human blood. Such a metabolism is often subjected to genetically determined interindividual differences. An enzyme polymorphism has been described for aromatic amines (Lewalter and Korallus 1985). Two distinct subpopulations of conjugators and non-conjugators have also been described for the enzymatic reaction of several methyl halides with glutathione in human erythrocytes (Peter et al 1989; Hallier et al 1990). The distribution and possible metabolism of ethylene oxide in human blood have however not yet been investigated with sensitive methods.

Methods

Blood samples of 6 volunteers (9 ml each) were drawn in syringes containing 1 ml 3.8% aqueous sodium citrate. Previously, these volunteers had been investigated for their ability to conjugate methyl halides with glutathione (GSH) in erythrocytes (conjugators) or not (non-conjugators) according to Peter et al (1989) and Hallier et al (1990). 9 ml of each individual blood sample was transferred to 22 ml head space vials and mixed with 200 μl of an aqueous solution of 1.2-^{14}C ethylene oxide (Amersham spec. act.: 510 MBq/mmol, 25 μCi/ml water). The samples were brought to and incubated at 37°C with constant rolling for four hours. At different intervals, 0.5 ml aliquots were drawn from each vial and centrifuged a 400 g for 5 min. The sedimented blood cells were resuspended in 350 μl physiological saline by vortexing and layered onto 1.5 ml lymphocyte separation medium. Lymphocytes were isolated after centrifugation at 400 g for 30 min, mixed with 200 μl saline and centrifuged at 10000 g for 10 min. The lymphocytes were resuspended in 200 μl saline and stored at −20 °C. The platelets were removed according to Strange et al (1982). The erythrocytes were washed twice and lysed, the cell membranes were sedimented by centrifugation at 16000 g for 30 min. The supernatant was transferred into Millipore Ultrafree Membrane centrifugation MC vials (cutoff: 10000 Da) and centrifuged at 16000 g for 2 h. The low molecular weight fraction was isolated and the proteins were washed with 300 μl water. Proteins on the membrane were suspended in 300 μl water and stored at −20 °C. Blood plasm was similarly divided into low and high molecular weight fractions. The protein content of blood plasm was determined according to Lowry et al (1951) and the hemoglobin content in erythrocyte cytoplasm by use of a standard test. Radioactivity was measured in a liquid scintillation counter.

Results

The amount of radioactivity in the different blood compartments after four hours of incubation and the initial velocity of the increase of radioactivity bound in the different blood compartments were the parameters chosen to describe the pharmacokinetic processes involved (Table 1). In three samples shown as group A, the radioactivity associated with the low molecular weight fraction of the erythrocyte cytoplasm was eight times higher after four hours incubation than in the other three samples, group B (highly significant in Student's t-test). At the same time, the opposite effect was observed in the high molecular weight fraction, the radioactivity in the samples of group B being twice as high (significant in t-test) as in the samples of group A (Fig. 1). No significant interindividual differences in the distribution of radioactivity were observed in lymphocytes. In the low molecular weight fraction of blood plasma, the amount of radioactivity was twice as high in the samples of group A as

Table 1. Initial velocity (v) (mean ± standard deviation, n=3) of the reaction of ethylene oxide with lymphocytes and low and high molecular weight components of erythrocyte cytoplasm and blood plasma in conjugators (group A) and non-conjugators (group B)

			Group A	Group B
			v $\frac{pmol}{l*min}$	v $\frac{pmol}{l*min}$
Blood plasma	low MW	GSH	18.5 ± 3.9	9.9 ± 2.9
	high MW	AAs in albumin	1.6 ± 0.6	1.0 ± 0.4
Lymphocytes		Bases in DNA	1.4 ± 0.4	1.2 ± 0.3
Erhythrocytes	low MW	GSH	16.0 ± 2.3	2.0 ± 0.4
	high MW	AAs in hemoglobin	3.8 ± 0.6	4.8 ± 0.6

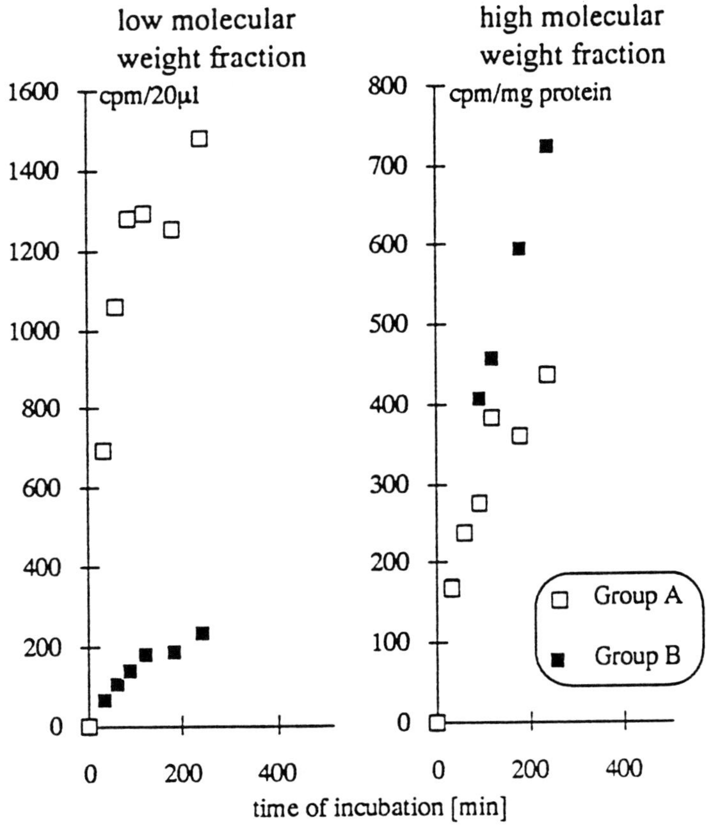

Fig. 1. Increase of radioactivity in low and high molecular weight fraction of the blood plasm of conjugators (group A) and non-conjugators (group B) after incubation with ^{14}C ethylene oxide

those in group B after four hours (Fig. 2): no significant differences between the groups were observed in the high molecular weight fraction. The initial velocity of the binding of radioactivity to the different blood compartments shows in general the same interindividual differences between the two groups as the amounts after four hours incubation. The velocity in the low molecular weight fraction of erythrocytes is eight times higher for group A than that for group B. In the other compartments, the differences are either less distinct or not detectable.

Discussion

The results show that ethylene oxide reacts with lymphocytes and components of high and low molecular weight in blood plasm and erythrocytes. However, marked interindividual differences were observed. These were highly significant in the low molecular weight fraction of erythrocyte cytoplasm. Two distinct groups of individuals (A & B) were found among the subjects investi-

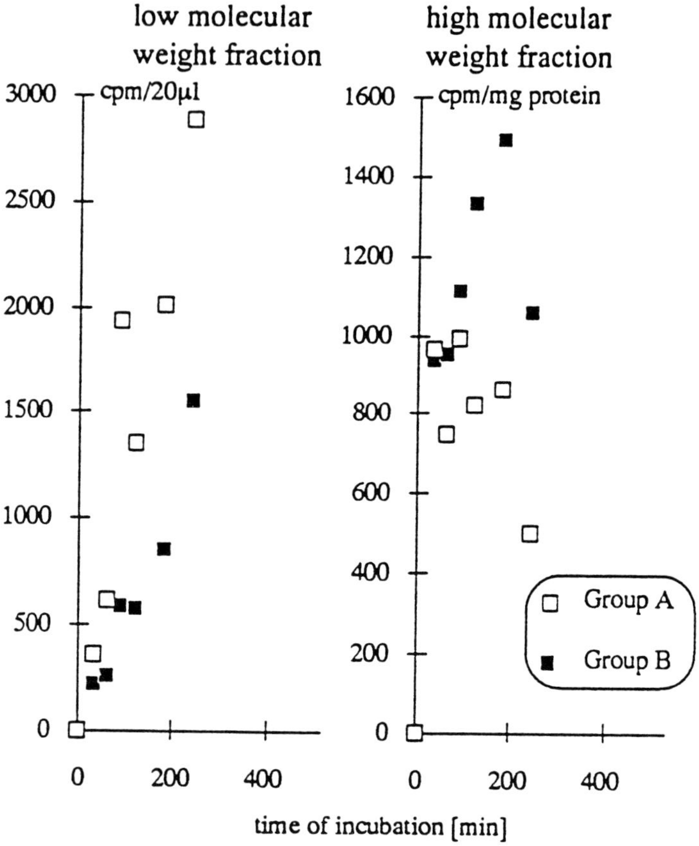

Fig. 2. Increase of radioactivity in low and high molecular weight fraction of the erythrocyte cytoplasm of conjugators (group A) and non-conjugators (group B) after incubation with ^{14}C ethylene oxide

gated. In a clinical study, hemodialysis patients showing immediate-type allergic reactions during treatment, the so-called reactors, had significantly higher levels of ethylene oxide adducts with serum albumin than those without allergic reactions (non-reactors) (Grammer et al 1984). From the data in Figure 2 and Table 1, a different binding of ethylene oxide to serum proteins in the high molecular weight fraction can be assumed between group A and group B. A statistical significance could, however, not be found due to the small number of samples investigated. As mentioned above, only a part of the human population are able to conjugate several methyl halides to glutathione in human erythrocytes (conjugators), the rest are not (non-conjugators) (Peter et al 1989; Hallier et al 1990). It was known beforehand, which of the individuals investigated in this study possessed this ability or not. Interestingly, the persons in group A were all conjugators for methyl halides and the persons in group B were non-conjugators. A common cause for this coincidence, such as enzyme polymorphism, must therefore be considered. Only 60% of the human population express the isozyme glutathione-S-transferase μ, which is responsible for conjugation of epoxides to glutathione in the liver (Warholm et al 1983); a similar phenomenon may be postulated for erythrocytes. Ethylene oxide is also produced endogenously in the liver; the above mentioned enzyme polymorphism may modify the effects of this phenomenon interindividually. Determination of mercapturic acids (Gérin and Tardif 1986) and binding products of ethylene oxide with macromolecules in blood have been discussed as biomonitors. An enzyme polymorphism in the liver and/or in erythrocytes must be taken into account for the production of metabolites or adduct formation. In the study presented here, the differences between group A and group B were found to be greatest in the erythrocyte cytoplasm and least distinct in lymphocytes and blood plasm. The formation of hemoglobin adducts may therefore vary interindividually, so that this parameter could be less suitable for biological monitoring than for example DNA adducts in lymphocytes.

References

Brugnone F, Perbellini L, Faccini GB, Pasini F, Bartolucci GB, DeRosa E (1986) Ethylene oxide exposure. Biological monitoring by analysis of alveolar air and blood. Int Arch Occup Environ Health 58:105–120

Brugnone F, Perbellini L, Faccini GB, Pasini F (1985) Concentration of Ethylene Oxide in the alveolar Air of Occupationally Exposed Workers. Am J Ind Med 8:67–72

DFG (1989) Deutsche Forschungsgemeinschaft. Maximum Concentrations at the Workplace and Biological Tolerance Values for Working Materials 1989. Report No XXV. VCH Verlagsges., Weinheim, FRG

ECETOC (1989) Monograph #13, DNA and Protein Adducts: Evaluation of their use in Exposure Monitoring and Risk Assessment. In: Stringer DA (ed). European Chemial Industry Ecology & Toxicology Centre, Brussels, Belgium

Farmer PB, Bailey E, Gorf SM (1986) Monitoring human exposure to ethylene oxide by the determination of haemoglobin adducts using gas chromatography-mass spectrometry. Carcinogenesis 7:637–640

Gérin M, Tardif R (1986) Urinary N-acetyl-S-hydroxyethyl-L-cysteine in rats as biological indicator of ethylene oxide. Fund Appl Toxicol 7:419–423

Grammer LC, Roberts M, Nicholls AJ, Plattz MM, Patterson R (1984) IgE against ethylene oxide-altered human serum albumin in patients who have had acute dialysis reactions. J Allergy Clin Immunol 74:544–549

Hallier E, Deutschmann S, Reichel C, Bolt HM, Peter H (1990) A comparative investigation of the metabolism of methyl bromide and methyl iodide in human erythrocytes. Intern Arch Occup Environ Health 62:221–225

IARC (1985). Ethylene oxide. In: IARC Monographs on the evaluation of the carcinogenic risk of chemicals to humans. Allyl compound, Aldehydes, Epoxides and Peroxides, Volume 36, pp. 189–226, Lyon, France

Lewalter J, Korallus U (1985) Blood protein conjugates and acetylation of aromatic amines. New findings on biological monitoring. Int Arch Occup Environ Health 56:179–196

Lowry OH, Rosebrough NJ, Farr AL, Randall RJ (1951) Protein measurement with the Folin phenol reagent. J Biol Chem 193:265–275

Peter H, Deutschmann S, Reichel C, Hallier E (1989) Metabolism of methyl chloride by human erythrocytes. Arch Toxicol 63:351–355

Strange RC, Johnson PH, Lawton A (1982) Studies on the variability of glutathione S-transferase from human erythrocytes. Clin Chim Acta 120:251–260

Warholm M, Guthenberg C, Mannervik B (1983) Molecular and Catalytic Properties of Glutathione Transferase from Human Liver: An Enzyme Efficiently Conjugating Epoxides. Biochemistry 22:3610–3617

Recent Developments in Toxicology:
Trends, Methods and Problems
Arch. Toxicol., Suppl. 14, 254–258 (1991)
© Springer-Verlag 1991

Distribution of Methylene Chloride in Human Blood

R. THIER, U. FOEST, S. DEUTSCHMANN, K. R. SCHROEDER, G. WESTPHAL, E. HALLIER, and H. PETER

Institut für Arbeitsphysiologie an der Universität Dortmund, Ardeystraße 67, 4600 Dortmund, FRG

Introduction

Methylene chloride (dichloromethane) is widely used as a solvent for stripping of paint, as industrial cleaning agent, for coating of pills in the pharmaceutical industry, and in the decaffeination of coffee. There is "sufficient evidence for the carcinogenicity" of methylene chloride in animals and "inadequate evidence for its carcinogenity in humans", according to IARC (IARC 1987; CEC 1990).

A glutathione-dependent metabolism has been held responsible for the tumours observed in mice (ECETOC 1989). According to the investigations of Ottenwälder et al (1989), the glutathione-dependent pathway plays a minor role in the metabolism of methylene chloride in mice, below inhalatory concentrations of 500 ppm, similar to other species including humans (ECETOC 1989).

Previous investigations of the metabolism of methylene chloride were based on *in vitro* experiments with liver fractions of mice, rats, hamsters and humans. A possible metabolism in blood constituents was not taken into consideration. In the case of the monohalogenated methanes methyl chloride, methyl bromide and methyl iodide, a glutathione-dependent metabolism in human erythrocytes has been described (Peter et al 1989; Hallier et al 1990).

The aim of the study presented here was to investigate the possibility of a similar pathway of methylene chloride as described for methyl chloride.

Methods

Into syringes containing 1 ml aqueous sodium citrate (3.8%), individual 9 ml blood samples of 10 volunteers were drawn. Previously, it had been investigated whether these persons possessed the ability to conjugate methyl halides

Abbreviations: LMW = low molecular weight, HMW = high molecular weight, GSH = Glutathione in erythrocytes with glutathione (GSH) ("conjugators") or not ("non-conjugators") according to Peter et al (1989) and Hallier et al (1990).

The individual blood samples were transferred into 22 ml head space vials and mixed with radiolabelled ^{14}C-methylene chloride, 5 samples at the high concentration (156 μml, spec. activity: 1.3 MBq/mmol, i.e.: 8.3×10^{-3} MBq per blood sample, diluted with 10 μl nonlabelled methylene chloride) and 5 samples at the low concentration (0.57 μmol, spec. activity: 925 MBq/mmol, i.e.: 1623 MBq per blood sample, resolved in 200 μl bidistilled water). The vials were kept at 37°C and incubated for eight hours under constant rolling. At different time intervals, 0.5 ml aliquots were taken and centrifuged at 400 g for 10 min. The further separation of the plasma and cellular fraction is summarized in Fig. 1. The platelets were removed according to Strange et al (1982).

The protein content of the high molecular weight fraction of the blood plasma was determined by the method of Lowry et al (1951). Radioactivity was measured for 10 min in a liquid scintillation counter.

Results

The distribution of radioactivity between the different blood compartments following incubation of methylene chloride with whole human blood is summarized in Table 1.

In all samples, only a minimum of the radioactivity was found in the low and high molecular weight fractions of erythrocyte cytoplasm. Likewise, no radioactivity was associated with the erythrocyte membranes. A minor radioactivity count (about 5%) was determined in the lymphocytes.

In contrast, a high radioactivity count was bound in blood plasma. Two distinct subpopulations were found in regard to the distribution of radioactivity in plasma, subgroup A and subgroup B. In 5 samples, subgroup A, practically no radioactivity (ca. 1%) was found in both fractions of blood plasma (Fig. 2). In the other 5 samples, subgroup B, the radioactivity increased over time and reached 30% in the low molecular weight fraction and 5% in the high molecular weight fraction after 8 hours.

Discussion

A glutathione-dependent metabolism in erythrocytes was discovered in previous experiments for the mono-halogenated methanes methyl chloride, methyl bromide and methyl iodide. Two distinct subpopulations of "conjugators" and "non-conjugators" were found. In the case of the dihalogenated compound methylene chloride, such a metabolism in erythrocytes is improbable, since only low amounts of radioactivity were associated with the fractions of the red blood cells.

In contrast, a major radioactivity count was found in the low molecular weight fraction of the blood plasma of a subgroup of the persons investigated, group B. This implies that albumin (molecular weight: 67,000) is not a primary

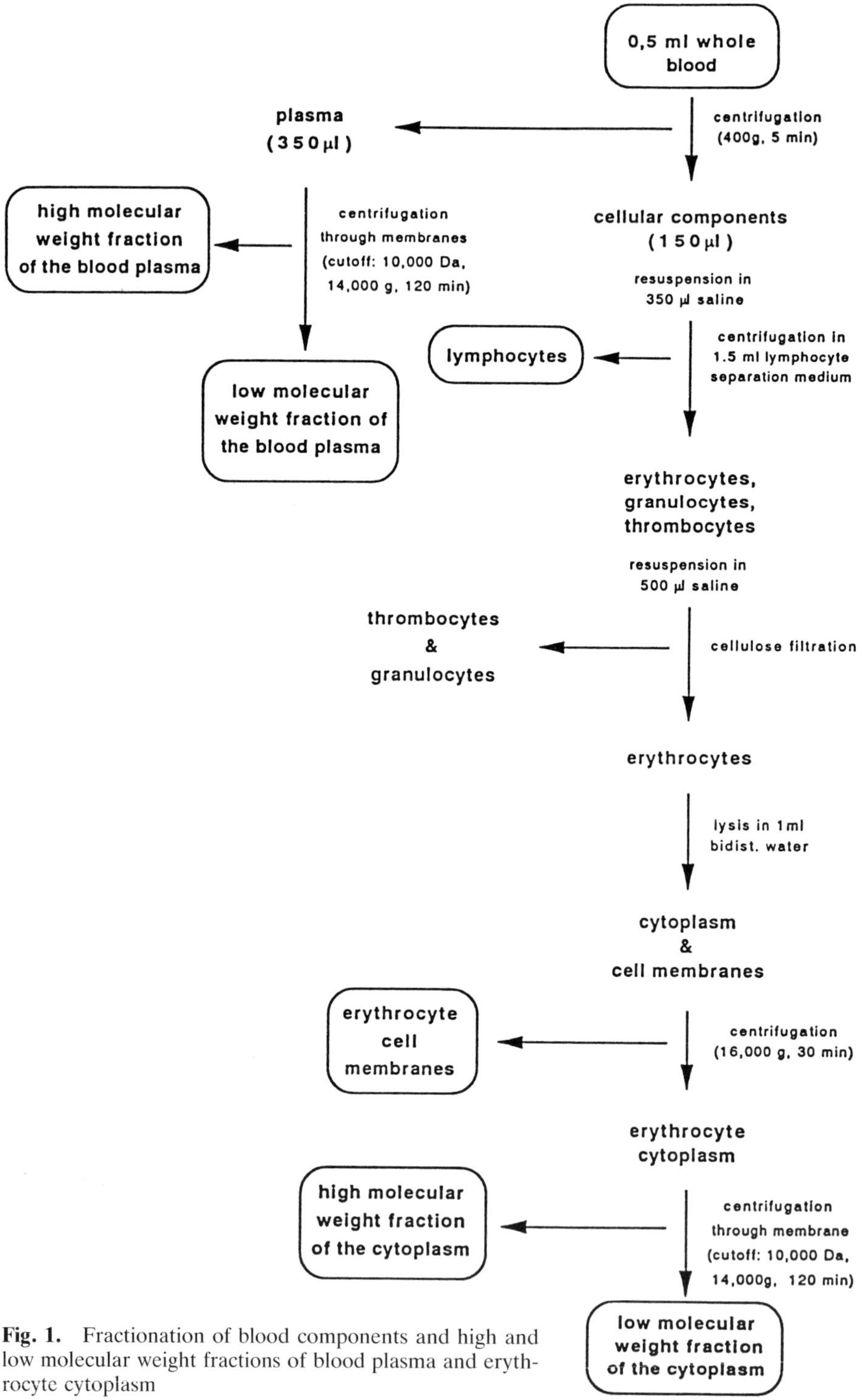

Fig. 1. Fractionation of blood components and high and low molecular weight fractions of blood plasma and erythrocyte cytoplasm

Table 1. Increase of radioactivity in the different blood compartments after incubation with ^{14}C-methylene chloride (% of target dose)

		Group A			Group B		
		2 h	4 h	8 h	2 h	4 h	8 h
Plasma	High molecular weight fraction	0.25	0.27	0.43	1.34	1.98	4.76
	Low molecular weight fraction	1.09	0.87	1.19	9.06	14.29	30.28
Erythrocytes	Membrane	0.02	0.03	0.02	0.02	0.04	0.06
	High molecular weight fraction	0.01	0.00	0.00	0.08	0.07	0.21
	Low molecular weight fraction	0.08	0.07	0.1	0.61	0.41	0.97
Lymphocytes		2.86	2.39	3.45	3.87	4.28	5.78

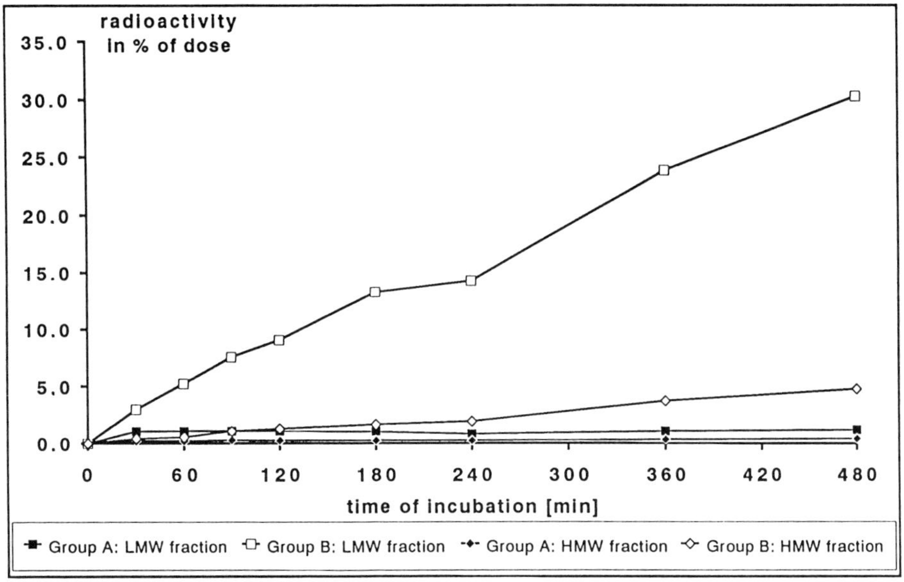

Fig. 2. Increase of radioactivity in low and high molecular weight fraction of the blood plasma after incubation with ^{14}C-methylene chloride (LMW = low molecular weight, HMW = high molecular weight)

target of the radiolabelled substance. It is likely that a metabolic transformation of methylene chloride had taken place in all plasma samples of subgroup B.

In the samples of subgroup A, only a minor radioactivity count was associated with both the high and the low molecular weight fraction of blood plasma, which might indicate interindividual differences in the transformation of methylene chloride in human blood.

According to previous data, only a part of the human population was able to conjugate monohalogenated methanes with glutathione in erythrocytes ("conjugators"), the other part was not ("non-conjugators"). The individuals investigated in this study had also been characterized in this respect. Interestingly, all the persons of subgroup B were "conjugators" for monohalogenated methanes and all persons in subgroup A were "non-conjugators" in this respect. As a reason for this coincidence, enzyme polymorphism should be considered.

Such an enzyme polymorphism with respect to metabolism of methylene chloride in man might have an impact on human risk evaluation.

References

CEC, Commission of the European Communities, Industrial Health Service (1990) The toxicology of chemicals: 1 Carcinogenity 2:67–74

ECETOC (1989) Methylene chloride (dichloromethane): An overview of experimental work investigating species, differences in carcinogenicity and their relevance to man. Technical Report No. 34

Hallier E, Deutschmann S, Reichel C, Bolt HM, Peter H (1990) A comparative investigation of the metabolism of methyl bromide and methyl iodide in human erythrocytes. Intern Arch Occup Environ Health 62:221–225

IARC, International Agency for Research on Cancer (1987) IARC Monographs on the evaluation of carcinogenic risks to humans, Suppl. 7, Overall evaluations of carcinogenicity: an updating of IARC Monographs. Volumes 1 to 42 Lyon, 194–195

Lowry OH, Rosenbrough NJ, Farr AL, Randall RJ (1951) Protein measurement with the Folin phenol reagent. J Biol Chem 193:265–275

Ottenwälder H, Jäger R, Thier R, Bolt HM (1989) Influence of cytochrome P-450 inhibitors on the inhalative uptake of methyl chloride and methylene chloride in male B6C3F1 mice. Arch Toxicol Suppl 13:258–261

Peter H, Deutschmann S, Reichel C, Hallier E (1989) Metabolism of methyl chloride by human erythrocytes. Arch Toxicol 63:351–355

Strange RC, Johnson PH, Lawton A (1982) Studies on the variability of glutathione S-transferase from human erythrocytes. Clin Chim Acta 120:251–260

Recent Developments in Toxicology:
Trends, Methods and Problems
Arch. Toxicol., Suppl. 14, 259–262 (1991)
© Springer-Verlag 1991

Acute Experimental Exposures to Acetone and Ethyl Acetate

R. R. Vangala, M. Blaszkewicz, H. M. Bolt, K. Golka, E. Kiesswetter, and A. Seeber

Institut für Arbeitsphysiologie an der Universität Dortmund, Abteilung für Toxikologie und Arbeitsmedizin, Ardeystraße 67, 4600 Dortmund 1, FRG

Introduction

Among the volatile organic solvents, acetone (AC) 54% and ethyl acetate (EA) 39% were found as major components after toluene 69% in 101 glues analyzed in Germany (Auffarth and Häger 1989). Pulmonary absorption, biotransformation and excretion of acetone and ethyl acetate, as single solvents, have been thoroughly studied in man (Fernández and Droz 1974; Nomiyama and Nomiyama 1974; Wigaeus et al 1981). As there is no evidence available on combined exposures, studies of acute exposures to solvents of AC and EA in combination at Maximum Allowable Concentration (MAC) were carried out on humans in an exposure chamber to monitor the neurobehavioral effects, uptake and excretion during and after the exposure.

Methods

The exposure of 16 healthy male volunteers was carried out in a closed room of 28 m^3 in volume. The exact amounts of acetone (Merck Germany, PA) (AC) and ethyl acetate (Merck Germany, PA) (EA) alone and a mixture of both were evaporated with a constantly maintained nitrogen stream in thermostatic special glass vessels and injected into the main stream of fresh filtered atmospheric air at constant temperature and humidity, flowing into the chamber continuously at a rate of 200 m^3/h (3 m/sec) with an air exchange of 7.5 times per hour. Solvent vapour concentrations (AC: 990 ppm; EA: 402 ppm alone and in mixture AC/EA: 497/205 ppm) in the room air were monitored continuously with the GC. The alveolar air and urine samples were collected before, during and after exposure and analyzed with GC and an automatic headspace sampling device.

Results

The EA alveolar air concentrations of 16% and 13% (mixture) of the exposure concentrations were obtained after 30 min whereas AC concentrations of 22% and 19% (mixture) were reached after 120 and 60 min. The relative EA vapour concentrations in alveolar air dropped rapidly to 2.0% and 0.5% (mixture) within 30 to 60 minutes after the cessation of exposure. On the other hand the reduction of AC to 8% and 7% (mixture) was obtained after 140 minutes which is in good agreement with the results of Wigaeus et al (1981), where they found 8% of AC in alveolar air with a half-time of 4 h at an exposure concentration of 540 ppm of AC. Ethanol as a metabolite was measured in alveolar air during the 4 h exposure at concentrations of 4.2 ppm and 3.6 ppm when the subjects were exposed to EA alone (402 ppm) and in combination (205 ppm) with AC (497 ppm) respectively (Fig. 1). The cumulative acetone excretion in urine 89.5 mg and 47.4 mg (Fig. 2) was first reached in 18 h after the cessation of exposure whereas ethyl acetate excretion 1.75 mg and 0.75 mg (Fig. 3) was reached within 2 h after exposure cessation. No further EA was excreted in the urine by the subjects.

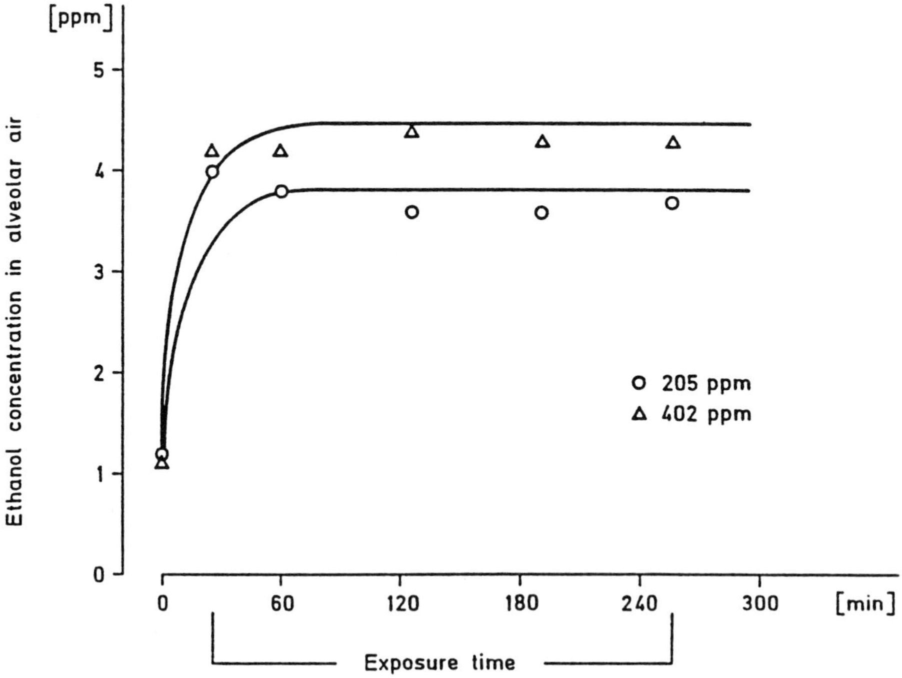

Fig. 1. Ethanol in alveolar air during the 4 h exposure of 16 subjects to ethyl acetate alone 402 ppm (△) and in combination 205 ppm (○) with acetone 497 ppm

Discussion

At a constant concentration of AC in the room air, a plateau of maximal alveolar air concentration was reached after 2 hrs. Accordingly, the concentration of AC in alveolar air is decreased very slowly at the end of the exposure. On the other hand, because of the quick transformation and metabolism of EA in the organism, almost the same amounts of ethanol (4.2/3.6 ppm) were released through the lung without any influence of the exposure nature (alone or mixture) and of exposure concentrations of EA (Fig. 1). It appears that the maximum amount of absorbed ethyl acetate was hydrolysed during the exposure. Further, the constant amount of ethanol in alveolar air reveals that there is an equilibrium between the formation of ethanol through hydrolysis of EA and its oxidative metabolism. The steady state of ethanol during the exposure is clearly seen as the metabolism of ethanol reached the saturation point very quickly and on the other hand the enzyme might be showing a higher activity. The steady increase of cumulative AC values of urine (Fig. 2) 18 h after the exposure cessation shows the slow excretion of AC in the urine. On the contrary, the value of EA in urine (Fig. 3) shows that EA is totally excreted within 2 h after the cessation of exposure. As there is no steady state any more after 4 h exposure, the reduction in concentration is first determined by the

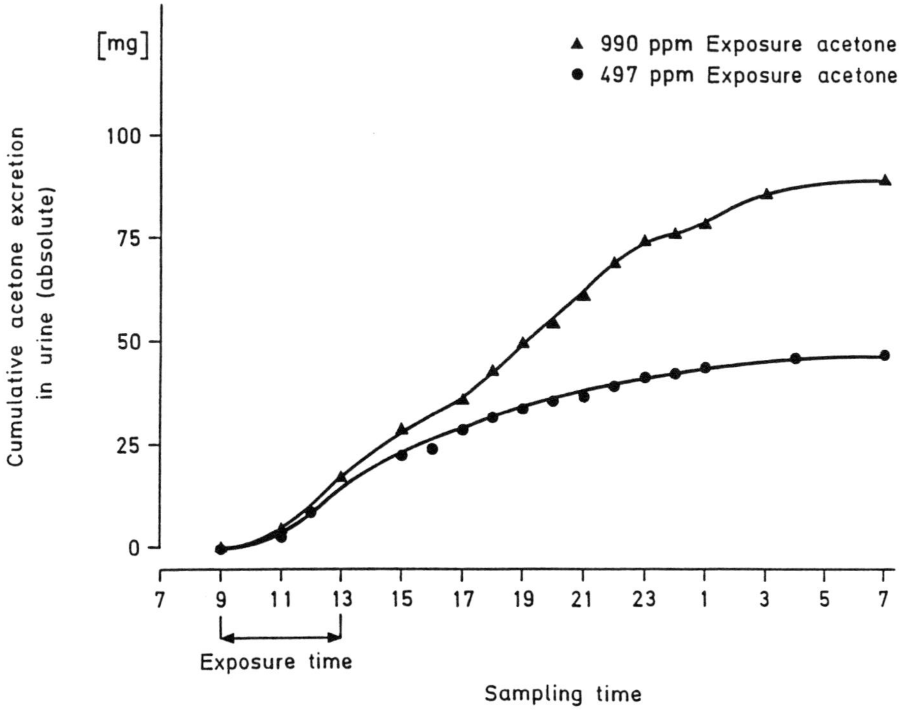

Fig. 2. Cumulative acetone excretion in total urine per person in 24 h during and after the exposure to 990 ppm acetone alone (▲) and in combination 497 ppm (●) with ethyl acetate 205 ppm

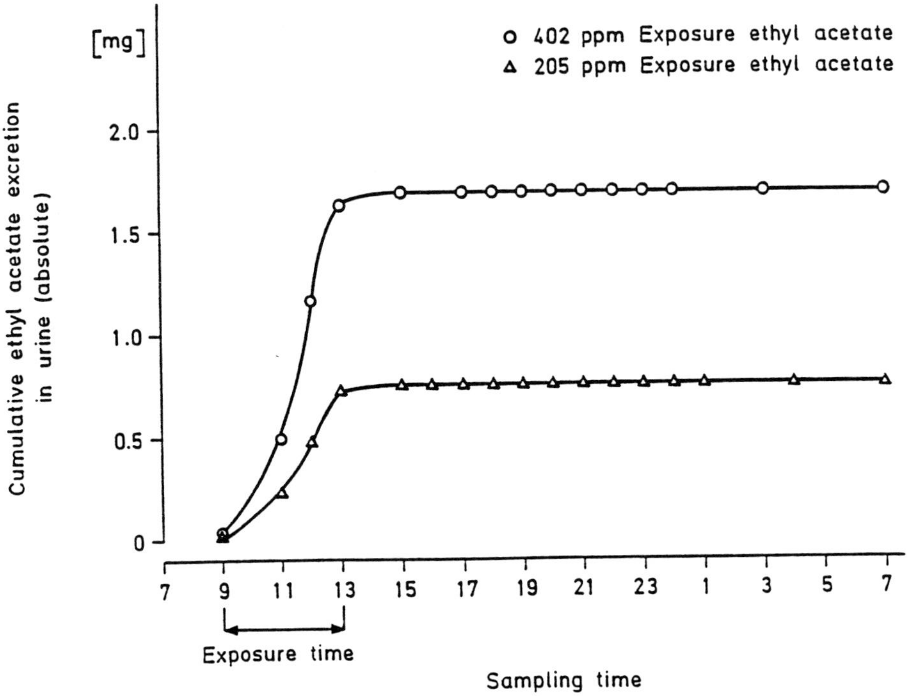

Fig. 3. Cumulative ethyl acetate excretion in total urine per person in 24 h during and after the exposure to 402 ppm ethyl acetate alone (○) and in combination 205 ppm (△) with acetone 497 ppm

distribution coefficients of AC and EA which markedly differ from each other. In comparison with AC, the rapid exhalation in the lung and the excretion of EA in urine can be explained by the low solubility of EA in water.

The possible combination effect observed by subjective symptoms should be confirmed by further pharmacokinetic studies.

Acknowledgement. The authors wish to thank Ms. I. Glaeser, B. Straka-Emden and Mr. Ch. Lippmann for their technical assistance.

References

Auffarth J, Häger J (1989) Exposition beim Umgang mit lösemittelhaltigen Zubereitungen. Herstellung von Lösemittelklebstoffen. Schriftenreihe der Bundesanstalt für Arbeitsschutz, Dortmund, Gefährliche Arbeitsstoffe – GA 30, Wirtschaftsverlag NW, Bremerhaven

Fernández J, Droz P (1974) Absorption et élimination pulmonaire de l'acétate d'éthyle. Étude expérimentale sur des sujets humains. Arch Mal Prof 35:953–961

Nomiyama K, Nomiyama H (1974) Respiratory retention, uptake and excretion of organic solvents in man. Int Arch Arbeitsmed 32:75–83

Wigaeus E, Holm S, Åstrand I (1981) Exposure to acetone. Uptake and elimination in man. Scand J Work Environ Health 7:84–94

Recent Developments in Toxicology:
Trends, Methods and Problems
Arch. Toxicol., Suppl. 14, 263–265 (1991)
© Springer-Verlag 1991

Toxic Effects of Carbon Tetrachloride During Short and Long Term Ethanol Intake in Rats

E. KNIEPERT[1], A. SIEGEMUND[2], M. ROSENKRANZ[3], and V. GÖRISCH[1]

[1] Institutes of Pharmacology and Toxicology,
[2] Clinical Chemistry and Laboratory Diagnostic, and
[3] Pathology, Department of Medicine, University Leipzig, Härtelstraße 16–18, 7010 Leipzig, FRG

Introduction

In continuing the investigations into the influence of ethanol pretreatment on the toxic effects of a single CCl_4 dose (Kniepert et al 1990), changes following the repeated simultaneous administration of ethanol and CCl_4 were studied.

Materials and Methods

Adult male Wistar rats (Prob:WIST, LVS Probstheida, Leipzig) weighing 150–200 g at the beginning of the treatment were kept on an ad libitum diet of commercial chow (pellets) and water or 10% (v/v) ethanol – pure – (VEB Likörfabrik Zahna) in a temperature controlled room (20 ± 2°C, relative humidity 60–80%).

The ethanol consumption, together with drinking water, extended for periods of 1, 10 (common water control group), and 52 weeks. After this pretreatment the animals were given ten injections (the 1st and 10th intraperitoneally, 2nd to 9th subcutaneously) of 0.16 g/kg body weight carbon tetrachloride – pure – (VEB Laborchemie Apolda) dissolved in sun flower oil every fortnight, while ethanol was given until the experiment was finished. One week (168 hours) before and 24, 48 hours following the last CCl_4 treatment, blood samples were taken by puncture of the retroorbital venous plexus. Serum enzyme activities, food intake, fluid intake and body weight were determined as described before (Kniepert et al 1990). From five rats of each group liver biopsies were taken 48 hours after the last CCl_4 treatment for microscopy (Kniepert and Seffner 1989).

Abbreviations: W = water; CCl_4 = carbon tetrachloride; ALAT = alanine aminotransferase; ASAT = aspartate aminotransferase; LDH = lactate dehydrogenase; AP = alkaline phosphatase

Table 1. Relative changes (%) of serum activities of ALAT, ASAT, LDH and AP, decrease in body weight (x ± s) and lethality (%) of rats – which received 10% (v/v) ethanol ad libitum for periods of 19, 28 and 70 weeks and ten injections of CCl_4 (0.16 g/kg body weight) every 14 days in the last 18 weeks – one week before and 24, 48 hours following the last CCl_4 dose

Time (hours before or after CCl_4)	Ethanol consumption (weeks)	Relative changes (%) of serum activities in comparison with W + CCl_4 (= 100%)				Decrease in body weight	Lethality
		ALAT	ASAT	LDH	AP	(g)	(%)
	1 + 18	192*	308**	410**	104	−20.8 ± 14.3	30
−168	10 + 18	147	215*	337*	95	−14.4 ± 10.2	20
	52 + 18	68**	63*	186	82	−11.2 ± 8.5	8
	1 + 18	717**	887**	599**	136*	19.0 ± 2.2**	40
24	10 + 18	99	89	69	110	5.0 ± 10.0	20
	52 + 18	161*	150	135	109	6.2 ± 4.8	17
	1 + 18	262*	230*	166	131*	28.0 ± 7.6**	50
48	10 + 18	51	94	326**	100	9.4 ± 9.4	20
	52 + 18	173	264**	308**	108	13.8 ± 4.8*	17

* $p \leq 0.05$ significant difference from animals receiving water + CCl_4
** $p \leq 0.01$

The statistical significance of differences in enzyme activities (geometric mean values) and body weight differences (arithmetic mean values) were calculated by Student's t-test or U-test of Wilcoxon, Mann and Whitney.

Results

Increase of serum enzyme activities, change in body weight and lethalities were higher in rats treated with 10% (v/v) ethanol for 1 + 18, than for 10+18, or for 52+18 weeks one week (after the 9th) before the 10th CCl_4 injection (see Table 1). The greatest alterations following the last CCl_4 dose (e.g. the most extensive fibrosis, too) were also seen in animals with 1 + 18 weeks of ethanol consumption. Many serum enzyme activities were decreased following 10 + 18 weeks of ethanol intake + 10 CCl_4 injections in comparison with water +CCl_4. However, the highest overall degree of histological damage was found in animals got 10 + 18 weeks of ethanol intake + 10 CCl_4 doses. Some elevated serum enzyme activities and decrease in body weight also were seen following 52 + 18 weeks of ethanol intake + 10 CCl_4 doses.

Discussion

In rats with short-term ethanol intake the CCl_4-toxicity was highest after ten injections every fortnight, similar to one CCl_4 injection. The decisive factor for the increase in toxicity was the overall duration of ethanol intake and not the pretreatment time as could be supposed after a single administration of CCl_4 (Kniepert et al 1990). Obviously after 1 to 20 weeks ethanol intake mechanisms of compensation or regeneration (see Bell et al 1988) will occur which may reduce the toxic effects of CCl_4 such as an increase in serum enzyme activities, a decrease in body weight, casualties. However the changes of the liver which can be histologically determined seem to be less evident for this hypothesis, although they are often considered more suitable for identifying chronic lesions (Zimmerman 1983).

References

Bell AN, Young RA, Lockard VG, Mehendale HM (1988) Protection of chlordecone-potentiated carbon tetrachloride hepatotoxicity and lethality by partial hepatectomy. Arch Toxicol 61:392–405

Kniepert E, Seffner W (1989) Histological and histometrical changes of liver damage in rats after treatment with ethanol and carbon tetrachloride. Exp Pathol 36:185–188

Kniepert E, Siegemund A, Görisch V (1990) Influence of ethanol pretreatment of differing duration on toxic effects of carbon tetrachloride in rats. Biomed Biochim Acta 49:1097–1102

Zimmerman HJ (1983) Hepatotoxic Evaluation Methodology. In: Diagnostic Procedures in the Evaluation of Hepatic Diseases. Ed. Beker S, New York, Alan R Liss, 159–194

Recent Developments in Toxicology:
Trends, Methods and Problems
Arch. Toxicol., Suppl. 14, 266–268 (1991)
© Springer-Verlag 1991

Effects of Pyridinium,1-{[(4-Carbamoyl-Pyridinio) Methoxy] Methyl}-2-(Hydroxyimino-Methyl) Dichloride Monohydrate (HI6) and Atropine on the Circulation and Respiration of Anesthetized Guinea Pigs

S. DE LA MOTTE, and L. SZINICZ

Institut für Pharmakologie und Toxikologie, Akademie des Sanitäts- und Gesundheitswesens der Bundeswehr, Ingolstädter Landstraße 98, D-8046 Garching-Hochbrück, FRG

Introduction

Atropine (AT) and HI6 are effective antidotes in organophosphate (OP) poisoning. However, AT may enhance oxime-toxicity (Boskovic et al 1984), and a small overdose of HI6 may impair circulation seriously (Lipp and Dola 1980).

Materials and Methods

In each dose group four female guinea pigs (Pirbright white), 270–320 g, from Lippische Versuchstierzucht (Extertal, FRG) were used. Urethane for anesthesia (1.7 g/kg ip) and atropine sulfate were purchased from Sigma (Munich, FRG). HI6 was received from JG Clement (DRES, Ralston, Canada). AT and HI6 were checked for purity by HPLC. Cardio-pulmonary parameters were measured as described by Molz and Szinicz (1988). AT (15 μmol/kg) and HI6 (30 or 100 μmol/kg), alone or in combination, were injected intravenously (iv) in 0.9% saline (volume 0.2% of body weight).

Results

AT, 15 μmol/kg, increased heart rate (HR). HI6, 100 μmol/kg, reduced HR to 90% of pretreatment values. In combination with AT, this reduction was lessened. Each drug decreased blood pressure (BP), AT to 70%, HI6 dose-depedently to 80% (30 μmol/kg), or to 65% (100 μmol/kg). The addition of AT to HI6, 30 μmol/kg, enhanced this decrease to 65%, but did not modify the effect of HI6, 100 μmol/kg. The effects on circulation reversed 20 min after

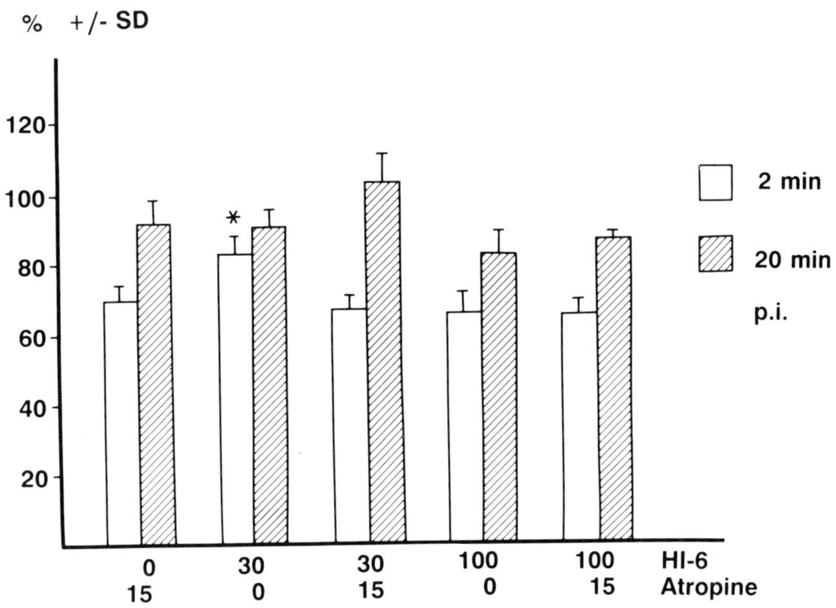

Fig. 1. Effects of μmol dosages of HI6 and atropine on blood pressure of anesthetized guinea pigs 2 min and 20 min after iv injection (PI), shown as percent of pretreatment values. n = 4. Values in dose group with HI6, 30 μmol/kg, are significantly greater ($p < 0.01$, U-test) than in other dose groups

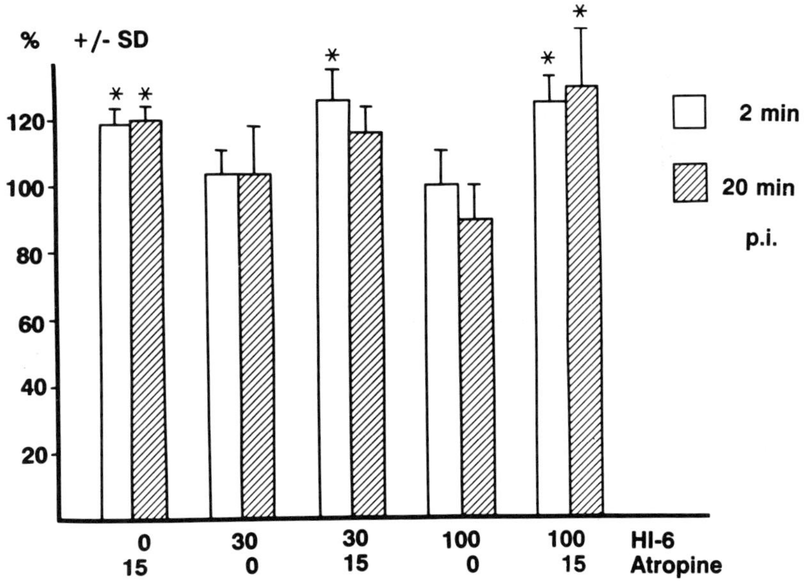

Fig. 2. Effects of HI6 and atropine on respiratory rate (see Fig. 1). n = 4. Values in dose groups with atropine are significantly greater ($p < 0.01$, U-test) than in groups with HI6 alone

drug injection. The median BP (Fig. 1) reflected systolic and diastolic. BP amplitude was not influenced.

Respiration was not altered by HI6. AT stimulated respiratory rate (RR) to 120%. The combination of AT with both doses of HI6 showed the same effect as AT alone (Fig. 2). The increase, which was accompanied by a rise of the respiratory minute volume, lasted longer than the effects on circulation and did not reverse after 20 minutes. No effect on tidal volume, compliance or airway resistance was observed.

Discussion

In preliminary experiments AT, 1.5 μmol/kg, increases HR considerably and does not influence BP, but 72 μmol/kg decreases both HR and BP. In guinea pigs 15 μmol/kg of AT may produce toxic effects different from the muscarine-blocking properties (Green et al 1977).

Circulation may be affected by HI6, while respiration seems intact (Klimmek and Eyer, 1986). HI6 and AT obviously act by different mechanisms. Interestingly, the combination only produces minor additive effects on BP. Moreover, adverse effects of HI6 on HR might be improved by AT. Perhaps the findings of Lipp and Dola (1980) are due to pyridostigmine.

The increase of RR by AT may be explained by central stimulation. Although HI6 penetrates the blood-brain-barrier (Klimmek and Eyer, 1986), it does not seem to alter the central effects of AT. This may be important in therapy of OP poisoning.

References

Boskovic B, Kovacevic V, Jovanovic D (1984) PAM-2 Cl, HI-6, and HGG-12 in Soman and Tabun Poisoning. Fundam Appl Toxicol 4:S106–S115

Green DM, Muir AW, Stratton JA, Inch TD (1977) Dual mechanism of the antidotal action of atropine-like drugs in poisoning by organophosphorus anticholinesterases. J Pharm Pharmacol 29:62–64

Klimmek R, Eyer P (1986) Pharmacokinetics and pharmacodynamics of the oxime HI6 in dogs. Arch Toxicol 59:272–278

Lipp J, Dola T (1980) Comparison of the Efficacy of HS-6 versus HI-6 when Combined with Atropine, Pyridostigmine and Clinazepam for Soman Poisoning in the Monkey. Arch Int Pharmacodyn 246:138–148

Molz K-H, Szinicz L (1988) Effect of Pyridinium, 1-{[(3-benzoylpyridinio)-methoxyl]-methyl}-2-[(hydroxyimino)methyl]-dichloride (HGG 12) on Various Respiratory and Circulatory Parameters of Anesthetized Guinea Pigs. In: The Target Organ and the Toxic Process. Arch Toxicol, Suppl 12:428–431

Recent Developments in Toxicology:
Trends, Methods and Problems
Arch. Toxicol., Suppl. 14, 269–271 (1991)
© Springer-Verlag 1991

The Mechanism of the Anaesthetic Action of Organic Solvents Studied with Different Membrane Models in Vitro

H. Tähti, and A. Valaja

Department of Biomedical Sciences, University of Tampere, P.O. Box 607, Tampere, Finland

Introduction

The structural organization of membrane proteins and their interaction with lipids renders them sensitive to perturbations of organic solvents. Some membrane proteins are inhibited up to 50 per cent at the anaesthetic concentration of many inhalation anaesthetics, aliphatic and aromatic hydrocarbons, alcohols, ketones, esters and alkanes (Franks and Lieb 1982). Hydrophilic pockets of proteins may bind anaesthetics with high selectivity and cause conformational changes. Changes of the allosteric properties of membrane proteins may provide a possible mechanism of anaesthesia. The aim of the present study was to evaluate whether peripheral erythrocyte and platelet membranes as well as cerebral synaptosome membranes can be used in determinations of the anaesthetic potency of organic solvents.

Materials and Methods

Human erythrocyte and platelet membranes and rat (males, Sprague Dawley, Department of Biomedical Sciences, University of Tampere) synaptosomal membranes were isolated (Korpela and Tähti 1988) and exposed to various concentrations (3 mM, 6 mM, 9 mM, 20 mM, 30 mM) of benzene, toluene, o-xylene and styrene (Chemicals, p.a. Merck) in the incubation mixtures *in vitro* by a method described previously (Korpela and Tähti 1986, 1987). After solvent exposure (30 minutes), the total ATPase (adenosine triphosphatase) activities were determined (Lowry and Lopez 1946). All measurements were made as triplicates. The effects of the solvent on platelet, erythrocyte and synaptosome membrane ATPase activity were compared. In statistical analysis, Student's t-test was used.

Abbreviations: ATPase = adenosine triphosphatase

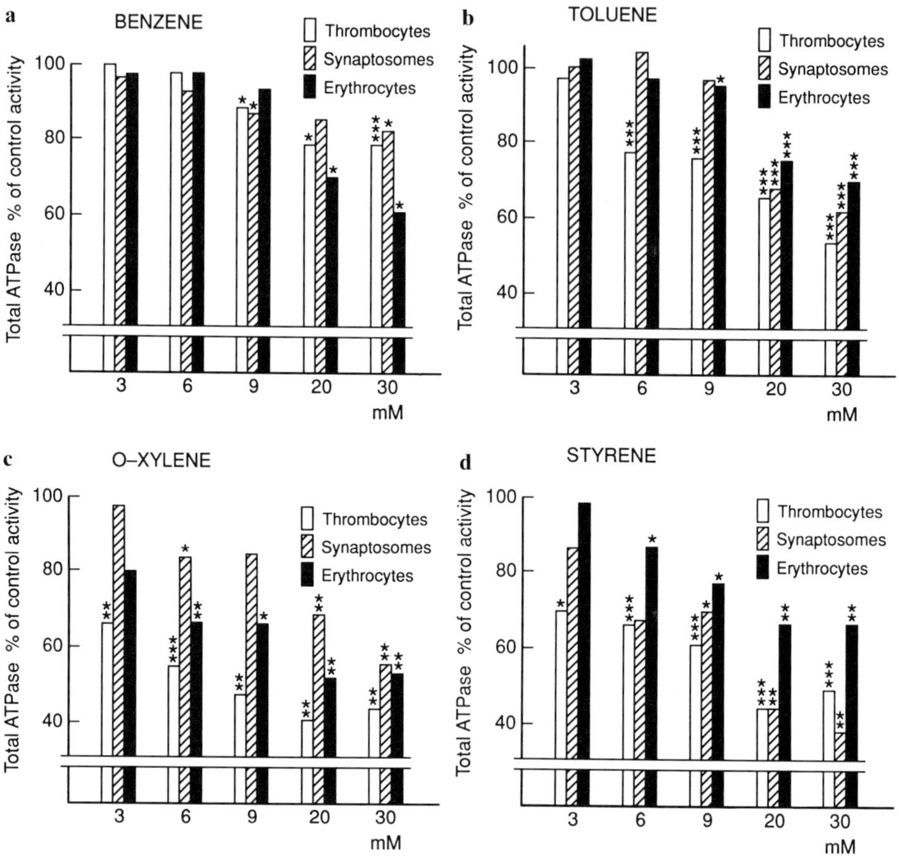

Fig. 1 a–d. The total ATPase activity in thrombocyte, synaptosome and erythrocyte membranes after various benzene (**a**), toluene (**b**), o-xylene (**c**), and styrene (**d**) treatments *in vitro*

Results

The enzyme inhibition varied after the various solvent treatments in increasing order as follows: benzene, toluene, styrene, o-xylene (Figs. 1 a–c). Toluene, styrene and o-xylene had a more intensive inhibiting effect on platelet ATPase than on erythrocyte and synaptosomal ATPase. Benzene inhibited more erythrocyte membrane ATPase than platelet and synaptosomal membrane ATPase. Xylene and styrene were more potent enzyme inhibitors.

Discussion

Organic solvents exert an anaesthetic action on the central nervous system. The mechanism of this action may be based on the changes in the cell membrane structural proteins, found in the present study and in previous studies (Korpela and Tähti 1986). Peripheral cell membranes are also targets for the anaesthetic effect exerted by solvents. Therefore, these membrane models may serve as test models when the occupational exposure to solvents is evaluated. The action of organic solvents may occur in lipid protein interfaces in the membranes. The differences in the enzyme-inhibiting potency of the solvents studied (benzene, toluene, o-xylene and styrene) may be caused by their different lipid solubilities and different molecular structure. In membrane models there are differences, because the ratio of lipid to protein and the composition of lipids in various types of membranes can vary. However, the effects of solvents in peripheral models are quite similar to those in a neural membrane *in vitro* measured as changes of ATPase activity.

Acknowledgements. This study was financially supported by the Finnish Ministry of Agriculture and Forestry.

References

Franks NP, Lieb WR (1982) Molecular mechanism of general anaesthesia. Nature, London 300:487–493

Korpela M, Tähti H (1986) The effect of selected organic solvents on intact human red cell membrane acetylcholinesterase *in vitro*. Toxicol Appl Pharmacol 85:257–262

Korpela M, Tähti H (1987) Effects of industrial solvents on human erythrocyte membrane adenosine triphosphatase activities *in vitro*. Scand J Work Environ Health 13:513–517

Korpela M, Tähti H (1988) The effect of *in vitro* and *in vivo* toluene exposure on rat erythrocyte and synaptosome membrane integral enzymes. Pharmacol Toxicol 63:30–32

Lowry OH, Lopez JA (1946) The determination of inorganic phosphate in the presence of labile phosphate esters. J Biol Chem 198:421–428

Recent Developments in Toxicology:
Trends, Methods and Problems
Arch. Toxicol., Suppl. 14, 272–275 (1991)
© Springer-Verlag 1991

An *in Vivo* Method to Assess the Local Tolerability of Intranasally Administered Drugs

D. Frattola, C. Sorini, V. Aliverti, and M. Losa

ISF-Laboratories for Biomedical Research, Via Leonardo da Vinci 1,
I-20090 Trezzano s/N, Milan, Italy

Introduction

The intranasal administration represents a useful alternative route to the oral one, particularly for drugs such as peptides which undergo to degradation processes in the gastro-intestinal tract. In order to set up an intranasal formulation it is essential to evaluate the local tolerability both of the active ingredient and of different excipients to be used for the pharmaceutical form.

Usually these studies are based on post-mortem observations or on *in vitro* (Batts et al 1989) and *ex-vivo* (Levrier et al 1989) models of cilia functionality. On the contrary, little information is available about the cellular reactions of the mucosa after intranasal drug administration using *in vivo* models.

The purpose of this study is to evaluate directly *in vivo,* the response of the nasal mucosa to the topically applied solution, using a simple method which allows in the rabbit repeated and non-traumatic mechanical cellular sampling under standardized conditions.

Materials and Methods

Cellular samples from the nasal mucosa of New Zealand White rabbits were taken and processed as described below. A total of 24 rabbits weighing 2.5–3.0 kg were used. Eight untreated animals were used as normal controls; eight animals treated intranasally with phosphate buffer saline (PBS) were used as negative standards, eight animals treated intranasally with 5% sodium taurocholate solution (Na-T, mucosa damaging agent) were used as positive standards.

The solutions were given by intranasal route using a suitable pump. Each animal received seven times in an hour two 200 microlitres puff for each nostril.

Seventy-two hours before and 4, 24, and 72 h after administration of the test compound, cell samples were collected from the nasal mucosa of unanaesthetized rabbits. A small nylon brush (0.4 × 1 cm), supplied by Oral-B, Cooper Care Inc., Palo Alto, California was introduced into each nostril and then immediately removed with a slight rotating movement. The brush was immediately shaken in a small plastic 1.5 ml tube containing 1 ml Hanks' balanced salt solution plus 0.1% bovine serum albumin (HBSS + BSA).

A 100 microliters aliquot of cell suspension was stained with Trypan Blue and the total number of cells (dead and alive) was calculated using a Bürker chamber.

The remaining suspension was centrifuged for 8 min at 1500 rpm and the pellet resuspended in 20 microlitres of HBSS (Gibco BRL) + BSA (Sigma). The resulting suspension was deposited on a glass slide and fixed with methyl alcohol. Differential count (leukocytes and epithelial cells) was performed on 200 cells of Giemsa stained preparations.

The split-plot design was employed for statistical analysis.

In addition, at each time interval, a careful macroscopical inspection of the nasal cavities was carried out.

Results

The macroscopical appearance of the nasal mucosa in untreated rabbits as well as in those treated with PBS was found to be normal throughout the study period. The animals treated with sodium taurocholate evidenced nasal bleeding from the first administration onward.

The material obtained using the brush technique consisted either of fully dispersed cells or, occasionally, included small aggregates of epithelial cells. The cells are generally well preserved with excellent structural details (Figs. 1, 2).

Table 1. Epithelial cell and leukocyte counts. Mean group ± S.E. values (expressed as number of cells, x 1000). Number of animals/group: 8

Time (h)		Group 1 (untreated)	Group 2 (PBS)	Group 3 (sodium taurocholate)
−72	Epithelial cells	609 ± 180	1078 ± 179	1512 ± 308
	Leukocytes	12 ± 2	22 ± 8	172 ± 117
4	Epithelial cells	424 ± 47	1677 ± 286[a]	1354 ± 214[a]
	Leukocytes	16 ± 3	1127 ± 282	9359 ± 1453[a,b]
24	Epithelial cells	982 ± 174	1921 ± 265	3869 ± 973[a,b]
	Leukocytes	178 ± 155	1402 ± 241	29928 ± 4808[a,b]
72	Epithelial cells	453 ± 68	1899 ± 266[a]	2213 ± 331[a]
	Leukocytes	67 ± 33	1057 ± 231	11382 ± 2519[a,b]

a = $p<0.01$, among treatments; b = $p<0.01$, among times

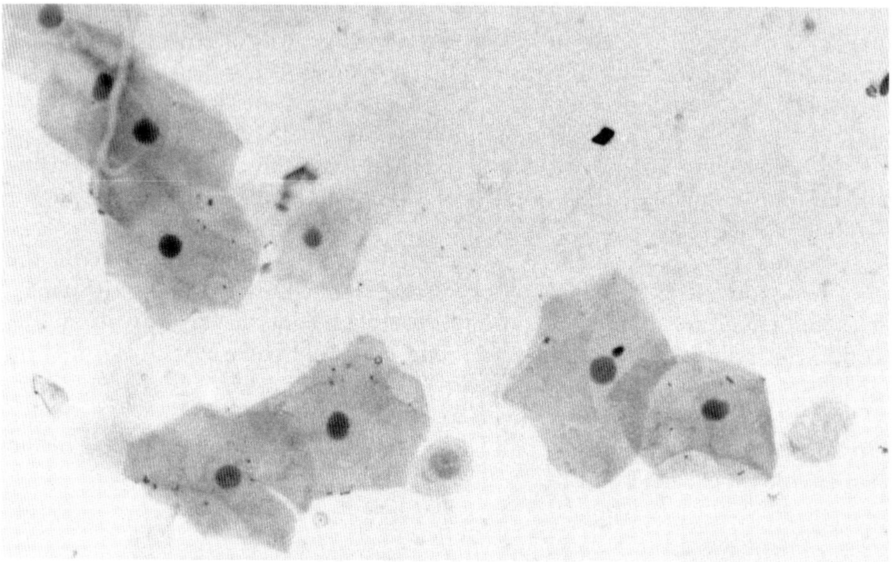

Fig. 1. A representative cell sample from PBS-treated animal showing only shedded epithelial cells. Giemsa, 300x

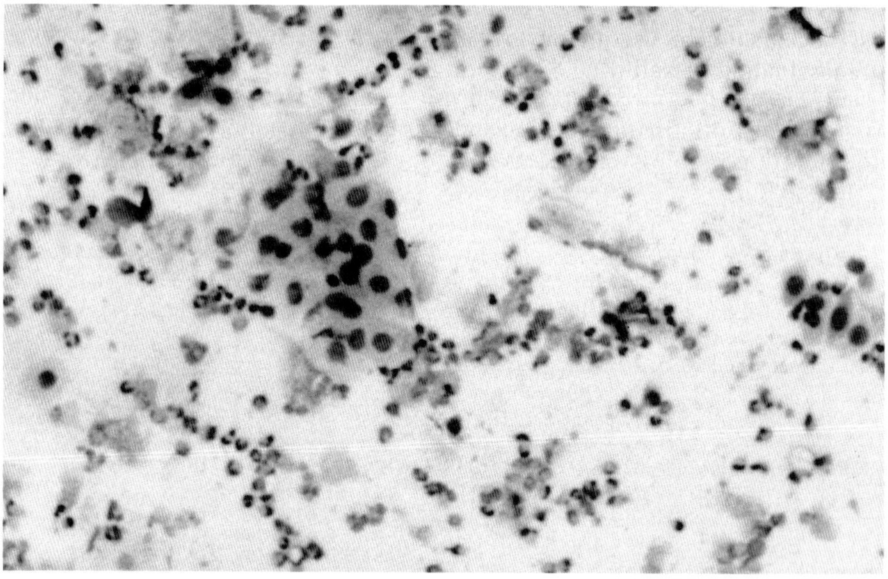

Fig. 2. A representative cell sample from sodium taurocholate-treated rabbit showing erythrocytes and leukocytes and a tissue fragment containing several epithelial cells. Giemsa, 200x

PBS-treatment caused only a slight increase of cell counts (Table 1). The cells were mainly epithelial elements derived from the exfoliation of the surface level which are in the process of being shed (Fig. 1).

On the contrary, Na-T caused a severe inflammatory reaction with deeper layer epithelial cell detachment and haemorrhage (Fig. 2). Differential count showed a marked, statistically significant, increase (maximum value at 24 h) of both leukocytes and epithelial cells (Table 1).

Discussion

The results obtained in the present study indicate that the brush technique can be used as a sensitive and reliable method for the evaluation of the nasal mucosa response to intranasally administered pharmaceutical preparations.

In fact, the brushing procedure did not induce significant mechanical irritation of the nasal mucosa as evidenced by the lack of any inflammatory reaction or epithelial cells detachment even after repeated application. This is important in order to evaluate the actual irritative properties of the test compound, avoiding background noise produced by the sampling technique.

Conversely, PBS and Na-T administration determined, respectively, a slight and severe inflammatory reaction with epithelial cell detachment.

Based on the morphological analysis of the exfoliated cellular population this method allows to evaluate the degree of the nasal mucosa injury from minimal to severe. In conclusion, the brush method, that can be easily applied in order to obtain repeated non-traumatic cell sampling from the nasal mucosa in rabbits, appears to be useful for the study of the local tolerability of a new intranasal administered drug. Furthermore, it is advisable that the same technique could be applied to human, allowing direct comparison between experimental and clinical observations.

References

Batts AH, Marriott C, Martin GP, Bond SV (1989) The effect of some preservatives used in nasal preparations on mucociliary clearance. J Pharm Pharmacol 41:156–159

Levrier J, Molon-Noblot S, Duval D, Lloyd KG (1989) A new *ex vivo* method for the study of nasal drops on ciliary function. Fundam Clin Pharmacol 3:471–482

Recent Developments in Toxicology:
Trends, Methods and Problems
Arch. Toxicol., Suppl. 14, 276–279 (1991)
© Springer-Verlag 1991

Study on the Penetration of Tetanus Toxoid Through the Nasal Mucosa

E. Spranz, B. Siegemund, J. Frevert, and H. E. Knoell

Department of Toxicology and Pharmacology, Battelle-Institut e.V., Frankfurt, Germany

Introduction

Certain infections and human diseases require immediate and often repeated application of drugs (e.g. tetanus, diabetes). Parenteral application causes pain due to the route of administration, and trained adults must always be present when drugs have to be administered to children or animals. Most of the vaccines developed during the last 80 years or so must be administered parenterally, which involves the problem of infections and hyperallergic reactions.

A more comfortable vaccination method with less risk of negative side effects is intranasal administration (instillation into the nasal cavity). However, this way of application may also involve some difficulties and/or disadvantages:
- Part of the dose may reach the gastro-intestinal tract, where the vaccine will be destroyed.
- Large molecules cannot penetrate the nasal mucosa
- and/or will be destroyed by enzymes in the mucosa.

The aim of this study was therefore to establish a successful primary vaccination by intranasal application of a newly developed tetanus toxoid formulation (Behring Werke, Marburg). In addition, the penetration rate of large molecules like tetanus toxoid (MW 150,000) was to be determined and the penetration mechanism through the nasal mucosa examined.

Methods

To judge the success of vaccination, nasal, i. e., applied onto the nostril, (tetanus toxoid) or intranasal instillation of tetanus toxoid formulation ($n = 8$) was compared with subcutaneous application (s.c.) ($n = 6$) in mice (female NMRI mice; BW: 22 ± 2 g; diet: Altromin R + M, Type 1324). The antibody response after primary vaccination was determined in serum by means of ELISAs.

To determine the penetration rates of tetanus toxoid or tetanus toxin in mice (control and s.c.: n = 5; nasal: n = 6; intranasal: n = 26; 6 samples/point of time; female NMRI mice; BW: 34–40 g; diet: Altromin R + M, Type 1324), a sandwich ELISA was used and the amount of toxoid was evaluated by computer.

The penetration rates after different applications were compared by evaluating the areas under curve (AUC). This evaluation was based on the determination of tetanus toxoid in the serum of mice after intravenous administration (n = 6).

Finally, the penetration mechanism was evaluated by means of various histological methods: micro-autoradiography, immunfluorescent techniques and aminoethyl carbazole techniques with slides.

Results

The test substances were examined by means of SDS gel electrophoresis before application. This confirmed that the vaccine did not consist of a single protein of MW 150,000. Small bands were observed on 48 kiloDalton (kD) and 95 kD, and a large band mostly between 140 and 200 kD.

The titres after nasal and subcutaneous vaccination were examined on day 50 (Fig. 1). Only the result after subcutaneous administration of 10 Lf (limit of flocculation) tetanus toxoid and nasal application of 100 Lf tetanus toxoid could be regarded as sufficient protection against tetanus infection. When 25 Lf tetanus toxoid without adjuvant was used for nasal application, only minor protection against tetanus was observed. When using the newly developed

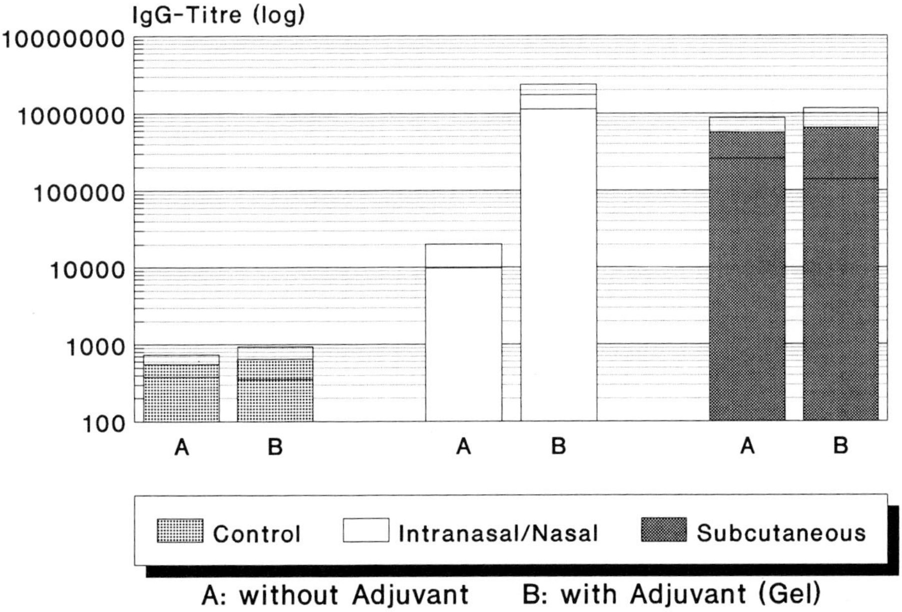

Fig. 1. IgG-titres 50 days after application of tetanus toxoid

Table 1. Penetration rates after 6 h (%)

i.v.	s.c.	n. toxoid	n. toxin	i.n. gel + toxoid
100	6.28	0.45	1.22	2.42

tetanus toxoid formulation (Behring Werke, Marburg) for nasal administration, the result achieved with 25 Lf was better than after subcutaneous administration of 10 Lf tetanus toxoid. The titres in the individual animals were very different, but in fact they all guaranteed efficient protection.

Intranasal vaccination with 25 Lf tetanus toxoid in this new gel-adjuvant formulation actually resulted in a higher IgG concentration than subcutaneous vaccination with 10 Lf tetanus toxoid.

The penetration rates after intravenous (24 μg tetanus toxoid), subcutaneous (48 μg tetanus toxoid), nasal (360 μg tetanus toxoid or tetanus toxin) and intranasal (24 μg tetanus toxoid) application were examined, too. It should be noted that the penetration after intranasal vaccination of the gel formulation lasted longer.

Comparing the penetration rates of all the administration routes employed, based on the formula of Igawa et al (1988), showed that intranasal vaccination with the new formulation was very successful in mice (Table 1, Fig. 2).

As regards the penetration mechanism, it has been found that transepithelial transport is an intracellular process. By immunohistological techniques it was possible to demonstrate that it is the ciliated cellsof the respiratory epithelium which take up and transport tetanus toxoid through the epithelial barrier of the nasal mucosa.

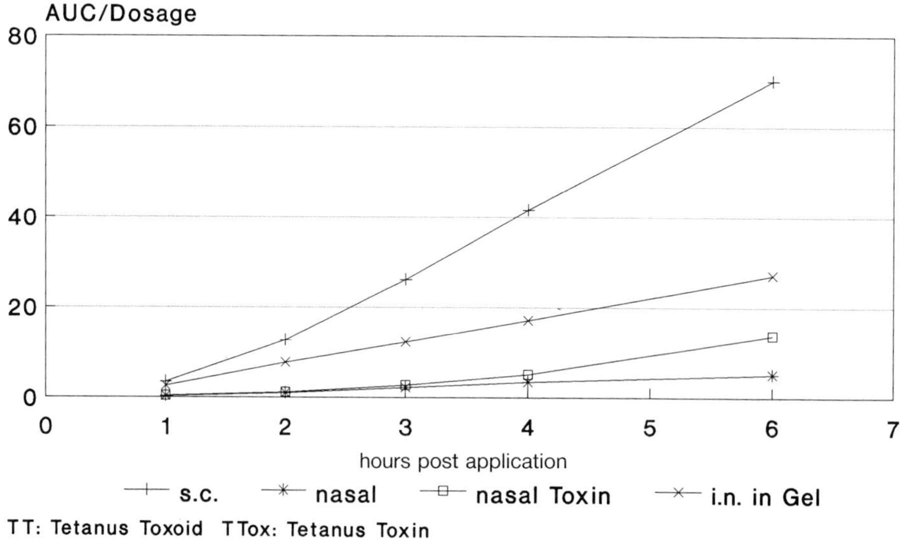

Fig. 2. Penetration of tetanus toxoid and tetanus toxin

Discussion

One of the most important points concerning successful nasal vaccination is the time of contact with the mucosa; this can definitely be prolonged by using the newly developed gel (Behring Werke, Marburg).

Besides, the new formulation permits more exact dosage in humans and large animals. The experimental results have shown that on the one hand the penetration rate of tetanus toxoid is increased and on the other hand the formulation itself works as an adjuvant. In addition this formulation (i.e. the gel) of the vaccine allows fast and easy administration even by untrained persons. Furthermore, the amount of tetanus toxoid to be administered for successful nasal vaccination can be lowered to the level that is necessary in the case of subcutaneous administration.

Acknowledgements. We wish to thank Dr. Bremecker of the University of Marburg and Dr. Hungerer of Behring Werke, Marburg, for making available to us both tetanus toxoid and the new tetanus toxoid formulation.

References

Bremecker KD (1986) Entwicklung eines Nasengels zur Primärimmunisierung. Sci Pharm 54:151

Bremecker KD (1987) Published Patent Application DE 3601923 A1, Date of application: 23. 1. 1986, Date of publication: 30. 7. 1987

Igawa T (1988) Enzyme immunoassay of human fibroblast interferon after intranasal administration with several excipients in rabbits. Chem Pharm Bull 36:3055–3059

Spranz E (1989) Studien zur Penetration von Tetanustoxoid an der Nasenschleimhaut von Mäusen. Vet med Diss (doctoral thesis), München; in print

Recent Developments in Toxicology:
Trends, Methods and Problems
Arch. Toxicol., Suppl. 14, 280–284 (1991)
© Springer-Verlag 1991

Stobadine Toxicity and Transplacental Movement

A. Krištofová, E. Ujházy, Š. Bezek, T. Balonová, and R. Nosáľ

Institute of Experimental Pharmacology Slovak Academy of Sciences Bratislava.

Introduction

Stobadine (STO) is a prospective antiarrhythmic and cardioprotective agent with antihypoxic effects on the myocardium (Beneš and Štolc 1989). This paper presents the most important findings in toxicological and transplacental transport studies, which were performed with STO in order to evaluate its safety.

Methods

STO, (-)-cis-2,8-dimethyl-2,3,4,4a,5,9b-hexahydro-1H-pyrido[4,3-b]indole, m.w. 202.30, as dipalmitate salt DP 1031 and dihydrochloride salt DH 1011, was synthetized at the Institute of Organic Chemistry and Biochemistry, CSAS, Prague. ^{3}H-STO (as DH 1011) was supplied by the Institute for Research, Production and Use of Radioisotopes, Prague (sp. activity 495 GBq/mM; radiochemical purity 95%, as determined by TLC).

In toxicity studies, adult SPF Wistar-strain rats, initial body weight 201–210 g (males) and 196–220 g (females) were administered STO by oral gavage. STO as a Methocell suspension (0.6%) was given once daily at dose levels of 13, 39 and 130 mg/kg (DP 1031) and as a water solution at dose levels of 5, 15 and 50 mg/kg (DH 1011) for 28 days. Mature, virgin SPF Wistar-strain rats, 3–4 months of age, 300 g mean weight at day 20 of gestation, were used in the transplacental study. ^{3}H-STO, as DH 1011, (5 mg/kg) was administered on day 20 of gestation orally.

In toxicity studies criteria evaluated for compound effect were mortality, body weight, clinical signs, clinical chemistry and hematology, organ weight, and gross and microscopic pathology. Evaluations were performed before the start, at the end of STO administration and after 4 weeks without medication. Decapitation was used to obtain blood from the rat for serum chemistry deter-

Abbreviations: STO – stobadine, DH 1011 – dihydrochloride salt of STO, DP 1031 – dipalmitate salt of STO minations. BIO-LA-CHEMA (CSFR) test kits were used to measure the following parameters: alkaline phosphatase, aspartate aminotransferase, alanine aminotransferase, lactate dehydrogenase, urea nitrogen, glucose, creatinine, total protein, albumin. Blood for hematology was obtained from the tail vein. Red and white cell counts, hemoglobin, hematocrit and mean cell volume were determined on a Coulter Counter ZF6 instrument. The leukocyte differential determinations were done manually using Pappenheim stain smears. In the transplacental study, levels of the parent drug and the total activity were determined in maternal plasma, placentas and foetuses 10, 60, 120 and 1440 min after STO administration.

Results

In the toxicity studies no clinical signs of adverse reactions were observed throughout the whole treatment period. Survival was unaffected by STO. The growth rate was slightly retarded in female rats of the 39 and 130 mg/kg dose groups and a slight increase in the glucose level (males) and albumin (females) was observed (Fig. 1). A small increase in the number of red blood cells,

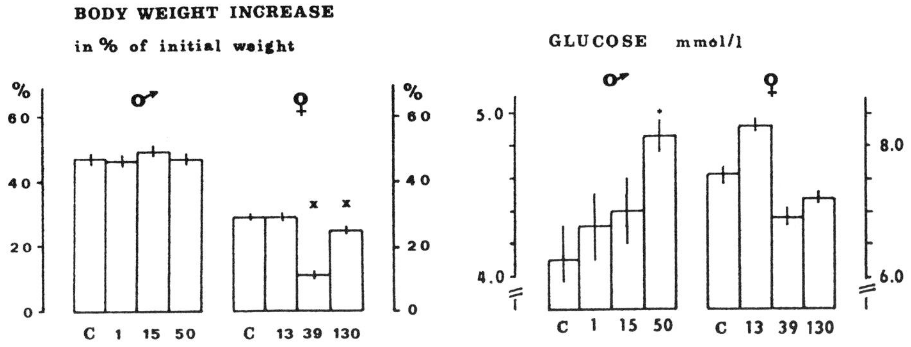

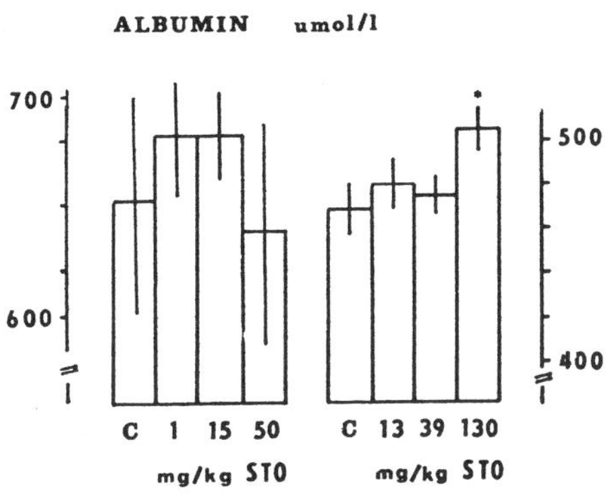

Fig. 1. Changes of body weight, serum glucose and albumin levels of rats treated with stobadine for 28 days. Mean ± SE, n = 8 (males), n = 6–10 (females. * Significantly different from control animals (C). Student'ts t-test, $p < 0.05$

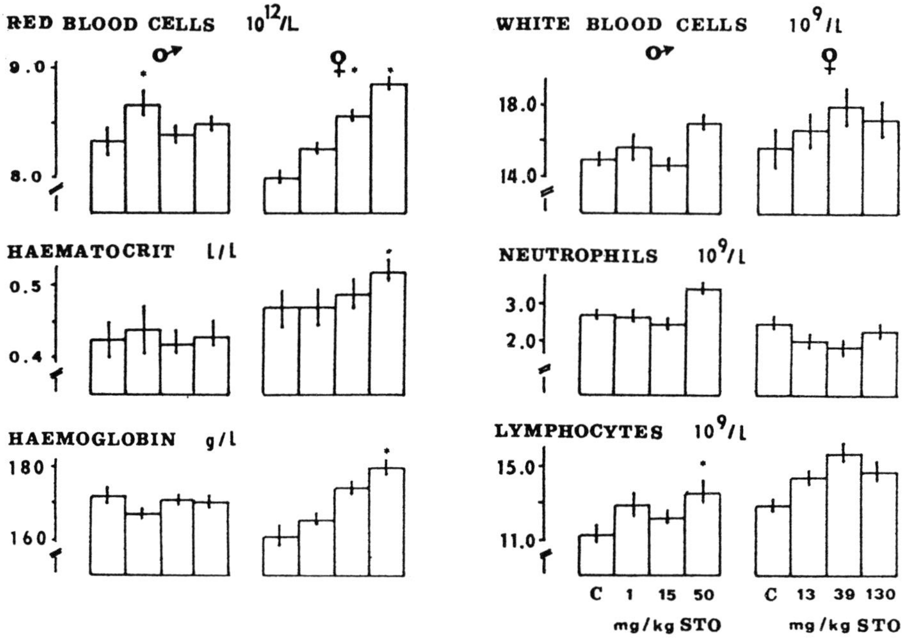

Fig. 2. Hematological parameters in rats given stobadine for 28 days. Mean ± SE, n = 8 (males), n = 14 (females). * Significantly different from control animals (C). Student's t-test, $p<0.05$

hematocrit and hemoglobin was obtained in the females in the highest dose group after 4 weeks of STO administration. A significant increase of the absolute lymphocyte count was observed only in the highest dose group in males (Fig. 2). There were no histopathological changes in organs and no differences in the relative organ weights between control and STO treated animals.

In the transplacental transport study transfer of ^{3}H-STO across rat placenta was observed. Peak concentrations in maternal plasma were reached within 60 min. A certain retention of ^{3}H-STO was observed at 60 and 120 min after administration of STO, giving higher concentrations in the placental and foetal tissues than in the maternal plasma (Fig. 3).

Discussion

In one-month toxicity studies in rats it was clearly shown that STO had no adverse effects at dose levels up to 15 mg/kg/day. The observed changes in the hematological parameters were small and therefore considered to be of minor

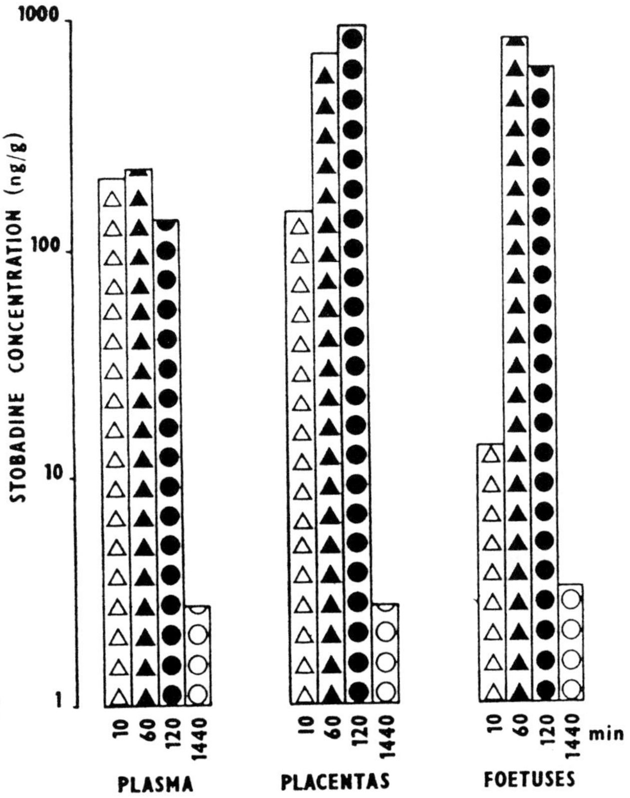

Fig. 3. The time course of stobadine concentration in rat plasma, placentas and foetuses

toxicological importance. Growth rate retardation has been observed in pregnant rats given STO (50 mg/kg/day) in a long term reproductive toxicity study (Balonová and Zeljenková 1985). In the transplacental study of STO no significant retention was observed in foetal and placental tissue. There was no evidence of adverse effects of STO on late chick embryogenesis (Ujházy et al. 1988).

References

Beneš L, Štolc S (1989) Stobadine. Drug Future 14:135–137

Balonová T, Zeljenková D (1985) Reproductive and teratological studies with a pyridoindole derivative. I. Effect on fertility and general reproductive performance of the rat. QSAR in toxicology and xenobiochemistry 8. Amsterdam. Elsevier, 1985, p. 373–378

Ujházy E, Zeljenková D, Ďurišová M, Končeková Z, Miček J (1988) Effect of Pyridoindole Derivative (DH 1011) on Late Chick Embryogenesis. Anat Anz Jena 167:113–117

Recent Developments in Toxicology:
Trends, Methods and Problems
Arch. Toxicol., Suppl. 14, 284–287 (1991)
© Springer-Verlag 1991

Some Effects of Furazolidone, Humic Acid and a Combination of Both on Male Reproduction in Rats

H. Zimmermann[1], S. Golbs[1], and G. Gutte[2]

[1] Institute of Vet.-Pharmacology, Pharmacy and Toxicology, and
[2] Institute of Vet.-Anatomy, Histology and Embryology of the Leipzig University, Zwickauer Straße 55, O-7010 Leipzig, Germany

Introduction

Many methods are used to investigate possible effects of chemicals on male reproduction. Special questions arise in connection with chronic exposure to a substance and the exposure to several compounds. Histological investigations of the testes provide good results but are relatively time consuming. Mating studies require many animals and are not very sensitive because male rats are still fertile if sperm numbers are considerably decreased (Ettlin et al 1984; Russell and Ettlin 1989). Therefore it seems to be better to assess the sperm number in the epididymis and the morphology of mature sperm in the ductus deferens. The aims of this study were to estimate

- the normal values for reproductive parameters of the inbreeding strain WIST/Lppt,
- the sensitivity of this strain to a well known spermatoxic substance, furazolidone,
- if a combination of humic acid with furazolidone would influence the reproductive effects of furazolidone.

Material and Methods

Adult Wistar rats of the inbred strain WIST/Lppt were kept under conventional conditions. Water and ground standard food pellets were given ad libitum. Humic acid can be used as a veterinary therapeutic agent for gastrointestinal disorders in young animals (calves, pigs, sheep, dogs, cats, and zoo animals). Humic acid concentrate (about 81.4% humic acid content, from Sächsische Pharma-Werke GmbH Weinböhla) was admixed to the food in a concentration of 5 g/kg, starting from 5 days before the first furazolidone dose till the 5th day after the last furazolidone dose. Furazolidone (about 97% purity, from Sächsische Pharma-Werke GmbH Weinböhla) was suspended in

Table 1. Dosages of the experimental groups (N = number of animals)

group	N	humic acid concentrate	furazolidone
OC	8	none	none
OF1	8	none	5 × 50 mg/kg bw
OF2	8	none	5 × 100 mg/kg bw
HC	8	5 g/kg food over 15 days	none
HF1	8	5 g/kg food over 15 days	5 × 50 mg/kg bw
HF2	8	5 g/kg food over 15 days	5 × 100 mg/kg bw

1% carmellose sodium solution (sodium carboxymethylcellulose content more than 90%, from Sächsische Pharma-Werke GmbH Weinböhla). It was administered by gavage tube into the distal part of the oesophagus for 5 days. All animals were killed between the 56th and 58th day after the first furazolidone administration. The dosages are shown in Table 1. The following parameters were estimated: body weight (bw), weight of both testes (tw), epididymis weight of the right epididymis (ew), weight of the seminal vesicles (svw), prostate weight (pw) number of spermatozoa in the head and the proximal part of the left body epididymis (s/h) and in the tail and the distal part of the body of the left epididymis (s/t), the percentage of intact sperm in the ductus deferens smears (is %), sperm head morphology in the ductus deferens smears and motility (gross estimation). Spermatozoa were counted in a Neubauer chamber after dividing the left epididymis at the middle of the body and using the maceration technique (Heywood and James 1978).

Statistics: Data obtained were statistically analysed employing paired Student's t-test and the level of significance was set at $p < 0.05$.

Results and Discussion

There were no significant differences between control (OC) and those fed humic acid (HC) and so these groups could be combined. Table 2 shows the normal values of the parameters tested in this study. The weights of the prostate and the seminal vesicle glands were not affected in all groups and the portion of abnormal sperm heads was always lower than 1%. About 30–50% of the spermatozoa in the head and the proximal part of the body of the epididymis showed forward movement in all groups. The number of motile sperms in the tail and the distal part of the body was about 70% in OC, HC, OF1 and

Table 2. Mean values (MV) and standard deviations (SD) of the measured parameters of control group (OC + HC)

	bw (g)	tw (mg)	ew (mg)	svw (mg)	pw (mg)	s/h (10^6)	s/t (10^6)	is % (%)
MV	437	3368	573	1328	822	56.4	140.7	95.2
SD	37	157	33	111	109	4.8	21.4	1.7

HF1 whereas only some spermatozoa of the high dose groups, OF2 + HF2, were motile. Fig. 1 shows the results of the other parameters in groups OF1, OF2 and HF2 related to C = OC + HC = 100%. The measured parameters of HF1 did not show a normal distribution. Five of eight animals were not affected in any of the measured parameters, two rats showed similar effects to OF1 and one male showed only slight effects.

Reasons for this could be 1) the adsorbing properties of humic acids, which could be distinctly marked because of different behaviour in food intake or 2) normal interindividual differences to the reproductive-toxic effects of furazolidone which have been described for other substances, e.g. for sulphasalazine in man (Birnie et al 1981). To clarify the results of HF1 further investigations are necessary. The inbred rat strain WIST/Lppt is sensitive to spermatoxic chemicals. Humic acid concentrate were shown to be ineffective in the prevention of sperm toxicity of furazolidone at dosages of 100 mg/kg bw over 5 days.

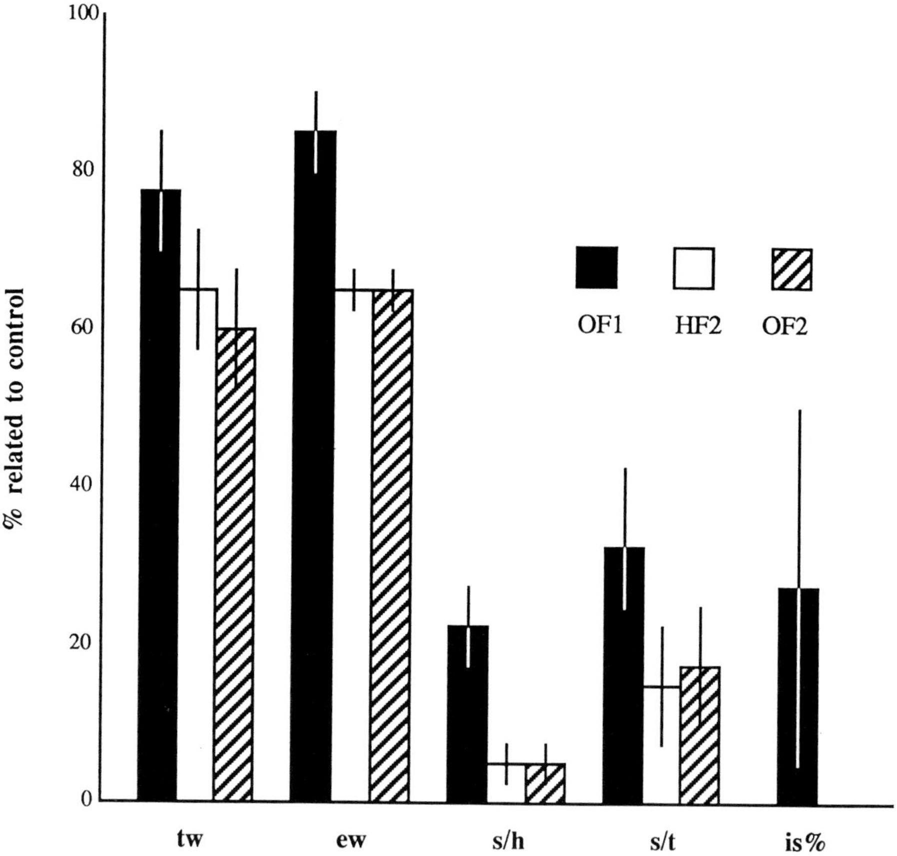

Fig. 1. Testes weight (tw), epididymis weight (ew), prostate weight (pw), number of sperms in the head and the proximal part of the body (s/h), in the tail and the distal part of the body of the epididymis (s/t) and percentage of intact sperms (is %) related to C = 100%. Values are means and standard deviations of 8 rats. In OF2 and HF2 no sperm were intact. All values shown are significant different from C at $p < 0.05$

References

Birnie G, McLeod T, Watkinson G (1981) Incidence of sulphasalazine-induced male infertility. Gut 22:452–455
Ettlin RA, Bechter R, Lee IP, Hodel C (1984) Aspects of testicular toxicity induced by anticancer drugs. Arch Toxicol Suppl 7:151–154
Heywood R, James RW (1978) Assessment of testicular toxicity in laboratory animals. Environ Health Perspect 24:73–80
Russell LD, Ettlin RA (1989) Histopathology of the testes. EST Training Session on Histological Evaluation of Spermatogenesis in Laboratory Animals 30 May 1989:17

Recent Developments in Toxicology:
Trends, Methods and Problems
Arch. Toxicol., Suppl. 14, 288–292 (1991)
© Springer-Verlag 1991

Residues of Carbadox Metabolites in Edible Pork Products

A. J. Baars, L. P. Jager, Th. J. Spierenberg, G. J. de Graaf, and J. W. Seinhorst

Central Veterinary Institute, P.O. Box 65, 8200 AB Lelystad, The Netherlands

Introduction

Carbadox (methyl-3-[2-quinoxalinylmethylene]carbazate-N1, N4-dioxide; CBX) is a widely used feed additive for young pigs (Trasher et al 1969), but because of its adrenal toxicity there is some risk in using it (Van der Molen et al 1985, 1986). CBX is rapidly metabolized; known biotransformation products involve the three N-O-reduced compounds (N1-, N4-, and N1,N4-desoxycarbadox; DCBX), and quinoxaline-2-carboxylic acid (QCA). QCA has been identified as the major residual metabolite in pigs (Federal Register 1972; Ferrando et al 1975). Since CBX and DCBX are suspected carcinogens, CBX may only be administered to swine under the age of 4 months, and a withdrawal period of 4 weeks has been imposed.

The regulatory residue limit for QCA is 30 ppb in edible products (Federal Register 1972). Ferrando et al (1975) used an analytical method with a detection limit of 30 ppb, and suggested that QCA has a biological half-life in pig liver of 4–5 days. McIntosh et al (1985) reported the depletion of CBX and DCBX from blood, muscle, and kidney within 72 hours; traces of DCBX were still detectable in liver (detection limit 2 ppb).

The aim of the present study was to investigate the depletion of CBX, DCBX, and QCA applying a feeding schedule as used in routine pig husbandry.

Material and Methods

Seventy-five pigs (3-week-old, females and castrated males) were allotted to 13 experimental and 2 control groups. The housing conditions were described earlier (Van der Molen et al 1989). The pigs were fed a commercial diet (Hope Farms, Woerden, The Netherlands) with 50 ppm CBX (Mecadox, Pfizer) or

Abbreviations: CBX: carbadox; DCBX: desoxycarbadox; QCA: quinoxaline-2-carboxylic acid; ppb: parts per billion (ng per g); ppm: parts per million (μg per g)

vehicle, for 94 days. After withdrawal from the drug, they received the same diet without CBX (verified by analysis). The pigs were then killed at regular intervals, and samples of blood, liver, kidney, and muscle were collected and stored at −20°C until analysis. CBX, DCBX, and QCA were analyzed as described by De Graaf and Spierenburg (1985), Aerts et al (1988), De Graaf et al (1990), Van Ginkel et al (1990), and Keukens et al (1990).

Results and Discussion

The growth of the pigs during and after the administration of CBX was fully normal. No differences were observed between the CBX-treated and the untreated pigs (Baars et al 1990).

Plasma concentrations of QCA during the CBX feeding period reached a maximum after one week, and then slowly decreased (Fig. 1). This finding confirms earlier results on the kinetics of CBX administration (De Graaf et al 1988), who suggested an immature biotransformation of CBX, or induction of one or more CBX-metabolizing enzymes as possible explanations. In the present study, however, the uptake of CBX per kg body weight slowly declined over time; thus, it seems likely that plasma concentrations of QCA decreased concommitantly as well.

After only one day of withdrawal of CBX from the feed, the parent compound was not detectable in blood, muscle tissue, liver, or kidney. This finding indicates that the amount of CBX in edible products derived from CBX-treated pigs is probably insignificant.

DCBX virtually disappeared from the muscles within 3 days, from the kidney within 7 days, and from the liver within 14 days (Fig. 2). Although the levels

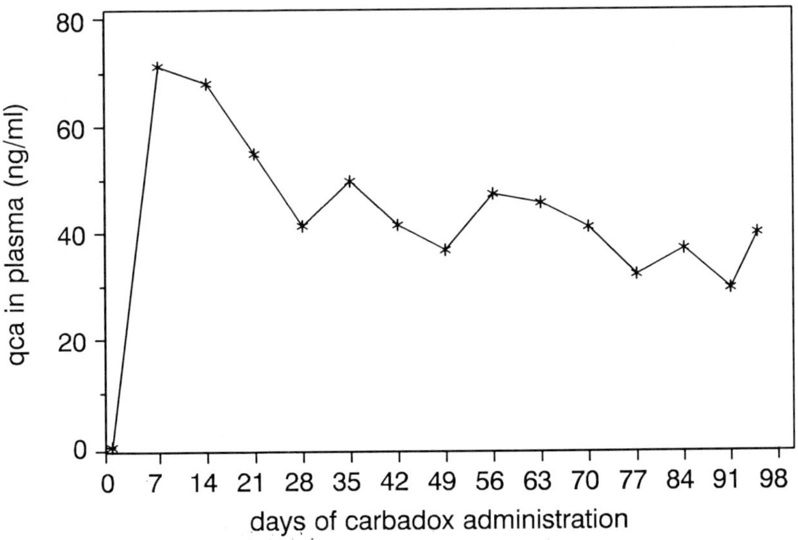

Fig. 1. Plasma levels of quinoxaline-2-carboxylic acid (QCA), the major residual metabolite of carbadox, during administration of carbadox to pigs

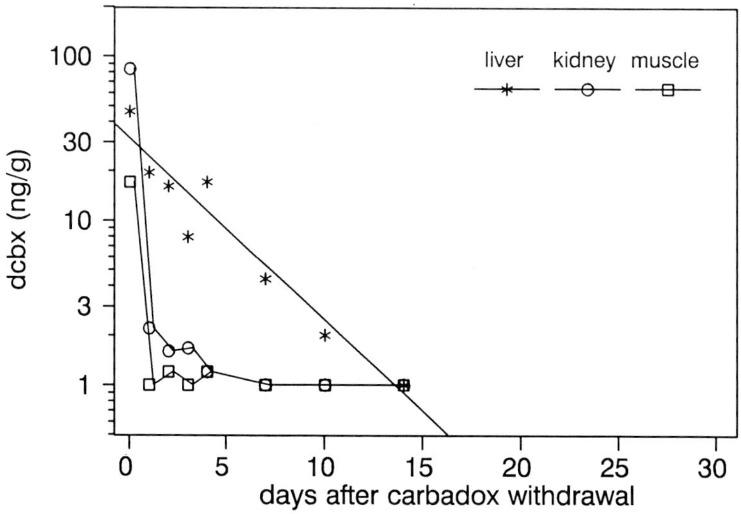

Fig. 2. Depletion of desoxycarbadox (DCBX), the primary metabolite of carbadox, from muscle, liver, and kidney of pigs after withdrawal of carbadox. Detection limit is 1 ppb

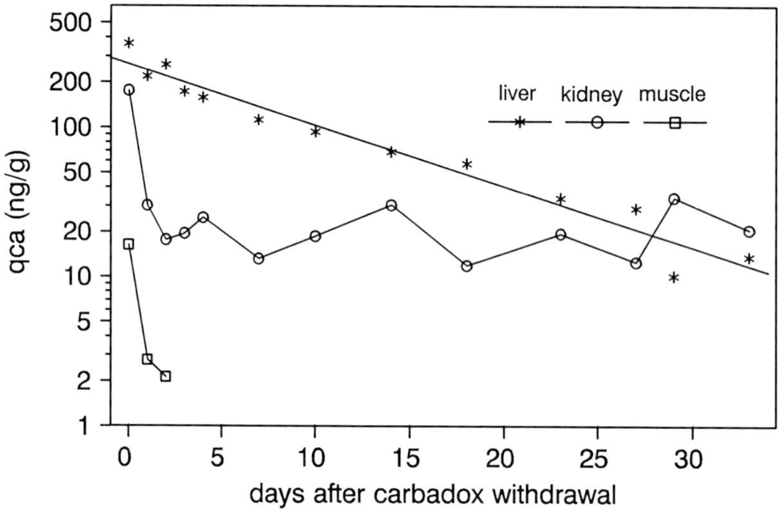

Fig. 3. Depletion of quinoxaline-2-carboxylic acid (QCA), the major residual metabolite of carbadox, from muscle, liver, and kidney of pigs after withdrawal of carbadox. Detection limit is 2 ppb

fell to the limit of detection (1 ppb) within a fortnight, residual amounts of DCBX at this level cannot be excluded.

Figure 3 shows the elimination of QCA. In muscle tissue the detection limit (2 ppb) was reached 3 days after CBX withdrawal. QCA disappeared from the liver and the kidney much more slowly, however. The kinetics of QCA in the liver suggest a linear elimination with an äpparant half-life of 7 days. In contrast, the decline of QCA in the kidney appeared to be biphasic. After decreasing rapidly for 2 days following CBX withdrawal, QCA concentrations remained fairly constant, indicating a relatively slow elimination process.

Analyzing identical samples by different methods, it was found that the alkaline extraction method (De Graaf et al 1990) detected 2–4 times more QCA than the subtilisin digestion method (Van Ginkel et al 1990). This suggests that one or more unknown metabolites of CBX exist. Because the two methods were essentially equal in detecting DCBX, the unknown metabolite(s) appear(s) to be (an) intermediate(s) of the pathway of DCBX to QCA. It seems likely that the enzymatic digestion method estimates the true amount of metabolically produced QCA, while the alkaline extraction method also estimates (part of) the intermediary products by converting them to QCA.

The regulatory limit of 30 ppb of CBX-related residues in edible pork products is met within 3 days for meat. However, 4 to 5 weeks are needed to reach this limit in the liver and kidney. Therefore, an added safety factor of 2 seems advisable in view of the possible presence of unknown metabolites. Consequently, a CBX-withdrawal period of 8 weeks is recommended to ensure that unwanted drug residues are well below the regulatory limits.

Acknowledgements. This study was part of a collaborative project of the Central Veterinary Institute, Lelystad; the National Institute of Public Health and Environmental Protection, Bilthoven (Dr. L. A. van Ginkel); and the State Institute for Quality Control of Agricultural Products, Wageningen (Drs. M. M. L. Aerts and H. J. Keukens). We thank the Ministry of Agriculture, Nature Management and Fisheries, The Hague, for financial support, and Pfizer (Groton, U.S.A.) for a gift of pure CBX and desoxymetabolites.

References

Aerts MML, Beek WMJ, Keukens HJ, Brinkman UATh (1988) Determination of residues of carbadox and some of its metabolites in swine tissues by HPLC using on-line pre-column enrichment and post-column derivatisation with UV/Vis detection. J Chromatogr 456:105–119

Baars AJ, Van Ginkel LA, Aerts MML, De Graaf GJ, Seinhorst WJ, Keukens HJ, Schwillens PLWJ, Spierenburg ThJ, Jager LP (1990) Kinetics of carbadox in pigs. In: Haagsma N, Ruiter A, Czedik-Eysenberg PB (eds) Residues of veterinary drugs in food. University of Utrecht, pp 110–113

De Graaf GJ, Jager LP, Baars AJ, Spierenburg ThJ (1988) Some pharmacokinetic observations of carbadox medication in pigs. Vet Quarterly 10:34–41

de Graaf GJ, Spierenburg ThJ (1985) Liquid chromatographic determination of carbadox and desoxycarbadox in medicated feeds and in porcine gastrointestinal tract. J Assoc Off Anal Chem 68:658–660

De Graaf GJ, Spierenburg ThJ, Baars AJ (1990) GC-MS determination of quinoxaline-2-carboxylic acid, the terminal metabolite of carbadox, in porcine meat, liver, and kidney. In: Haagsma N, Ruiter A, Czedik-Eysenberg PB (eds) Residues of veterinary drugs in food. University of Urecht, pp 202–205

Federal Register (1972) Vol 37, No 192, 20683, October 3

Ferrando R, Truhaut R, Raynaud JP, Spanoghe JP (1975) Toxicity by relay. II. Method for the assessment of safety to human consumers of carbadox, a growth-promoting additive to the feed of slaughter pigs. Toxicology 3:369–398

Keukens HJ, Binnendijk GM, Beek WMJ, Aerts MML (1990) Determination of residues of carbadox and metabolites in edible tissues – in vitro stability studies. In: Haagsma N,

Ruiter A, Czedik-Eysenberg PB (eds) Residues of veterinary drugs in food. University of Utrecht, pp 235–239
MacIntosh AI, Lauriault G, Neville GA (1985) Liquid chromatographic monitoring of the depletion of carbadox and its metabolite desoxycarbadox in swine tissues. J Assoc Off Anal Chem 68:665–671
Trasher GW, Shively JE, Askelson CE, Babock WE, Chalquest RR (1969) Effects of feeding carbadox upon the growth and performance of young pigs. J Anim Sci 28:208–215
Van der Molen EJ, Baars AJ, De Graaf GJ, Jager LP (1989) Comparative study of the effect of carbadox, olaquindox and cyadox on aldosterone, sodium and potassium plasma levels in weaned pigs. Res Vet Sci 47:11–16
Van der Molen EJ, De Graaf GJ, Baars AJ, Schopman W (1986) Carbadox induced changes in aldosterone, sodium and potassium levels in the blood of weaned pigs. J Vet Med (A) 33:617–623
Van der Molen EJ, Nabuurs MJA, Jager LP (1985) Clinical and pathological changes related to toxicity of carbadox in weaned pigs. J Vet Med (A) 32:540–550
Van Ginkel LA, Schwillens PLWJ, Jacquemeyns M, Zomer G (1990) The detection and identification of quinoxaline-2-carboxylic acid, a major metabolite of carbadox, in swine tissue. In: Haagsma N, Ruiter A, Czedik-Eysenberg PB (eds) Residues of veterinary drugs in food. University of Utrecht, pp 189–195

Recent Developments in Toxicology:
Trends, Methods and Problems
Arch. Toxicol., Suppl. 14, 293–298 (1991)
© Springer-Verlag 1991

Pharmacological-Toxicological Investigations with a Furan Derivative Produced from Sugar-Cane on Laboratory Animals

S. Golbs, M. Kuehnert, H. Zimmermann, and J. A. P. Donato

University of Leipzig, Institute of Vet.-Pharmacology, Pharmacy and Toxicology, Zwickauer Str. 55, D-7010 Leipzig, FRG

Introduction

The most important danger for health and productivity of the cattle population in Cuba are protozoan infections like anaplasmosis, piroplasmosis, babesiosis and eperitrozoonosis. The most efficient way for the treatment of these diseases is to attack the ticks which are responsible for transferring the infections of these diseases. Thus the pharmacological research of Cuba had the task to find domestic products against ticks. Furan derivatives which could be used as antiparasitic drugs after intensive pharmacological-toxicological tests rise as secondary products from the extracted sugarcane. The compound UC-244 (Fig. 1) is such a secondary product of sugar industry. It showed a good antiparasitic effect in concentrations from 0.3% on (Donato 1988).

Material and Methods

The pharmacological-toxicological tests were made on Wistar rats (WIST/Lppt) and on rabbits (Table 1). The determination of the acute toxicity of UC-244 from Santa Clara, Cuba, produced in the university laboratories based on the method of Litchfield and Wilcoxon (1949). In the acute experiment the substance was given once in the distal part of the oesophagus with the help of a metal probe. UC-244 was dissolved in peanut oil (Oleum Arachidis).

In subchronic experiments (administration of UC-244 in food over 90 days) a daily dose of 44.8 mg/kg body weight (group 1 = 10% of the LD_{50}), 89.6 mg/kg body weight (group 2 = 20% of the LD_{50}) and 179.2 mg/kg body weight (group 3 = 40% of the LD_{50}) was given.

In the third experiment the length of the hexobarbital-induced sleep as a parameter of the activity of cytochrome P 450-system was evaluated. Hexobarbital (AWD Dresden) 80 mg/kg body weight was given intraperitoneal.

In experiment 4 the damage and the primary acute irritation effect on eyes were estimated according to the method of Draize (1944). In the first part of

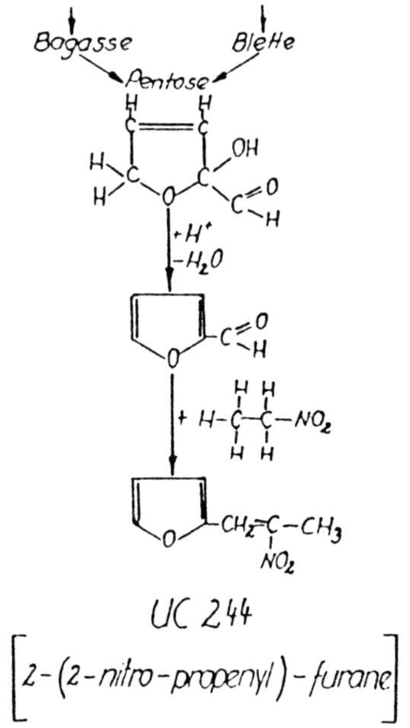

Fig. 1. Synthesis of UC-244 as a secondary product of sugar industry

Table 1. Scheme of pharmacological-toxicological experiments on Wistar rats and rabbits

Experiment	Number of animals	Species of animals	Sex	Generation	Age in days
1. Oral acute LD_{50}	60	rats	♀	73[iii]2	42 ± 7
2. Subchronic experiment	136	rats	♀	73[iii]3	42 ± 7
3. Hexobarbital sleeping time	77	rats	♀	73[iii]3	42 ± 7
4. Draize test	60	rabbit	♀/♂	EG-12	adult

experiment 100 mg per eye were administered in pure powder form and in second part the substance was given as a 0.3% solution in Tween-80.

Results and Discussion

The acute LD_{50} of oral administered UC-244 is 561.6 mg/kg body weight for male rats and 448.0 mg/kg body weight for female ones. The highest mortality appeared during the first 24 hours after dosing (75% of males and 72.3% of females died in this time). Rats which showed clear intoxication symptoms and which survived the first 48 hours recovered from intoxication after 72 hours. This reaction may be due to a fast biotransformation of the furan derivative.

In the acute experiment, a significant increase of the activity of the enzymes aspartate aminotransferase (AsAT), alanine aminotransferase (ALAT) and glutamate dehydrogenase (GLDH) was found 24 hours after administration (Fig. 2).

These effects correspond with the patho-morphological changes and they point to cellular damage in the liver (Schmidl 1981; Roth et al 1985).

In the subchronic experiment a significant increase of the enzymes ALAT and alkaline phosphatase (AP) could be observed with a daily dosage of UC-244 89.6 mg/kg body weight after 90 days of administration. The enzymes AsAT and leucine aminopeptidase (LAP) increased significantly with a daily dosage of 179.2 mg/kg body weight after 90 days of administration (Fig. 3).

No cellular changes could be found by means of pathomorphological investigations. The reason for this could be that the furan derivative UC-244 in long-term experiments causes only an unimportant cellular membrane damage without degeneration in liver and kidney cells.

The investigation of the haematological parameters resulted in an increase of total protein, of leucocytes and of mature segmented granulocytes in the acute experiment. These results point out a cellular reaction of immune system caused by UC-244. In contrast to these results in the subchronic experiment a decrease of the number of erythrocytes and an increase of haemoglobin were measured. The values of the calculated parameters MCV (mean corpuscular volume of an erythrocyte), MCHC (mean cellular haemoglobin concentration of erythrocytes) and MCH (mean content of haemoglobin of an erythrocyte)

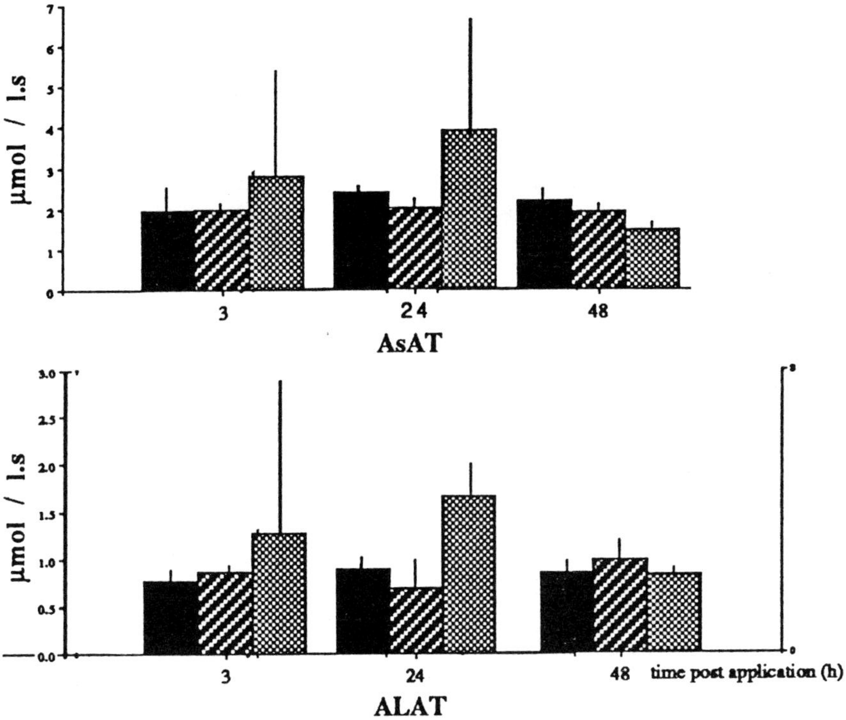

Fig. 2 Activity of AsAT and ALAT in female rats (acute test) negative control group ■; positive control group (peanut oil) ▨; 448 mg UC-244/kg body weight ▩

increased in comparison with the control group. Therefore it can be concluded that the furan derivative might cause a weak hyperchromic, normocytical anaemia after long-term administration (Mandell et al 1979; Kucers et al 1979).

The daily oral administration of 89.6 or 179.2 mg/kg body weight over 30 days causes a reduction of the hexobarbital sleeping time by 21.4 or 28.4%. This shortening has probably a strong connection to an induction of microsomal enzymes (Gillet et al 1978; Loomis 1982).

After a daily administration of 89.6 and 179.2 mg UC-244 per kg body weight over 90 days a significant increase of creatinine in the blood plasma could be found. This may be due to an disturbance of metabolism of the kidneys. The kidney function is obviously influenced by UC-244.

The glutathione (GSH) content of the liver was decreased from 5.40 μmol/g liver to 0.82 μmol/g and 0.45 μmol/g in group 1 (62.5 mg UC-244/kg b.w.) and group 2 (125.0 mg UC-244/kg b.w.) 90 minutes after administration. The strong reduction of the glutathione content in the liver verifys the fact of a liver damage (Scheler 1989; Bakke et al 1984; Koch et al 1986).

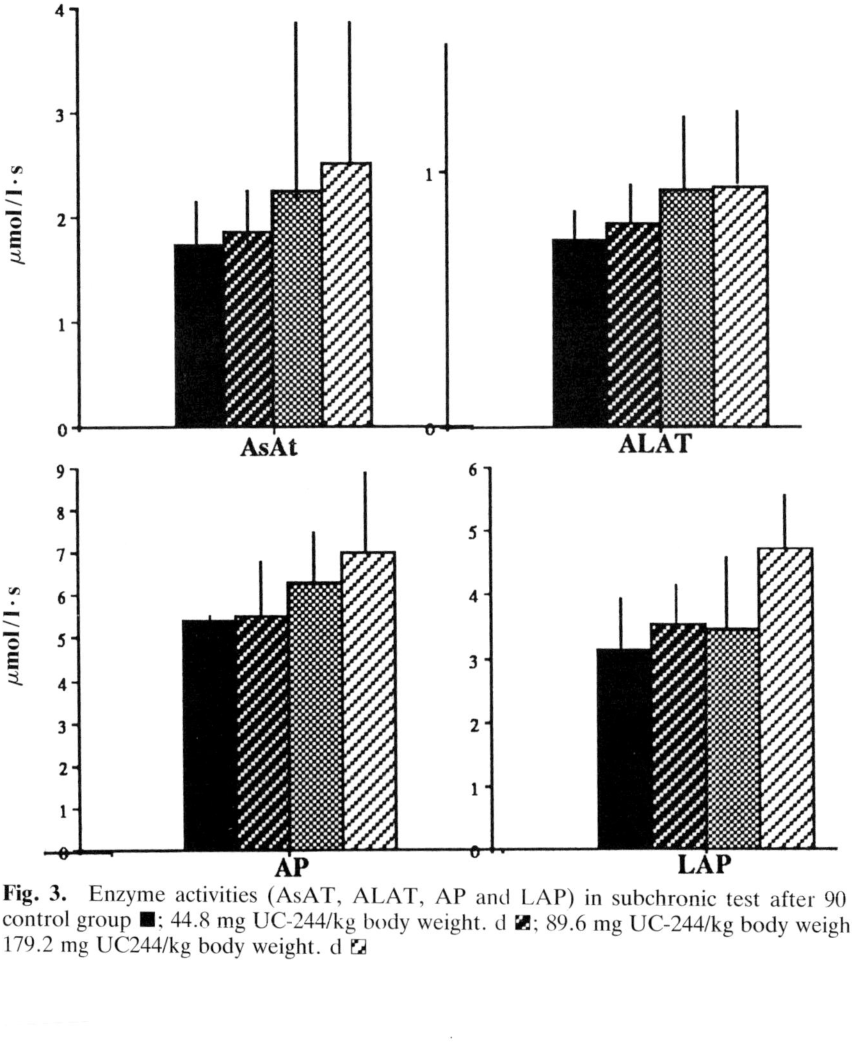

Fig. 3. Enzyme activities (AsAT, ALAT, AP and LAP) in subchronic test after 90 days control group ■; 44.8 mg UC-244/kg body weight. d ▨; 89.6 mg UC-244/kg body weight. ▩; 179.2 mg UC244/kg body weight. d ▨

Table 2. Content of reduced glutathione (GSH) in liver 90 minutes after administration

Group	Dose (mg/kg bodyweight)	μmol GSH/g liver
control	0	5.40 ± 0.56
1	62.5	0.82 ± 0.36
2	125.0	0.45 ± 0.13

The results of the examination on the eyes of rabbits were that the substance UC-244 produces strong irritations on iris, conjunctiva and cornea. Therefore care is necessary when working with the substance (Rodriguez 1986).

Conclusions

The results show that the furan derivative UC-244 is a moderately toxic substance. Structural and functional changes are concentrated in the liver in the acute experiment. From these results it can be concluded that by using the compound UC-244 in antiparasitic dosage no or only minimal local irritation effects for warm blooded animals are to be expected. Based on these results it seems to be possible that UC-244 could be used against ticks in subtropical and tropical countries in the future. This would imply that the imports of such antiparasitic drugs could be reduced or relieved. But more pharmacological and toxicological studies are necessary to increase the safety of this compound.

References

Bakke J, Gustafsson JA (1984) Mercapturicacid pathways metabolites of xenobiotics: generation of potentially toxic metabolites during enterohepatic circulation. TIPS, 5:517–521

Donato JAP (1988) Pharmakologisch-toxikologische Untersuchungen mit dem Furanderivat UC-244 an der Laboratoriumsratte und am Kaninchen. Dissertation A, Karl-Marx-Universität, Fakultät für Agrarwissenschaften

Draize JH, Woodward G, Calvery HO (1944) Methods for the study of irritation and toxicity of substances applied topically to the skin and mucus membranes. J Pharmcol Exp Therap 82:377–390

Gillet AP, Andrews JM (1978) Nitrofurantoin. Laboratory methods in antimicrobial chemotherapy, 265–268

Koch NP, Ritschel WA (1986) Synopsis der Biopharmazie und Pharmakokinetik. Landsberg, München: ecomed, 173–204

Kucers A, Bennett NM (1979) The use of antibiotics. William Heinemann Medical books London, 749–760

Lichfield JR, Wilcoxon F (1949) A simplified method of evaluating dose effect experiments. J Pharmcol Exp Therap 96:99–115

Loomis TA (1982) Fundaments de Toxicologie. Editorial Acriba Zagoreza Espana

Mandell DL, Douglas RC, Bennett IN (1979) Principles and practice of infectious diseases. John Wiley and Sons

Rodriguez S, Gruz R, Morales A (1986) Determinatione de la irritabilitad del UC-244 applicado a nivel ocular en conejos. V. Internationale Konferenz der Landwirtschaft. UCLV, Santa Clara, 46

Roth HI, Braun R, Froline D (1985) Pharmazeutisches Taschenbuch. Wissenschaftliche Verlagsgesellschaft mbH Stuttgart, 25–26

Scheler W (1989) Grundlagen der allgemeinen Pharmakologie. Gustav Fischer Verlag

Schmidl M (1981) Laboruntersuchungen für die Diagnose und Verlaufskontrolle. Boehringer Mannheim GmbH, Mannheim

Recent Developments in Toxicology:
Trends, Methods and Problems
Arch. Toxicol., Suppl. 14, 298–302 (1991)
© Springer-Verlag 1991

Influence of the Degree, Position and Nature of the Halide Substitution on the Rate of Photobleaching of Xanthene Dyes

W. LEGRUM

Department of Pharmacology and Toxicology, Philipps-University, Lahnberge, D-3550 Marburg, FRG

Introduction

Aqueous solutions of xanthene dyes irridiated with visible light generate singlet oxygen in various quantum yields and are therefore used as sensitizers for photochemical reactions (Lamberts and Neckers 1984). Xanthene dyes might be suitable for the elucidation of photodegradation of pharmacologically active drugs, e.g. amiodarone and thyroxine. These compounds contain ortho-iodinated phenoxy groups as structural elements and easily are decomposed. Relatively high photobleaching is disadvantageous. Kinetic analysis of the rates of photobleaching of ten differentially substituted xanthene dyes served to investigate the influence of the degree, the position and the nature of the halide substitution on the photostability.

Materials and Methods

Chemicals: The following xanthene dyes were dissolved in water as sodium salts: fluorescein (FL), Riedel de Haën (Seelze, FRG); erythrosin (ERY = 4',5'-diiodo-FL), Aldrich-Chemie (Steinheim, FRG); erythrosin B (ERB = 2',4',5',7'-tetraiodo-FL), Serva-Feinbiochemica (Heidelberg, FRG); rose bengal (RB = 2',4',5',7'-tetraiodo-3,4,5,6-tetrachloro-FL), Serva; 4',5'-dibromofluorescein (DB), Sigma-Chemie (Steinheim, FRG); eosine B (EB = 4',5'-dibromo-2', 7'-dinitro-FL), E. Merck (Darmstadt, FRG); eosine Y (EY = 2',4',5',7'-tetrabromo-FL), Serva; phloxin B (PB = 2',4',5',7'-tetrabromo-3,4,5,6-tetrachloro-FL), Serva; 2',7'-dichlorofluorescein (DC), Sigma; merbromin (MB = 2',7'-dibromo-4'-hydroxymercuri-FL), Serva. The dye solutions tested (20, 10 and 4 μM) had a pH of 6.5. Sodium azide (N_3) was from E. Merck (Darmstadt, FRG) and imidazole (IZ) from Fluka (Ulm, FRG).

Bleaching experiments: An universal-white fluorescent tube (40 W) served for the irradiation of the xanthene dyes in covered plastic Petri dishes ($\varnothing$ = 85

mm) placed 6 cm above the tube on a plexiglas platform. Bleaching was measured in water at room temperature by reading the absorbance every 10 min for 1 hour keeping constant the volume irradiated. The measurements were carried out with an Uvikon 810 spectrophotometer (Kontron) in cuvettes of 1 cm light path setting the wavelength to be maximum of absorbance. The coefficients of absorbance were determined accordingly.

Kinetic analysis of the bleaching process reflects that the decrease of the concentration per time is proportional to the light absorbed by the molecules (first law of photochemistry) and that the intensity of the light decreases while passing through the dye solution (Lambert-Beer's law). Both functions are combined in the following differential equation: $dc/dt = -I_0\ (1-e^{-\varepsilon' cs})\ k$ for a constant initial light intensity where: c = concentration, $\varepsilon' = \varepsilon \cdot \ln 10$ = absorption coefficient, I = intensity of light, k = bleaching coefficient, s = light path, t = time. Integration over time leads to functions fitting the experimental time courses of absorption (Legrum et al 1989). The parameter "initial bleaching rate" (v_0) expressed in "pmoles of dye destroyed per second" is obtained by extrapolating the bleaching rate of a 10 micromolar dye solution towards a time and a light path of zero.

Results and Discussion

The time courses of absorption, i.e. "bleaching curves" of erythrosin (ERY) in aqueous solutions saturated with oxygen are shown in Fig. 1 for three different initial concentrations (4, 10 and 20 μM). Increasing the volume of the dye solution irradiated (i.e. longer light path) lowers the virtual bleaching. However, bleaching is the same for identical amounts of dye (identical number of absorption centres) irrespective of the degree of dilution as visualized by the percentages on the curves. Kinetic analysis of the six bleaching curves leads as described above to the determination of the initial bleaching rate of ERY and of the other xanthene dyes.

Table 1 combines the parameter v_0 of the different dyes investigated together with some of their physicochemical properties. The data reveal:

1. Photobleaching and the degree of substitution or the molecular mass are independent from each other.
2. Photobleaching and the yield of singlet oxygen formation are independent from each other. In some cases it seems they behave antiparallely to each other.
3. Substitution of the xanthene dyes with heavier atoms (I > Br > H) renders the dyes more susceptible towards photobleaching (ERY > DB; ERB > EY; RB > PB > FL). This is also true for a substitution with one hydroxymercuri-group or with two nitro-groups (MB>EB>EY).
4. Substitution of the xanthene dyes at the positions 2' and 7' stabilizes the molecule against photobleaching (FL<DC; ERB<ERY; EY<DB).

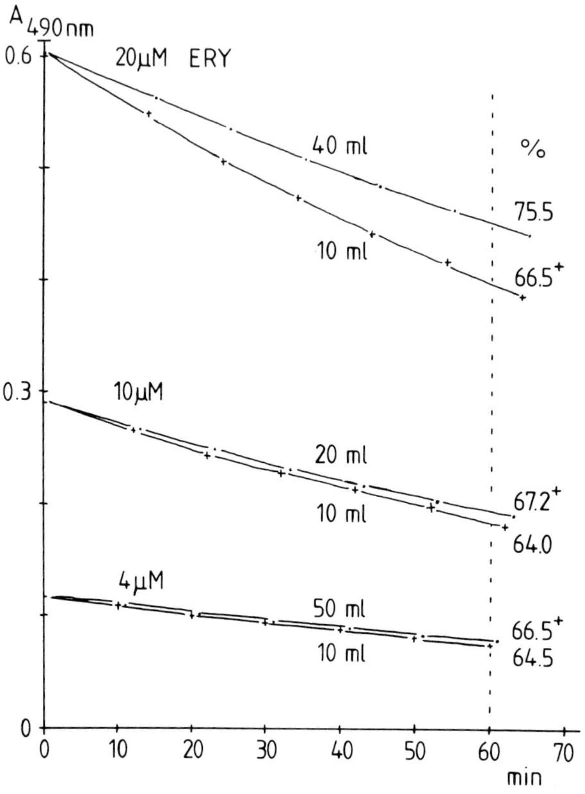

Fig. 1. Time course of the photobleaching of erythrosine (ERY). The remaining dye after a period of 60 minutes of irradiation with white light is indicated for each curve in per cent of its initial value. +denotes that identical amounts of absorption centres were irradiated

Table 1. The abbreviations of the dyes investigated are explained in Materials and Methods. The parameters Φ_t and Φ^1O_2 were taken from Gandin et al (1983), – denotes data not available. MB carries only one HgOH substituent at position 4' of the xanthene. BGR means bromopyrogallol red which carries a $-SO_3H$ instead of a $-COOH$ group; it is not really a xanthene dye. v_o is the initial photobleaching rate in pmol/sec. M_r gives the molecular weight of the free acids of the dyes. 'A' symbolizes the positions 4' and 5' at the xanthene structure (see formula in Fig. 2), 'B' the positions 2' and 7', while 'C' summarizes the positions 4, 5, 6 and 7 at the phenyl ring

dye	Φ_t	Φ^1O_2	v_o pmol/s	M_r g/mol	A	B substituents	C
ERY	0.33	0.48	12.5	584.1	I	H	H
ERB	0.69	0.63	10.0	835.9	I	I	H
RB	0.86	0.75	7.37	973.7	I	I	Cl
DB	0.29	0.42	8.43	490.1	Br	H	H
MB	–	–	7.20	706.7	HgOH	Br	H
EB	0.37	0.52	6.50	780.0	Br	NO_2	H
EY	0.32	0.57	5.34	647.9	Br	Br	H
PB	0.40	0.65	2.84	785.6	Br	Br	Cl
BGR	–	–	0	558.2	OH	Br	H
DC	0.04	0.07	0.86	401.2	H	Cl	H
FL	0.03	0.03	1.67	332.3	H	H	H

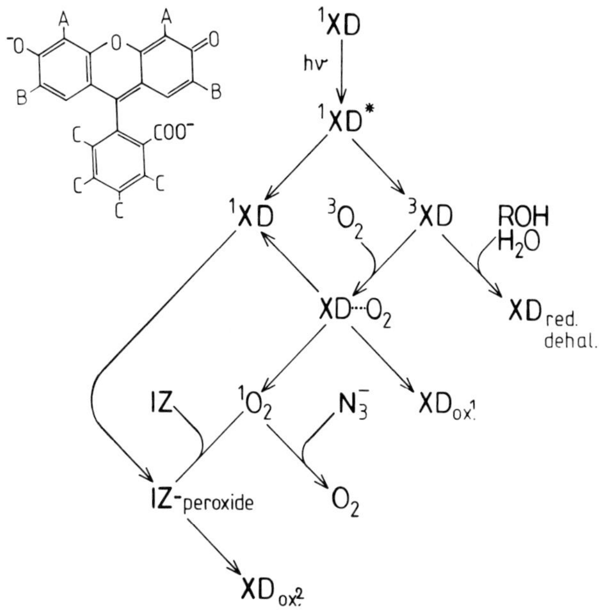

Fig. 2. Structural formula of a xanthene dye and scheme of reactions during the processes of bleaching. In the formula 'A' (4' and 5'), 'B' (2' and 7') and 'C' (4, 5, 6 and 7) indicate the positions of the substituents mentioned in table 1. The abbreviations used in the scheme are: hμ photoenergy, ^{1}XD* excited molecule of a xanthene dye, ^{1}XD xanthene dye singlet state, $^1O_2$ singlet oxygen, ^{3}XD xanthene dye triplet state, $^3O_2$ oxygen, XD···O_2 complex of the ^{3}XD with $^3O_2$ during energy transfer, IZ imidazole, N_3 azide-anion, $XD_{o\ x}^{1}$ and $XD_{o\ x}^{2}$ different oxidized and $XD_{red/dehal}$ reduced and dehalogenated xanthene dyes

5. Additional substitution at the positions 4, 5, 6 and 7 furthermore protects xanthene dyes against photobleaching (RB<ERB; PB<EY).
6. Substitution at the positions 4' and 5' only, with heavy atoms (I, Br and Hg) strongly favors the photobleaching (ERY, DB and MB).

Photobleaching of xanthene dyes is rather a direct reaction of the photoexcited dye with other molecules than a reaction with singlet oxygen via a type II mechanism (Imamura and Koizumi 1955; Wilson and Heitz 1984). This is confirmed by experiments using sodium azide as a quencher for singlet oxygen (Fig. 2). When present during irradiation at a concentration of 20 mM, azide does not influence the rate of photobleaching. On the other hand 10 mM imidazole which is converted by $^1O_2$ into a highly oxidizing transannular peroxide (Kraljić and El Mohsni 1978) increases the degradation of the dyes depending on the singlet oxygen formation. Sodium azide totally abolishes this additional degrading effect of imidazole.

References

Gandin E, Lion Y, Van de Vorst A (1983) Quantum yield of singlet oxygen production by xanthene derivatives. Photochem Photobiol 37:271–278

Imamura M, Koizumi M (1955) Irreversible photobleaching of the solution of fluorescent dyes. I. Kinetic studies on the primary process. Bull Chem Soc Jpn 28:117–124

Kraljić I, El Mohsni S (1978) A new method for the detection of singlet oxygen in aqueous solutions. Photochem Photobiol 28:577–581

Lamberts JJM, Neckers DC (1984) Rose bengal and non-polar derivatives: the birth of dye sensitizers for photooxidation. Z Naturforsch 39b:474–484

Legrum W, Bannenberg G, Hogardt M, Zenner G (1989) Kinetics of the photobleaching of halogenated xanthene dyes. Naunyn-Schmiedeberg's Arch Pharmacol 340 Suppl:R33

Wilson WW, Heitz JR (1984) Oxygen consumption during photobleaching of aqueous solutions of rose bengal. J Agric Food Chem 32:615–617

Recent Developments in Toxicology:
Trends, Methods and Problems
Arch. Toxicol., Suppl. 14, 303–306 (1991)
© Springer-Verlag 1991

Toxicological Investigations of Waste Products from the Plasma Etching Process in the Semiconductor Industry

S. Bauer[1], I. Wolff[1], N. Werner[1], P. Hoffmann[1], R. Hillig[2], R. Voigt[2], B. Merten[3], K. Oemus, and F.-W. Rath

[1] Institute of Industrial Toxicology, Martin Luther University, Franzosenweg 1a, O-4020 Halle, Germany
[2] Microelectronic Industry, Rudolfstraße 10, O-5010 Erfurt, Germany
[3] Research Centre of Microelectronics, Grenzstraße 28, O-8800 Dresden, and Institute of Pathology, Martin Luther University, Leninallee 4, O-4020 Halle, Germany

Introduction

Plasma etching processes in the semiconductor industry are accompanied by the formation of waste products with potential toxicity. There is only limited knowledge about the occupational and environmental risks of these substances.

Material and Methods

Characterization of the sample: The solid waste represents a complex mixture of inorganic and organic compounds. Dissolving in 10% hot ethanol resulted in an ethanolic solution (ES) containing 62% of the waste and a liquid residue (LR).

Acute oral toxicity: In a pilot experiment male specific pathogen-free rats (Shoe: Wistar, n = 5) aged about ten weeks were given 1.9 g/kg LR p.o. after overnight fasting. Observation of clinical and behavioral signs of toxicity for 14 days was followed by macroscopic pathology. In a continuing test 5.0 g/kg LR p.o. were given to male (n = 5) and female (n = 5) rats. The investigations were performed 48 h after LR administration: haematology, serum clinical chemistry (liver enzymes, creatinine [CRE], BUN, electrolytes), blood pressure, ECG, gross and histopathology. Standard pellet diet and tap water were available ad libitum.

Genotoxicity: Chromosome aberration test: Bone marrow cells of the rats were analyzed 48 h after oral administration of 5 g/kg LR or 15 ml/kg ES (1.23 g/kg soluble substance) (Werner et al 1985).

Somatic mutation and recombination test (SMART) with Drosophila melanogaster: Flies in the third larval stage (72 ± 2 h) were fed with LR- or ES-containing standard medium (0.01%–0.06% substance) up to the end of larval development. Because the mwh-flr-system was used, wings of freshly hatched adults were prepared for microscopical analyses of single and twin spots in the wing tissue (Graf et al 1983, 1984).

Results

ES contains inorganic compounds (mainly boric acid) and traces of perhalogenated hydrocarbons, whereas LR is mainly composed of perhalogenated hydrocarbons as analyzed by GC-MS. Neither deaths nor signs of behavioral toxicity were observe in the acute oral toxicity testing.

The exposure to 1.9 g/kg LR resulted in a significant decrease in body weight compared with the control animals 14 days after administration (249 ± 10 g vs. 354 ± 13 g, $p<0.05$, $n = 5$). The food consumption was decreased in the first week after administration and the water intake was increased in the second week. Signs of a forced diuresis were observed during this period. The relative weights of liver (86%), suprarenal glands (75%), stomach (39%), and lung (31%) were increased significantly.

Blood analyses 48 h after administration of 5.0 g/kg LR showed only mild biochemical and haematological changes. A significant increase in erythrocyte count (ERY), haemoglobin concentration (Hb), liver enzymes (ALAT, LDH) as well as a decrease in electrolytes (Na^+, Ca^{++}) were seen in females (Table 1).

Heart rates in both sexes and blood pressure of males were decreased (Table 2).

Histopathological examinations revealed changes in the liver which could be attributed to a slight acute toxic hepatosis. Chromosome analyses did not show an increase in the aberration rate in LR or ES-treated animals (data not

Table 1. Haematological and biochemical parameters 48 h after the administration of 5 g/kg LR (mean ± SE, n = 5)

		Male		Female	
		Control	Exposed	Control	Exposed
ERY	($x10^6/\mu l$)	7.2 ± 0.2	7.6 ± 0.3	7.0 ± 0.4	7.9 ± 0.3*
Hb	mval/l)	10.4 ± 0.1	10.4 ± 0.2	9.8 ± 0.4	10.6 ± 0.2*
ALAT	(nkat/l)	630 ± 30	880 ± 170	520 ± 30	640 ± 40*
LDH	(nkat/l)	3100 ± 161	3400 ± 678	2080 ± 467	4140 ± 425*
BUN	(mg/dl)	2.38 ± 0.16	2.45 ± 0.22	2.46 ± 0.13	2.78 ± 0.19
CRE	(μmol/l)	114 ± 3.6	121 ± 4.2	129 ± 9.2	144 ± 11.0
Na^+	(mmol/l)	142 ± 2	137 ± 1	137 ± 2	127 ± 3*
Ca^{++}	(mmol/l)	2.13 ± 0.10	2.08 ± 0.12	2.22 ± 0.04	2.13 ± 02.*

* $p<0.05$: compared with control

Table 2. Mean blood pressure (mmHg) and heart rate (min^{-1}) in urethane-anesthetized rats 48 h after the administration of 5 g/kg LR (mean ± SE, n = 5)

	Male		Female	
	Control	Exposed	Control	Exposed
Blood pressure	140 ± 8	109 ± 6*	141 ± 9	118 ± 8
Heart rate	500 ± 18	292 ± 11*	498 ± 27	348 ± 23*

* $p<0.05$: compared with control

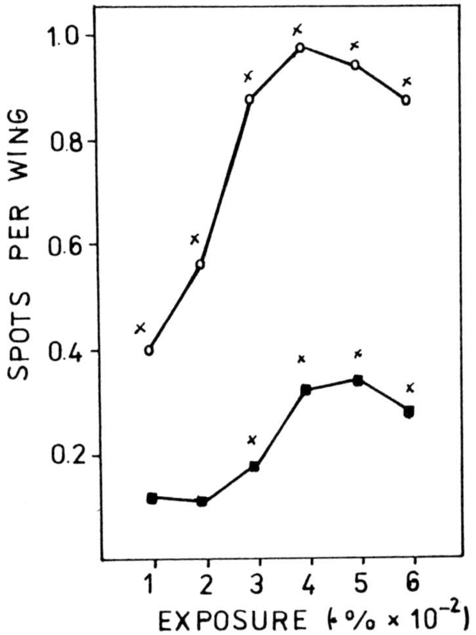

Fig. 1. Exposure effect-curves for the induction of wing spots by ES. O–O: mitotic recombinations, ■–■: gene mutations-/deletions; * $p<0.05$ compared with control (0.1 and 0.05 spots per wing, Kastenbaum-Bowman-test)

shown). Examination of wing spot distribution in SMART revealed well defined exposure effects. An increase in the mitotic recombination rate (LR and ES) and gene mutation frequency (ES) was observed (Fig. 1).

Conclusions

The results demonstrate only a slight acute oral toxicity of waste products from the plasma etching process. There was no increased chromosome aberration rate in the rats, but genotoxic effects of LR and ES in mammals cannot be excluded. The data of the SMART with Drosophila can be considered as a reference to possible genotoxic effects concerning different genetic end points.

References

Graf U, Juon H, Katz AJ, Frei HJ, Würgler FE (1983) A pilot study on a new Drosophila spot test. Mut Res 120:233–239

Graf U, Würgler FE, Katz AJ, Frei H, Juon H, Hall CB, Kale PG (1984) Somatic mutation and recombination test in Drosophila melanogaster. Invironm Mutag 6:153–188

Werner N, Freye H-A, Scheufler H (1985) Die kombinierte Prüfung auf Mutagenität und Teratogenität an der Labormaus – Versuche mit Cyclophosphamid und Ethylmethansulphonat. Biol Rdsch 23:251–254

Recent Developments in Toxicology:
Trends, Methods and Problems
Arch. Toxicol., Suppl. 14, 307–310 (1991)
© Springer-Verlag 1991

Sister Chromatid Exchanges in Furniture Workers Exposed to Unsaturated Polyester Resins

A. E. Karakaya, S. Şardaş, and M. Sun

Department of Toxicology, Gazi University, Faculty of Pharmacy, 06330 Ankara, Turkey

Introduction

In recent years it has become evident that occupational exposure to certain industrial chemicals may induce chromosomal damage (Das 1988). Unsaturated polyester resins have a wide commercial use in the furniture industry for surface coating. They are complex mixtures and made in several steps. A saturated dihydric alcohol (glycol) is generally condensed with both a saturated and an unsaturated dicarboxylic acid. Polyhydric alcohols are sometimes used to provide strength and chemical resistance. Maleic acid (anhydride) and fumaric acid are the usual unsaturated acids although itaconic or mesaconic acid may be used to give flexibility.

Phthalic anhydride is the most widely used saturated acid component. This prepolymer is dissolved in a vinyl monomer, usually styrene with appropriate inhibitors (Montgomery 1982). It has been estimated that 35 to 50 percent of the major component of the polyester resins employed in the workshops where the study was made contained styrene monomer. The present study evaluates the genotoxic risk of unsaturated polyester resins on the basis of two main criteria;

(a) the task of the exposed furniture workers with no protective measures and the lack of human studies giving information about the mixture under chronic exposure
(b) there is some experimental evidence of genotoxicity about the component of this mixture, (Vainio et al 1976; de Meester 1977).

The sister chromatid exchange (SCE) test was choosen, since it was considered the most sensitive test for detecting low concentrations of genotoxic agents (Kato 1974; Latt et al 1981, Gomez-Arroyo and Sauza 1985) and at the same time to be applicable to the monitoring of human populations (Latt et al 1981; Evans 1982; Wolff 1982).

Methods

A cytogenetic analysis was done among 30 workers exposed to polyester resins in three different workshops where they coat the furniture surface with this mixture. The men, aged 20 to 50 years, had been employed in polishing work from one year to 27 years. Each workers general health was examined at the time the blood sample was taken, and a history of exposure to any agent known to induce chromosomal damage was carefully recorded. A control group of 30 healthy men with no history of exposure to this mixture or any agent with known clastogenic activity was selected by using a questionnaire in order to enable matching the age and the amount of cigarettes before monitoring. Seventeen of the furniture workers were smokers and the other 13 were non-smokers. All smoked more than 10 cigarettes a day and the number of years of habitual smoking was more than ten years for all subjects. The number of years of occupational exposure were classified as over 10 years and under 10 years, since all had long occupational history (except one who started the job a year ago). SCE analysis were performed according to the method of Perry and Wolff (1974). SCE frequency was determined in at least 25 well spread cells from each subject. SCE frequencies were expressed on the number of SCE per cell.

Results

Data on the mean frequency of SCE per cell are presented in Table 1. Furniture workers as a group had a mean SCE of 7.50 ± 1.72, which was significantly higher than the mean value of 5.70 ± 1.95 seen in controls ($P < 0.001$). The mean SCE frequency in 17 who smoked more than 10 cigarettes per day, irrespective of the duration, was 8.35 ± 1.27: while that in 13 workers who never smoked was 6.38 ± 1.60. This difference was highly significant ($P < 0.001$). The mean SCE in 17 smoker controls, who had matched smoking habits was 7.11 ± 1.16, as against a mean of 3.85 ± 0.89 in 13 non smoker controls. This difference was also significant ($P < 0.001$). Duration of occupational

Table 1. SCE frequency in peripheral lymphocytes from a group of furniture workers exposed to unsaturated polyester resins in furniture industry, and controls

	SCE per cell ± SD in exposed	SCE per cell ± SD in controls	p-value
Smokers	8.35 ± 1.27 (n=17)	7.11 ± 1.16 (n=17)	p<0.01
Non-smokers	6.38 ± 1.60 (n=13)	3.85 ± 0.89 (n=13)	p<0.001
All subjects	7.50 ± 1.72 (n=30)	5.70 ± 1.95 (n=30)	p<0.001

Mean ± standard deviation (n = number of subjects) Comparison of exposed and control subjects by the t-test. The difference between smoker and non-smoker workers was also significant (p<0.001)

Table 2. Mean frequencies of SCE for duration of exposure in exposed workers

Furniture workers	Number of subjects	Duration of exposure	SCE per cell Mean ± SD
Smokers	9	over 10 years	9.22 ± 0.66
Smokers	8	under 10 years	7.37 ± 1.06
Non-smokers	6	over 10 years	7.67 ± 0.81
Non-smokers	7	under 10 years	5.29 ± 1.25

Difference between > 10 years exposed and <10 years exposed smoker furniture workers are significant ($p < 0.001$) by the t-test. Difference between > 10 years exposed and < 10 years exposed non-smokers are significant ($p < 0.01$) by the t-test

exposure, and the number of SCE was also studied in occupationally exposed smoker and non-smoker workers. The mean SCE frequency in those (Table 2) who smoked and worked more than 10 years was 9.22 ± 0.66: while in those who smoked but worked less than 10 years was 7.37 ± 1.06. This difference was highly significant ($P < 0.001$). But the mean SCE in non-smokers who worked more than 10 years was 7.67 ± 0.81, while it was 5.29 ± 1.25 in non-smokers who worked less than 10 years ($P < 0.01$).

Discussion

The cytogenetic observations made in this study shows that, of the workers in the furniture industry, those exposed to unsaturated polyester resins exhibited a statistically significant increase in the number of SCE compared with nonexposed controls. A statistically significant relationship between smoking and increased frequency of SCE was observed both in workers and in controls. However, this is in qualitative and quantitative agreement with the observations of other authors (Lambert et al 1978; Husgafvel-Pursiainen et al 1980; Livingston et al 1983) that there is a correlation between cigarette consumption and increase in SCE. The study also reveals a statistically significant association between the occupational duration of exposure to unsaturated polyester resins and chromosomal damage. Since these workers are exposed not to a singular unique exposure but to a complex mixture, the results obtained so far indicate a necessity to perform a more thorough study on the genotoxicity of polyester resins. Therefore it should be emphasized once more that the cytogenetic damage should be considered for all resins not just single substances. As indicated the major compound of the mixture was styrene monomer and the correlation between exposure to styrene and the frequency of SCE and chromosomal aberrations in lymphocytes of exposed subjects are discussed by previous reports (Uggla et al 1980; Watanable et al 1983; Camurri et al 1983; IARC Monographs 1987). These workers spend a working period of 8 h in the same work area as furniture surface coaters and do have direct contact with polyester resins. Although the polyester resin concentration in the working atmosphere was not determined it most probably exceeded the maximum allowed limits. In none of the studied areas did the workers employ any protection measures for preventing inhalation (wearing masks, gloves, appropriate ventilation). Finally in the light of positive induction of SCE in workers exposed to polyester resins and because of the risk of using mixtures, it is necessary to improve the working conditions. The results also show that smoking increases significantly the SCE frequencies in people exposed to several chemicals.

References

Camurri L, Codeluppi S, Ledroni C, Scarduelli L (1983) Chromosomal aberrations and sister chromatid exchanges in workers exposed to styrene. Mutat Res 119:361–363

Das BC (1988) Factors that influence formation of sister chromatid exchanges in human blood lymphocytes. CRC Critical Reviews in Toxicology 19:43–86

Evans HJ (1982) Chromosomal mutations in human populations. Cytoget Cell Genet 33:48–56

Gomez AS, Souza V (1985) In vitro and occupational induction of sister chromatid exchange in human lymphocytes with furfuryl alcohol and furfural. Mutat Res 156:233–238

Husgafvel-Pursianen K, Maki-Paakkanen J, Norppa H, Sorsa M (1980) Smoking and sister chromatid exchange. Heretidas 92:247–250

IARC (1987) IARC Monographs on the Evaluation of Carcinogenic Risk to Humans Suppl No. 7, IARC, Lyon, France, pp. 345–347

Kato H (1974) Induction of sister chromatid exchange by chemical mutagens and its possible revelance to DNA repair. Exp Cell Res 85:239–249

Lambert B, Lindblad A, Nordenskjöld M, Werilius B (1978) Increased frequency of sister chromatid exchanges in cigarette smokers. Heretidas 88:147–149

Latt SA, Allen J, Bloom S, Carrano A, Falke E, Kram D, Schneider E, Schreck R, Tice R, Witfield B, Wolff S (1981) Sister chromatid exchanges: a report of the Gene-Tox program. Mutat Res 99:257–271

Livingston GK, Fineman RM (1983) Correlation of human lymphocytes with smoking history. Mutat Res 119:59–64

de Meester C, Poncelet F, Roberfroid M, Ronde let J, Mercier M (1977) Mutagenicity of styrene and styrene oxide. Mutat Res 56:147–152

Montgomery RR (1982) Polymers. In: Cralley LV, Cralley LJ (eds) Patty's Industrial Hygiene and Toxicology. Vol IIc. A Wiley Interscience Publication USA, pp. 4209–4526

Perry P, Wolff S (1974) New Giemsa method for the differential staining of sister chromatids. Nature 251:156–158

Uggla AH, Anderson HC, Transberg EA, Zetterberg LG (1980) Correlation between exposure to styrene and the frequency of chromosomal aberrations and sister chromatid exchanges in lymphocytes of workers in a plastic boat factory. Mutat Res 74:199–201

Watanable T, Endo A, Kumai M, Ikeda M (1983) Chromosome aberrations and sister chromatid exchanges in styrene exposed workers with reference to their smoking habits. Environ Mutagen 5:299–303

Wolff S (1982) Difficulties in assessing the human health effect on mutagenic carcinogens by cytogenetic analysis. Cytogenet Cell Genet 33:7–13

Vainio H, Paakkönen R, Rönnholm K, Raunio V, Pelkonen O (1976) A study on the mutagenic activity of styrene and styrene oxide. Scand J Work Env Health 3:147–151

Recent Developments in Toxicology:
Trends, Methods and Problems
Arch. Toxicol., Suppl. 14, 311–315 (1991)
© Springer-Verlag 1991

Acute Intoxication as a Cause of Wild Bird Mortality in the Netherlands from 1975–1989

A. J. Baars, P. E. F. Zoun, Th. J. Spierenburg, and Th. Smit

Central Veterinary Institute, P.O. Box 65, 8200 AB Lelystad, The Netherlands

Introduction

The Working Group on Wild Bird Mortality was founded in 1975 as a cooperative effort of the Central Veterinary Institute, the Research Institute for Nature Management, the Institute of Taxonomical Zoology (University of Amsterdam), the State Forestry Commission, and the Netherlands Society for the Protection of Birds. The Working Group studies causes and consequences of diseases and mortality among wild birds in the Netherlands, and issues directives to prevent undesirable mortality.

Amateur and professional ornithologists in the Netherlands are organized in about 200 bird study groups. Thousands of bird watchers browse the fields during the week-ends. These observers, together with employees of the State Forestry Commission, wardens of nature sanctuaries, police officers, etc., act as the "eyes" of the Working Group in the field. If bird mortality is observed, they collect samples, report the situation on a standard form, and send these to the CDI.

A preliminary diagnosis is made on the basis of a pathological examination, and appropriate samples are examined by specialists to confirm the diagnosis. The ultimate diagnosis is reported to the observer, with a copy to the Bird Protection Society. The findings of the Working Group are reported annually to the Ministry of Agriculture, Nature Management and Fisheries. Recently the overall results of the period from 1975 to 1989 were published (Baars and Over 1989).

Abbreviations: CDI: Central Veterinary Institute; AAS: atomic absorption spectrometry; GLC: gas-liquid chromatography; HPLC: high performance liquid chromatography; TLC: thin layer chromatography

Materials and Methods

Upon arrival, birds were examined for gross pathological lesions. An X-ray was used to detect bullets or leadshot. If anamnesis and autopsy suggested that some intoxication was involved, selected samples were subjected to chemical-toxicological analysis, sometimes preceeded by a biological toxicity test using daphniae and guppies.

Unless anamnesis suggested otherwise, samples were routinely subjected to TLC for analysis of cholinesterase-inhibiting pesticides (Zoun and Spierenburg 1988). Samples were further analyzed if TLC results were negative but other pesticides were suspected. Strychnine and nicotine were detected by potassium iodoplatinate (Drost et al 1967), following TLC. Analysis of aldicarb (a carbamoyloxime pesticide) was done by HPLC (Spierenburg et al 1987). Halogenated hydrocarbons were detected by GLC with electron capture detection, using a modified method for endosulfan analysis (Zoun et al 1987). Alphachloralose was analyzed by GLC (Electron capture detection) after decomposition to chloralhydrate, or by GLC (mass spectrometric detection) after derivatization to the trimethylsilyl derivative. Lead intoxications were diagnosed by AAS of livers and kidneys (Van Beek et al 1987).

Results and Discussion

From 1975 until 1989, over 15,000 birds were examined for the cause of death. Various diseases, including parasitological infections, accounted for 46% of the total mortality (Fig. 1). Acute intoxication was the cause of death for 21% of the birds. Fourteen percent died as a result of some kind of mechanical

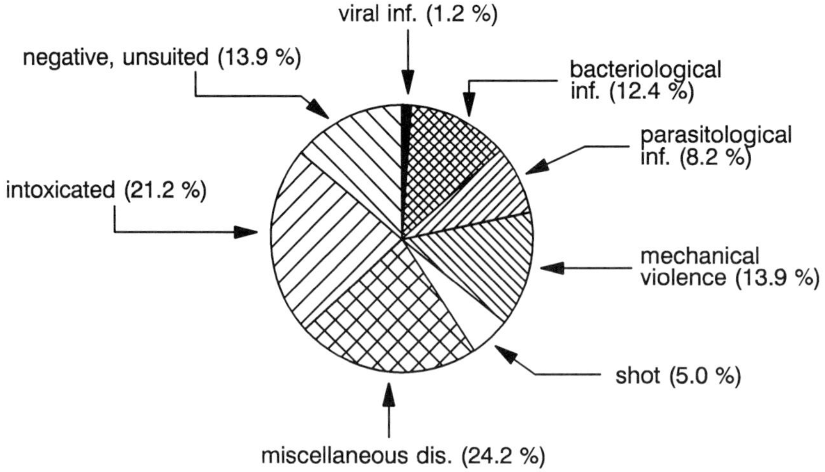

Fig. 1. Distribution of diagnoses of dead birds in the Netherlands, investigated between 1975 and 1989 (inf.: Infections; dis.: Diseases)

Table 1. Birds and related samples examined and percentage intoxicated between 1975 and 1989

Species	Number examined	Number poisoned	Percentage poisoned
Birds of prey	2702	621	23.0
Gulls and terns	2268	148	6.5
Geese, swans, ducks	2128	409	19.2
Crows	774	357	46.1
Owls	772	4	0.5
Moorhen, coot	523	59	11.3
Waders and stelts	468	10	2.1
Pigeons and doves	464	154	33.2
Finches	436	56	12.8
Starlings	370	110	29.7
Sparrows	168	72	42.9
Pheasants and partridges	137	69	50.4
Feral birds, crossbred ducks	2373	880	37.1
Other songbirds	817	53	6.5
Other birds	484	4	0.8
Soil, water and food	255	106	41.6
Seed and grain	81	61	75.3
Eggs and decoy	56	39	69.6
Mammals and fish	166	65	39.1
Total	15442	3277	21.2

violence. Another 14% was negative or unsuited for investigation, while 5% appeared to have died some time after being shot.

Different species of birds were intoxicated in widely varying percentages (Table 1). Illegal use of pesticides was the probable cause of death in about 55% of all intoxications. Only 7% of the intoxications appeared to be the result of legal pesticide applications, and most of these unintentional side effects occurred in the early years of the period under study.

These results have generally been confirmed by the British monitoring program (Fletcher et al 1989; Greig-Smith et al 1989; Environmental Panel, Advisory Committee on Pesticides 1990). In the period 1985–1989, these authors investigated 694 incidents of bird mortality and found that 39% appeared to have been poisoned by pesticides.

The results of the Working Group demonstrate that organophosphorus compounds are responsible for almost 80% of all bird intoxications diagnosed (Table 2). Organophosphorus pesticides, which are cheap, easy to obtain, and highly toxic, seem to be preferred by those who poison birds intentionally. In recent years, however, parathion intoxications have tended to decrease, while alphachloralose and carbamoyloximes (especially aldicarb) intoxications have increased.

This was also noted in the United Kingdom, where alphachloralose and mevinphos (parathion is not approved for use in the U.K.) are the main

Table 2. Toxic compounds detected in bird mortalities in The Netherlands, 1975 to 1989

Compound	Bird intoxications	
	Number	Percentage
Parathion	2393	73.0
Diazinon	87	2.7
Mevinphos	53	1.6
Misc. organosphosphorus pesticides	69	2.1
Carbamates	40	1.2
Carbamoyloximes	60	1.8
Alphachloralose	197	6.0
Strychnine	145	4.4
Crimidine	9	0.3
Nicotine	4	0.1
Endosulfan	55	1.7
Misc. chlorinated hydrocarbons	17	0.5
Lead	143	4.4
Other poisons	5	0.2

diagnosed toxicants (Fletcher et al 1989; Greig-Smith et al 1989; Environmental Panel, Advisory Committee on Pesticides 1990).

Most of the intoxications summarized in this report were caused by human abuse or even misuse of pesticides. The reasons to kill birds are numerous. Farmers sometimes sow poisoned grain to protect their crops against marauding birds. Some hunters may protect the pheasants they hunt by using poisoned bait to kill hawks. Trophy hunters poison birds of prey, have them stuffed, and sell them as a curiosity. However, the extensive information by e.g. the Bird Protection Society has resulted in a gradual reduction of intentional bird poisoning in recent years.

References

Baars AJ, Over HJ (eds) (1989) Wild bird mortality in The Netherlands 1975–1989. Netherlands Society for the Protection of Birds, Zeist, and Central Veterinary Institute, Lelystad, pp. 57

Drost RH, Reith JF (1967) Identificatie van stoffen in de toxicologische analyse met behulp van het extractiesysteem van Feldstein, dunnelaagchromatografie en UV-absorptiespectrofotometrie. I. Basische stoffen. Pharmac Weekbl 102:1379–1386

Environmental Panel, Advisory Committee on Pesticides (1990) Pesticide poisoning of animals 1988. Ministry of Agriculture, Fisheries and Food, London, pp. 20

Fletcher MR, Hunter K, Greig-Smith PW, Ruthven AD (1989) Investigations of suspected poisoning of animals by pesticides in Great Britain, October 1985 to December 1986. In: Environmental Panel, Advisory Committee on Pesticides (ed) Investigations of suspected poisoning of animals by pesticides in Great Britain, 1985–1987. Ministry of Agriculture, Fisheries and Food, London, pp. 7–22

Greig-Smith PW, Fletcher MR, Hunter K, Quick MP, Ruthven AD, Shaw IC (1989) Investigations of suspected poisoning of animals by pesticides in Great Britain in 1987. In: Environmental Panel, Advisory Committee on Pesticides (ed) Investigations of suspected poisoning of animals by pesticides in Great Britain, 1985–1987. Ministry of Agriculture, Fisheries and Food, London, pp. 23–44

Spierenburg ThJ, Kemmeren-Van Dijk MBH, Zoun PEF (1987) High-performance liquid chromatographic analysis of aldicarb in the stomach contents of birds of prey. J Chromatogr 393:137–139

Van Beek H, Greefkes HCA, Baars AJ (1987) Determination of copper, iron, manganese, lead and cadmium in automatically wet-digested animal tissue by graphite-furnace atomic absorption spectrometry with Zeeman background correction. Talanta 34:580–582

Zoun PEF, Spierenburg ThJ (1989) Determination of cholinesterase-inhibiting pesticides and some of their metabolites in cases of animal poisoning using thin-layer chromatography. J Chromatogr 462:448–453

Zoun PEF, Spierenburg ThJ, Baars AJ (1987) Gaschromatographic determination of endosulfan in fish and water samples. J Chromatogr 393:133–136

Author index

Subject Index